ANNUAL PLANT REVIEWS VOLUME 46

ANNUAL PLANT REVIEWS VOLUME 46

Plant Nuclear Structure, Genome Architecture and Gene Regulation

Edited by

David E. Evans

Department of Biological and Medical Sciences
Oxford Brookes University, Oxford, UK

Katja Graumann

Department of Biological and Medical Sciences
Oxford Brookes University, Oxford, UK

John A. Bryant

Biosciences, University of Exeter, Exeter, UK

Annual Plant Reviews

WILEY-BLACKWELL

A John Wiley & Sons, Ltd., Publication

Wiley-Blackwell is an imprint of John Wiley & Sons, formed by the merger of Wiley's global Scientific, Technical and Medical business with Blackwell Publishing.

Registered office: John Wiley & Sons, Ltd, The Atrium, Southern Gate, Chichester, West Sussex, PO19 8SQ, UK

Editorial offices: 9600 Garsington Road, Oxford, OX4 2DQ, UK
The Atrium, Southern Gate, Chichester, West Sussex, PO19 8SQ, UK
111 River Street, Hoboken, NJ 07030-5774, USA

For details of our global editorial offices, for customer services and for information about how to apply for permission to reuse the copyright material in this book please see our website at www.wiley.com/wiley-blackwell.

Library of Congress Cataloging-in-Publication Data has been applied for

ISBN 978-1-1184-7245-3 (hardback)

A catalogue record for this book is available from the British Library.

Wiley also publishes its books in a variety of electronic formats. Some content that appears in print may not be available in electronic books.

Cover image: Images supplied by Katja Graumann
Cover design by www.hisandhersdesign.co.uk

Set in 10/12pt Palatino by Aptara® Inc., New Delhi, India
Printed and bound in Singapore by Markono Print Media Pte Ltd

First Impression 2013

Annual Plant Reviews

A series for researchers and postgraduates in the plant sciences. Each volume in this series focuses on a theme of topical importance and emphasis is placed on rapid publication.

Titles in the series:

1. **Arabidopsis**
   Edited by M. Anderson and J.A. Roberts
2. **Biochemistry of Plant Secondary Metabolism**
   Edited by M. Wink
3. **Functions of Plant Secondary Metabolites and their Exploitation in Biotechnology**
   Edited by M. Wink
4. **Molecular Plant Pathology**
   Edited by M. Dickinson and J. Beynon
5. **Vacuolar Compartments**
   Edited by D.G. Robinson and J.C. Rogers
6. **Plant Reproduction**
   Edited by S.D. O'Neill and J.A. Roberts
7. **Protein–Protein Interactions in Plant Biology**
   Edited by M.T. McManus, W.A. Laing and A.C. Allan
8. **The Plant CellWall**
   Edited by J.K.C. Rose
9. **The Golgi Apparatus and the Plant Secretory Pathway**
   Edited by D.G. Robinson
10. **The Plant Cytoskeleton in Cell Differentiation and Development**
    Edited by P.J. Hussey
11. **Plant–Pathogen Interactions**
    Edited by N.J. Talbot
12. **Polarity in Plants**
    Edited by K. Lindsey
13. **Plastids**
    Edited by S.G. Moller
14. **Plant Pigments and their Manipulation**
    Edited by K. Davies

15. **Membrane Transport in Plants**
    Edited by M.R. Blatt
16. **Intercellular Communication in Plants**
    Edited by A.J. Fleming
17. **Plant Architecture and Its Manipulation**
    Edited by C.G.N. Turnbull
18. **Plasmodeomata**
    Edited by K.J. Oparka
19. **Plant Epigenetics**
    Edited by P. Meyer
20. **Flowering and Its Manipulation**
    Edited by C. Ainsworth
21. **Endogenous Plant Rhythms**
    Edited by A. Hall and H. McWatters
22. **Control of Primary Metabolism in Plants**
    Edited by W.C. Plaxton and M.T. McManus
23. **Biology of the Plant Cuticle**
    Edited by M. Riederer
24. **Plant Hormone Signaling**
    Edited by P. Hadden and S.G. Thomas
25. **Plant Cell Separation and Adhesion**
    Edited by J.R. Roberts and Z. Gonzalez-Carranza
26. **Senescence Processes in Plants**
    Edited by S. Gan
27. **Seed Development, Dormancy and Germination**
    Edited by K.J. Bradford and H. Nonogaki
28. **Plant Proteomics**
    Edited by C. Finnie
29. **Regulation of Transcription in Plants**
    Edited by K. Grasser
30. **Light and Plant Development**
    Edited by G. Whitelam
31. **Plant Mitochondria**
    Edited by D.C. Logan
32. **Cell Cycle Control and Plant Development**
    Edited by D. Inzé
33. **Intracellular Signaling in Plants**
    Edited by Z. Yang
34. **Molecular Aspects of Plant Disease Resistance**
    Edited by J. Parker
35. **Plant Systems Biology**
    Edited by G.M. Coruzzi and R.A. Gutiérrez
36. **The Moss *Physcomitrella patens***
    Edited by C.D. Knight, P.-F. Perroud and D.J. Cove
37. **Root Development**
    Edited by T. Beeckman
38. **Fruit Development and Seed Dispersal**
    Edited by L. Østergaard

39. **Function and Biotechnology of Plant Secondary Metabolites**
    Edited by M. Wink
40. **Biochemistry of Plant Secondary Metabolism**
    Edited by M. Wink
41. **Plant Polysaccharides**
    Edited by P. Ulvskov
42. **Nitrogen Metabolism in Plants in the Post-genomic Era**
    Edited by C. Foyer and H. Zhang
43. **Biology of Plant Metabolomics**
    Edited by R.D. Hall
44. **The Plant Hormone Ethylene**
    Edited by M.T. McManus
45. **The Evolution of Plant Form**
    Edited by B.A. Ambrose and M.D. Purugganan

CONTENTS

LIST OF CONTRIBUTORS

Susan J. Armstrong
School of Biosciences
University of Birmingham
B15 2TT
UK

Andreas Binder
Faculty of Biology, Genetics, University of Munich (LMU)
Großhaderner Straße 4
82152 Martinsried
Germany

Joanna Boruc
Department of Molecular Genetics
The Ohio State University
Columbus
OH 43210
USA

John A. Bryant
Biosciences
University of Exeter
Exeter
EX4 4QD, UK

David E. Evans
Department of Biological and Medical Sciences
Faculty of Health and Life Sciences
Oxford Brookes University
Headington Campus
Oxford
OX3 0BP
UK

F. Chris H. Franklin
School of Biosciences
University of Birmingham
B15 2TT
UK

Martin W. Goldberg
Department of Biological and Biomedical Sciences
University of Durham
Durham
DH1 3LE
UK

Katja Graumann
Department of Biological and Medical Sciences
Faculty of Health and Life Sciences
Oxford Brookes University
Headington Campus
Oxford
OX3 0BP
UK

Iris Meier
Department of Molecular Genetics
The Ohio State University
Columbus
OH 43210
USA

Eugenio Sanchez-Moran
School of Biosciences
University of Birmingham
B15 2TT
UK

Kim Osman
School of Biosciences
University of Birmingham
B15 2TT
UK

Martin Parniske
Faculty of Biology, Genetics
University of Munich (LMU)
Großhaderner Straße 4
82152 Martinsried
Germany

Monica Pradillo
Departmento de Genetica
Facultad de Biologia
Universidad Complutense
Madrid
28040 Spain

Nicola Y. Roberts
School of Biosciences
University of Birmingham
B15 2TT
UK

Juan L. Santos
Departmento de Genetica
Facultad de Biologia
Universidad Complutense
Madrid
28040 Spain

Christophe Tatout
GReD Laboratory
UMR CNRS 6293
INSERM U1103
University Blaise Pascal
Aubière
France

Mélanie Thomas
GReD Laboratory
UMR CNRS 6293
INSERM U1103
University Blaise Pascal
Aubière
France

Emmanuel Vanrobays
GReD Laboratory
UMR CNRS 6293
INSERM U1103
University Blaise Pascal
Aubière
France

Javier Varas
Departmento de Genetica
Facultad de Biologia
Universidad Complutense
Madrid
28040 Spain

Xiao Zhou
Department of Molecular Genetics
The Ohio State University
Columbus
OH 43210
USA

PREFACE

This volume was conceived to bring together reviews describing recent advances in knowledge and understanding of plant nuclear structures and functions, including that of the nuclear envelope. The book is particularly timely in that recent progress has been rapid in key areas including description and characterization of proteins of the nuclear envelope and nuclear pore complex, novel insights into nucleoskeletal structures, as well as developments related to chromatin organization, function and gene expression. Together these advances provide a framework for comparative understanding of nuclear envelope structure and function in a range of organisms and for understanding its evolution.

Current knowledge of the dynamic structure of plant DNA and chromatin is discussed by Sanchez-Moran in Chapter 5. Despite intensive study of histones and other chromosome-associated proteins, interactions to achieve the complex structures required both in interphase and during cell division remain poorly understood. The structures require several levels of organization, the first being the nucleosomal fibre comprising DNA wrapped around a core of histones. This is a dynamic structure and mechanisms for its remodelling are described. The nucleosomal fibre is then wound into a structure termed the chromatin fibre, which is arranged in loops associated with a multi-protein chromosome scaffold, the third level of structure. This interphase structure undergoes rapid dynamic change in mitosis with further condensation for replication and division. The importance of the structural organization of chromatin for processes such as transcription, replication, repair, recombination, condensation and segregation is also discussed. As in metazoans, plant chromatin is organized into regions of hetero- and euchromatin with heterochromatin adjacent to the NE.

Recent advances in understanding heterochromatin structure are presented in Chapter 6 by Vanrobays *et al.* Heterochromatin, originally thought to be condensed, gene-poor and 'silent', is now known often to be preferentially localized to the nuclear envelope and nucleolus and its significance is becoming clear as an epigenetic state required for many functions of the genome, including gene regulation, segregation of chromosomes and maintaining stability of the genome. Despite limited knowledge of it, in most species heterochromatin is the main form of chromatin and key questions remain to be answered. How is spatial organization of heterochromatin maintained through the cell cycle as DNA is replicated, chromatin condensed, and the nuclear envelope disrupted and reformed? Interactions between nuclear envelope, nucleoskeleton and chromatin are likely to be very significant and are discussed together with other theories for heterochromatin positioning.

Plant genomes vary greatly in size but in all cases the genome is contained within a very small compartment and it is clear that complex three-dimensional organization is needed in order for the many processes required for function. This three-dimensional structure requires interactions between chromatin, the envelope, the nuclear pores and the rest of the cell. In Chapter 4, Goldberg discusses from an ultrastructural and biochemical perspective the presence of an equivalent of the highly ordered lamina and nucleoskeleton described in metazoans. Such a structure appears to be required for nuclear function, but until recently its protein composition has eluded plant scientists. Plant cells have no proteins homologous to the lamins or other intermediate filament protein. Recent electron microscope studies in Goldberg's laboratory of the inner face of the plant nuclear envelope reveal a filamentous structure interconnecting the NPCs. This appears to be organized similarly to the lamina of *Xenopus* oocytes. Protein candidates for a plant nucleoskeleton have recently been suggested from a number of approaches; these long coiled-coil nuclear-localized proteins show some similarities to nucleoskeletal proteins of the metazoans and Goldberg presents the growing, but as yet incomplete, evidence for their role. The likely (direct or indirect) interactions of these proteins with the proteins of the nuclear envelope via a 'Linker of Nucleoskeleton and Cytoskeleton' complex is also considered in Chapter 2 by Graumann and Evans. Therein, the authors describe that, in common with metazoans, plants have one key family of proteins that in other kingdoms constitutes the inner nuclear envelope component of this bridging complex, namely the Sad1/Unc84 (SUN)-domain protein family. Absence of a variety of other inner nuclear envelope components involved in nuclear envelope-chromatin interactions in other kingdoms suggests that the SUN- domain proteins play a particularly significant and broader role in plants. In many respects however, the higher plant SUN-domain proteins show remarkable conservation in structure to those of other organisms. They are smaller than their metazoan counterparts, being closest in size to the yeast homologue Sad1. In addition, the authors discuss first evidence of proteins interacting with SUN-domain proteins in plants that show similarity in structure and mechanism to the Klarsicht/Anc-1/Syne Homology (KASH)-domain proteins of other kingdoms, which complete the nucleo-cytoskeletal bridging complexes. Chapter 2 also focuses on other protein components of the plant nuclear envelope as well as its lipid composition and highlights many of the cellular and nuclear processes in which the plant nuclear envelope plays key roles.

Structure and position of chromosomes must be achieved both for successful mitosis and meiosis. Evidence for SUN- domain protein involvement in the breakdown and reformation of the nuclear envelope in plant mitosis is presented in Chapter 2 together with suggestions of conserved mechanisms between kingdoms. Meiosis, while more complex, has received considerable attention and the role of telomeres is presented by Roberts *et al.* in Chapter 7, who reveal emerging evidence for their role in early events in the movement and synapsis of homologous chromosomes. Studies in Arabidopsis suggest that paired telomeres loosely cluster at the nuclear periphery in meiotic prophase 1; it is suggested that this facilitates chromosome alignment and synapsis. The proteins involved in the attachment of telomeres to the nuclear envelope remain elusive; however, in common with the yeast and metazoans, a role for SUN-domain proteins is suggested. Exploration of the structural protein interactions in meiosis is being vigorously pursued.

Recent characterization of proteins of the plant nuclear pore complex (NPC) has revealed that the structure more closely resembles those of vertebrates than yeast or fungi. In Chapter 3, Zhou *et al.* describe the significant progress made recently in identifying 30 constituent proteins of the plant NPC as well as characterizing plant NPC structure. While the overall architecture of NPCs is conserved in eukaryotes, the plant NPC are set apart by several unique features and absence of a number of vertebrate nucleoporins. Significantly, the anchorage of RanGAP, involved in the generation of the RanGTP/GDP gradient required for nuclear import and export (a mechanism conserved between kingdoms) has been shown to differ significantly between plants and other organisms. In mammals, for instance, RanGAP is anchored to the pore complex by sumoylation. In plants, this function is taken over by interaction with proteins associated with the nuclear pore complex termed the WPP (tryptophan proline proline) interacting proteins (WIPs) and WPP interacting tail-anchored proteins (WITs). Apart from structural differences, the authors also discuss plant-specific functions and non-trafficking processes that plant nucleoporins are involved in, including mitotic functions, plant development, hormone and abiotic stress responses and plant-microbe interactions. The latter topic is the primary focus of Chapter 8 by Binder and Parniske. Using *Lotus japonicus* as a model system, loss of function mutants of several nucleoporins result in impaired mycorrhizal association as well as root-nodule symbiosis linked to failure of nuclear calcium signalling. In *Arabidopsis thaliana,* nucleoporins have been shown to be required for the two major forms of response to fungal pathogens, namely pathogen-associated molecular pattern (PAMP)-triggered and disease resistance (R) gene-mediated defence signalling. This is presented in the context of expanding knowledge of the nuclear pore complex and other proteins of the nuclear envelope and suggests important targets for attention in relation to the introduction of nitrogen fixation into cereals and in the development of crops showing enhanced resistance to fungi. The authors also focus on the challenges of correlating specific functions with individual nucleoporins due to the complexity of interactions and functions of NPC components and functional redundancies.

It is evident that exploration of plant nuclear structure, genome architecture and gene regulation has widespread implications for crop improvement and food security. Movement of the nucleus occurring as stress and developmental responses are presented in Chapter 2 by Graumann and Evans and include movement in intense light, due to touch and viral and fungal infection. Such movements are likely to be significant in plant tolerance to stress and infection and to involve nucleo-cytoskeletal bridging complexes at the nuclear envelope. The positional effects of chromatin structure and the structure of the nucleus on gene expression, discussed in Chapter 6 by Vanrobays *et al.*, suggest an area with considerable potential for exploration as tools to study gene targeting to subnuclear localizations become available. It has yet to be established whether localization to the nuclear periphery, pore complex or other regions of the nucleus induces a repressive or activation effect in respect to gene expression. Such effects, if reproducible, have considerable potential for development. Perhaps the most comprehensively studied role for the plant nuclear structures with widespread significance concerns the role of the nuclear pore complex in fungal pathogenesis and symbiosis.

There is a very clear need to expand knowledge of protein interaction networks at the nuclear envelope involving cytoskeleton, nucleoskeleton and chromatin components. Study of the nuclear envelope proteome has been held back by a combination of limited interest by researchers and the technical difficulties of isolating and analysing it. Recent advances – the identification of SUN domain proteins and first evidence for a linker of nucleoskeleton and cytoskeleton complex, the characterization of more than 30 nucleoporins and increasing functional evidence and the tentative characterization of a plant lamina – all provide a framework for rapid advances coupled with increased understanding of chromatin structure and function. Given the outstanding importance of the nucleus and of epigenetic factors, we anticipate that the study of plant nuclear structure, genome architecture and gene regulation will play a very significant role in the near future.

As knowledge and understanding of the structure and properties of the nucleus and nuclear envelope expand, we come tantalizingly closer to understanding the origins of the structures of the eukaryotic cell. John Bryant (Chapter 1) uses the information presented together with knowledge of replication of nuclear DNA and the import of the replication proteins to present and develop current theories of the origins of the nucleus and its envelope. The early presence of the nucleus and nucleoskeleton, predating the arrival of chloroplasts and mitochondria in the proto-eukaryotic cell and the probable formation of the nuclear envelope from invaginations of the plasma membrane are discussed in the light of the development of key features of the higher plant nucleus. Just as we hope that presenting advances in understanding the structure and function of the plant nucleus will stimulate research in this field, it is equally our hope these advances will result in better appreciation of their origins not only in plants but across the orders of living things.

ACKNOWLEDGEMENTS

The editors wish to acknowledge the contribution of Professor J.S. (Pat) Heslop-Harrison and Dr Trude Schwarzacher, Department of Biology, University of Leicester, UK to the inception and development of this volume. They also wish to recognize the contribution of colleagues working on the nuclear envelope and on nuclear structure and function whose research provided inspiration for the work presented here. DE and KG acknowledge the support of the Leverhulme Trust under grant F/00 382/H and for an Early Career Fellowship for KG.

Annual Plant Reviews (2013) **46**, 1–18
doi: 10.1002/9781118472507.ch1 http://onlinelibrary.wiley.com

Chapter 1

INTRODUCTION: MYSTERIES, MOLECULES AND MECHANISMS

John A. Bryant
Biosciences, University of Exeter, Exeter, UK

Abstract: This brief chapter mentions the main structural and functional features of plant nuclei and in doing so, provides a very general introduction to other chapters in the book. It also covers aspects that are not featured elsewhere, especially the replication of nuclear DNA and the import of the replication proteins. Throughout the chapter there is an underlying theme of evolution, relating both to the similarities to and differences from the Archaea and to the possible evolutionary origins of the nucleus.

Keywords: Archaea; DNA replication; evolution; nuclear envelope; nuclear localization signal; origin; protein import

1.1 Darwin and Margulis revisited

In a famous letter sent in July 1879 to Joseph Hooker, the Director of Kew Gardens, Charles Darwin described the origin of the flowering plants as 'an abominable mystery'. Over 130 years later, the mystery seems to be solved, if not in detail, at least in general terms. It is now thought that flowering plants diverged from a lineage of seed ferns (now a totally extinct group) in the late Jurassic or early Cretaceous period (Doyle, 2006, 2008). Based on extensive phylogenetic analysis, the living plant that most resembles the earliest angiosperms (i.e. which is at the base of the angiosperm phylogenetic tree) is *Amborella trichopoda*, a semi-climbing shrub only found in the rain forests of New Caledonia. So, while a solution to that mystery has been found, a further, and perhaps more fundamental mystery remains. It is a

Annual Plant Reviews Volume 46: Plant Nuclear Structure, Genome Architecture and Gene Regulation, First Edition. Edited by David E. Evans, Katja Graumann and John A. Bryant.

mystery that involves not just flowering plants but all eukaryotes and at the beginning of the 21st century it is still not completely solved. That mystery is the origin of the nucleus, the organelle that is the subject of this book. As is evident in subsequent chapters, we have extensive knowledge of its structure and activities. It is a truly beautiful organelle – one that induces in many of us a sense of wonder. However, we are not at all sure where it came from although, as will become clear later in the chapter, a few hypotheses are beginning to emerge as front runners.

On the quest to solve the puzzle, one factor to consider is the origin of eukaryotes. It is now accepted that the two other major membrane-bound organelles, mitochondria and chloroplasts, have evolved from bacterial symbionts that invaded or were engulfed by what we could call proto-eukaryotes (as originally proposed by Margulis, 1971a, b, 1981). This idea has been extensively confirmed by genomic and proteomic studies, which also suggest strongly that those proto-eukaryotic host cells were derived from the Archaea and, in terms of energy metabolism, were using a form of glycolysis[1]. Further, it is clear that following the endosymbiotic events, transfer of genes from both the non-photosynthetic (i.e. mitochondrial) and the photosynthetic (chloroplastic) endosymbionts to the host's genome occurred on a large scale. Indeed, that the process is still going on (Huang *et al.*, 2004, Rousseau-Gueutin *et al.*, 2011, Wang *et al.*, 2012). But where, and in what state were the genomes of those proto-eukaryotic host organisms?

It was thought for several years that relevant information could be obtained by study of amitochondrial eukaryotes, eukaryotes presumed to date back to before the first endosymbiotic event. However, it is now known that these are secondarily amitochondrial, as revealed by the presence of endosymbiont-derived genes in the nucleus and the vestiges of a mitochondrion (e.g. van der Giezen and Tozar, 2005; Minge *et al.*, 2009). So, these cells cannot tell us what the proto-eukaryote looked like. Nevertheless, it is clear that in more recent instances of gene transfer (as mentioned above), the organelle gene has been integrated into a typical eukaryotic nuclear genome located in a typical eukaryotic nucleus. These structures are no hindrance to gene transfer. Further, the use of bioinformatics coupled with comparative cell physiology and biochemistry in attempts to 'root' the eukaryotic phylogenetic tree all lead to the conclusion that most of the approximately 60 differences between eukaryotes and prokaryotes were developed or developing before the first symbiotic event, the acquisition of mitochondria (de Duve, 2007; Margulis *et al.*, 2007; Cavalier-Smith, 2009).

The eukaryotic features possessed by the proto-eukaryotes are thought to have included the possession of a nucleus, nucleoskeleton and cytoskeleton (Margulis *et al.*, 2007; de Duve, 2007; Cavalier-Smith, 2009). Looking at the

[1]But note that in modern Archaea there are several variants of the 'conventional' glycolysis pathway (Sato and Atomi, 2011).

first two of these, these data do not provide any clear clues about where the nucleus came from and there are also questions about the nature of the nucleoskeleton in the earliest eukaryotes. Focussing specifically on this problem, we note that after the first symbiotic event (acquisition of mitochondria), the eukaryotic lineage split into two major branches (Cavalier-Smith, 2002), the unikonts (with one flagellum) that gave rise to, amongst other things, fungi and Metazoa, and the bikonts (with two flagella), one lineage of which became plants by the acquisition of chloroplasts (as mentioned above; see also Keeling, 2010).

Turning now to look at extant lineages, as is shown in Chapters 2 and 4, part of the nucleoskeleton in animals is the prominent lamina, consisting mainly of proteins known as lamins. However, plants lack lamins but do possess a lamina-like structure that has been called the 'plamina' (Fiserova *et al.*, 2009), consisting of plant-specific proteins that are functional analogues of lamins. Finally, in fungi, at least as represented by yeasts, the nucleoskeleton does not have any form of lamina. So, based on the origins of these groups, it is suggested that the proto-eukaryotic nucleoskeleton lacked a lamina and that this has developed subsequent to the uni-kont/bikont split. This gives us a little more information on the early nucleus, but the question of its origin remains.

At this point further specific discussion of the origin of nucleus is deferred to the end of the chapter, although it will appear more indirectly from time to time in the next three sections. Attention is now turned to the genome itself. Particular focus will be placed on the general structure of the genome, on its replication and on the implications for the latter process of enclosing the genome in an organelle.

1.2 Nuclei – general features

In plant cells that are not extensively vacuolated, the nucleus is the largest and usually the most obvious organelle. Even in mature cells with large vacuoles, the nucleus is usually clearly visible within the cytoplasm. It is the organelle that contains the bulk of the cell's DNA, the nuclear genome. Indeed, chromatin (Chapters 5 and 6), consisting mainly of a complex of DNA and proteins, is usually the most obvious component of the nucleus. The chromatin is attached *via* scaffold- or matrix-associated regions (SARs/MARs) to the nuclear matrix/scaffold/nucleoskeleton (Chapters 4 to 6). Within chromatin, the highly repeated genes encoding the major ribosomal RNAs (rRNAs) are looped out in structures called nucleoli. The fibrillar centres of the nucleoli are the sites of transcription of these genes and the transcripts are processed in the outer regions of the nucleoli.

The nucleus is bounded by the nuclear envelope or NE (Chapters 2 and 3), which consists, in effect, of three membranous components (shown diagrammatically in the cartoon in Figure 1.1). Firstly, the outer envelope is connected

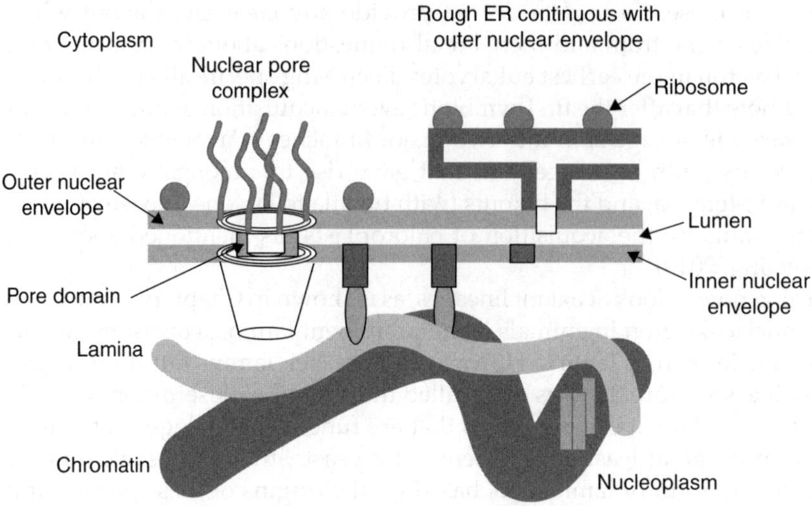

Figure 1.1 Diagrammatic cartoon of the nuclear envelope and nuclear pore complex. (From Evans *et al.*, 2004.) Reproduced by permission of the Society for Experimental Biology.

to the ER and the lipids and proteins of the outer NE are similar to those of the rough ER. Further, as with the rough ER, ribosomes are often present on the outer NE. So, the outer NE may be a site of protein synthesis and is certainly a part of the cell's endomembrane system. Secondly, there is the inner NE separated from the outer NE by the lumen, which is about 30 nm across. The inner surface of the inner NE is closely associated with the nuclear lamina, a structure consisting of filamentous proteins and which forms the main component of the nuclear matrix or nucleo-skeleton. Thirdly there is the pore membrane, which links the inner and outer NEs and forms part of the nuclear pore complex or NPC (Chapters 2, 4 and 8).

The containment of chromatin within its own membrane-bound organelle has major implications for the life of the cell. Amongst other things, it permits precise and complex regulation of gene activity and DNA replication 'protected' from more general aspects of cellular metabolism. However, it also imposes constraints. The nucleus does not contain protein-synthesizing machinery, even though proteins may be made on the surface of the outer NE. All the enzymes, together with structural and regulatory proteins necessary for the activities and components of the nucleus, over 1000 proteins in all (Nuclear Protein Database: http://npd.hgu.mrc.ac.uk/), must be able to get in from the outside. At the same time, several thousand more proteins, those that are not involved in the life of the nucleus, are kept out. There are also proteins that shuttle between the nucleus and the cytosol. Finally, all the different RNAs that function in the cytosol must leave the nucleus (in the form of nucleoprotein complexes). The NPCs have a major role in the control of entry into and exit from the nucleus (Chapters 4 and 8), along with specific signalling and transport mechanisms. This provides one more level of regulation of chromatin-associated biochemical activity.

1.3 The plant nuclear genome

1.3.1 General features

A recent review by Heslop-Harrison and Schwarzacher (2011) gives a wealth of information about plant nuclear genomes, while Chapters 5 and 6 in this volume deal with specific aspects of chromatin organization. In this chapter, the focus is on those features related to genome evolution and replication. Higher plant genomes vary enormously in overall size, ranging over three orders of magnitude. There is some correlation between genome size and nuclear size so that, in general, plants with large genomes have larger nuclei than plants with small genomes. Some of the differences in genome size have arisen by duplication of individual genes and of whole genomes (polyploidy). Within individual genomes, much of the DNA does not code for proteins or RNA. Comparison between closely related species (see, e.g. Bryant and Hughes, 2011) that have differing amounts of nuclear DNA show that most of the variation can be accounted for by repeated DNA sequences. Some of the variation is in the number of copies of repeated genes, such as those coding for rRNA but most of it is accounted for by variations in non-coding DNA. This includes highly repeated 'satellite' DNA of around 180 base pairs per repeat (Sibson *et al.*, 1991; Round *et al.*, 1997), simple sequence repeats and retrotransposons. 'Satellite' DNA sequences are concentrated at the centromeres where they appear to be essential for centromere function (Nagaki *et al.*, 2003 and Chapter 6, this volume). Retrotransposons or retro-elements include LINES (long interspersed sequences) and various highly repetitive sequences of different sizes and copy number. Taken together these retro-elements can make a very large proportion of the genome (for details, see Bryant and Hughes, 2011; Heslop-Harrison and Schwarzacher, 2011). There are also 'fossil' coding sequences or pseudogenes, some of which seem to have been derived from other species. All these types of sequence are present to some extent in all eukaryotes and have possibly been features of chromatin since the first appearance of the Eukarya. However, over the period in which higher plants have been evolving, their nuclear genomes have been and continue to be amongst the most dynamic.

As in all eukaryotes, the DNA itself exists as linear molecules, one long double helix per chromosome. In the chromosomes, the DNA is complexed with proteins, mainly the basic proteins known as histones, to form chromatin, as described in more detail in Chapters 5 and 6. Some chromatin (heterochromatin) remains condensed and therefore clearly visible throughout the cell cycle. As described in much more detail in Chapter 6, much of the heterochromatin is located at the centromeres (and thus involves 'satellite' DNA, mentioned above) and at the telomeres (ends of chromosomes). By contrast with heterochromatin, the majority of the chromatin, known as euchromatin, decondenses as mitosis is completed. The significance of these two patterns of behaviour and of the distribution heterochromatin is discussed in Chapter 6.

The linear structure of eukaryotic DNA molecules has caused some to question the origins of eukaryotic genomes (see Section 1.5). The consensus remains that the original proto-eukaryotic host cell was derived from the archaean lineage and yet amongst extant members of the Archaea, we do not yet know of any that have linear chromosomes. Nevertheless, DNA in many Archaea is complexed with histone-like proteins to form features that resemble the nucleosomes of eukaryotic chromatin (Pereira and Reeve, 1998; Sandman *et al.*, 2001), albeit that these nucleosomes contain only 80 base pairs of DNA wrapped round four, not eight histone molecules (see Chapter 5 for a detailed description of eukaryotic chromatin)[2]. Further, several eubacteria are known that have linear chromosomes, and some species are able to maintain both linear and circular DNA molecules (Volff and Altenbuchner, 2000; Lin and Moret, 2011), a situation that has also been described for mitochondria in some lower and higher plants and fungi (Borza *et al.*, 2009; Lo *et al.*, 2011). The absence of linear DNA molecules does not therefore rule out Archaea as being progenitors of or a sister group to the proto-eukaryotes.

1.3.2 Replication of the nuclear genome

The general features of eukaryotic genomes raise several problems for DNA replication. These have been discussed more fully in an earlier publication (Bryant, 2010) but need to be mentioned briefly here. The first is that the complex of DNA and protein in chromatin (which is of course common to all eukaryotes) means that copying the DNA is slower than in prokaryotes. Taking this together with the length of eukaryotic DNA molecules, especially in some plant genomes, has led to the evolution of multiple replication origins (points at which replication may start) along the axis of each DNA molecule. The nature of these replication origins in relation to DNA structure has been a matter for debate for many years (see e.g. Hernández *et al.*, 1988; Van't Hof and Lamm, 1992; Bryant and Francis, 2008; Bryant, 2010; Lee *et al.*, 2010; see also Berbenetz *et al.*, 2010) and it is still not clear whether or not specific sequences are involved. What is clear, however, is that origins are AT-rich and are therefore more prone to transient, localized short-range strand separation

[2]It must also be noted that many, if not all, Archaea possess a different type of DNA-binding protein, known as Alba, which is also able to generate a form of chromatin in which the DNA is inside the protein (Tanaka *et al.*, 2012) – i.e. very different from the nucleosome structure.

known as 'breathing'. It is also clear that the timing and order in which the origins are active ensures that the DNA is replicated within an S-phase that is completed within a few hours. There are again links with the Archaea in that several species have more than one replication origin (usually two or three: Lundgren and Bernander, 2005; Robinson and Bell, 2005), which seem to be attached to specific locations at the cell's periphery (Gristwood *et al.*, 2012). Like the replication origins of plants (and other eukaryotes), archaean origins are AT-rich and, in this group, specific sequence is important for correct function (Majernik and Chong, 2008).

Finally, the uni-directional (5′ to 3′) nature of DNA replication, coupled with the inability of DNA polymerases to initiate replication without a primer (see Bryant, 2010) causes problems at the ends of molecules. This has led to the development during evolution of specialized structures called telomeres at the ends of chromosomal DNA molecules, with an associated enzyme, telomerase[3]. As discussed in Chapter 7, the ends are protected from degradation and from being inappropriately targeted by the DNA repair machinery (see Chapter 5) because of the telomere/telomerase combination.

The enzymes and other proteins which carry out replication of nuclear DNA in plants have been described in some detail in two recent papers (Schultz *et al.*, 2007; Bryant, 2010). Here the focus is on a selection of those aspects that provide clues about evolution. Looking first at pre-replication events (see e.g. Aves, 2009; Bryant and Aves, 2011), it is clear that in plants, as in other eukaryotes, replication origins are bound and therefore 'marked' by a complex of six proteins, the Origin-Recognition Complex (ORC) (Collinge, *et al.*, 2004; Mori *et al.*, 2005; Shultz *et al.*, 2007; Bryant 2010). A protein known as CDC6, along with CDT1, then facilitates the loading of the CMG complex consisting of the GINS hetero-tetramer, MCMs 2-7 (the helicase that separates the two strands of DNA at the replication fork) and the protein known as CDC45, which will later facilitate loading of the initiating DNA polymerase. Looking at the Archaea, it is clear that both recognition of replication origins and the first stage in their activation are carried out by a single protein that is both similar to and fulfils the functions of the ORC and CDC6 (reviewed by Bryant and Aves, 2011). The function of the GINS complex is carried out either by a homo-tetramer or tetramer consisting of two different homo-dimers (Yoshimochi *et al.*, 2008). Sequence similarity to eukaryotic GINS proteins is limited but the proteins interact to form a complex of similar architecture to the eukaryotic complex. With respect to MCMs, most Archaea have just one, which forms a homohexamer as compared with the eukaryotic heterohexamer. Hints of multiple MCMs are seen in *Thermococcus kodakarensis* (Pan *et al.*, 2011), which has three. Both MCMs 2 and 3 can form homohexamers but only MCM3 is essential for DNA replication.

[3]See also the discussion on linear DNA molecules in mitochondria (e.g. Valach *et al.*, 2011).

Table 1.1 Comparison between Archaea and Eukarya in respect of proteins involved in the initiation of DNA replication

Function	Proteins in Archaea	Proteins in Eukarya
Origin recognition	Single Origin-Recognition Protein/CDC6	Complex of six different Origin-Recognition Proteins
Loading of pre-replicative complex onto origin	The same ORP/CDC6	CDC6 and CDT1
Helicase accessory proteins (GINS complex)	Either a homo-tetramer or two homo-dimers make up the complex	A hetero-tetramer
Replicative DNA helicase	A homo-hexamer of one type of MCM protein	A hetero-hexamer of six different MCM proteins
Polymerase loading factor	Member of the Rec-J protein family	CDC45

Finally, CDC45 is represented in Archaea by a member of the Rec J protein family, many of which have nuclease activity. What emerges from this is that heteromeric protein complexes involved in plant DNA replication are represented by homo-polymers in Archaea, or, as expressed by Bryant and Aves (2011), '*The proteins themselves represent the essential core, compared with the eukaryotic plenitude.*' Thus, individual proteins in Archaea are now represented by multiple versions in plants and other eukaryotes. Using the MCMs as an example, a detailed bioinformatic study (Lui *et al.*, 2009) indicates that the different MCMs have arisen by a series of gene duplication events and, by analogy, we can envisage a similar series of gene duplications giving rise to six ORC proteins plus CDC6 (Table 1.1). Further, the analysis carried out by Lui *et al.* (2009) suggests that these gene duplications are very ancient and were already present in the last common ancestor of all eukaryotes. Further, without going into detail here, similar conclusions have been reached from a study of transcription factors (Bell *et al.*, 2001).

After preparation of the DNA for replication, synthesis itself is initiated by DNA primase and the primers are then extended by DNA polymerase-α. This is the only DNA polymerase that can work with primase and, indeed, the two enymes form a complex, as described in previous publications (Bryant *et al.*, 1992; Bryant *et al.*, 2001; Bryant, 2010). Replication within each replicon (i.e. the tract of DNA replicated from one origin) is then completed by DNA polymerase-δ and DNA polymerase-ε as described in much greater detail in Bryant (2010). This also involves a nuclease to remove the primers (either a 'flap nuclease' such as Fen 1 or ribonuclease-H) and DNA ligase to join pieces of newly synthesized DNA (Bryant, 2010). Focussing specifically on the three replicative polymerases, these are all members of the B family of DNA polymerases and perhaps by now it is not surprising to learn that the replicative DNA polymerase of Archaea is also a member of this family.

For a much fuller description of DNA replication, of the enzymes and other proteins involved in replication and of the regulation of replication, the reader is referred to the comprehensive papers mentioned above. At this point in this chapter, it is time to move on to consider the implications for DNA replication and gene function of enclosing the genetic material in a sub-cellular compartment.

1.4 DNA inside, ribosomes outside

The enclosure of the genetic material inside a membrane-bound organelle has many advantages. It allows localized control mechanisms to operate inside the organelle, with that control being partly exerted by the inside-outside segregation itself. It frees the cell's gene expression and gene regulation machinery from the possibility of unwanted cross-reactivity with other aspects of metabolism. The division of labour between the sub-cellular organelles, including the nucleus, is held to have made multi-cellularity possible. However, this compartmentation (and here, obviously, focus is on just one of the organelles, the nucleus) also raises clear difficulties and especially that the enzymes and other proteins required for chromatin structure, DNA replication, gene expression and so on are made on ribosomes located in the cytosol. Further, once they have been synthesized by transcription from the relevant genes, RNA molecules must be able to fulfil their various functions in protein synthesis. These RNA molecules do not just diffuse out of the nucleus but are transported out in the form of nucleo-protein complexes. The proteins involved in this are thus able to both enter and leave the nucleus (and this is true of some proteins involved in DNA replication, in signalling and in control of gene expression).

The need to enable specific proteins to enter and leave the nucleus has led to the evolution of a sorting mechanism in which proteins destined for the inside of the nucleus contain within their amino acid sequence a label, the nuclear localization signal or NLS. In proteins that are required to come out again there is also a nuclear export signal or NES. In nuclei as they are now constituted, the gatekeeper between the cytosol and the nucleus is the nuclear pore complex (NPC) with import machinery based on importins and a RAN-GTPase (as described in Chapters 3 and 4). We have no clear picture of what a 'proto-nucleus' might have looked like in respect of NPCs. However, as indicated earlier, it is widely held that the engulfment/symbiosis that led to the development of mitochondria involved a host cell with many of the special eukaryotic features, including some form of nucleo-skeleton. This is somewhat confirmed by results from a genomic analysis of 19 (out of a total of about 30) nuclear pore complex proteins (Nups) across 60 different eukaryotes from a wide range of groups. The analysis indicates that all the major sub-complexes of proteins in the NPC are traceable as far back as the Last Eukaryotic Common Ancestor or LECA (Neumann *et al.*, 2010).

Clues to the development of these, perhaps in a lineage leading to the LECA, comes from a study of protein architecture and folding in Nups, leading to the suggestion that extensive gene duplication and motif duplication within genes has led to the development of the range of Nups from a small number of precursor proteins (Devos *et al.*, 2006).

At this point, attention is focussed purely on proteins imported into the nucleus (the same principles apply to those that must later leave the nucleus). In a wide-ranging study of proteins that are required to enter nuclei (Dingwall and Laskey, 1991), several different sequences were identified that were involved in nuclear uptake. Three of these are found in plants, namely:

- the SV-40 virus-type monopartite NLS, which consists of a run of five basic amino acids (named for the virus in which it was first found);
- the yeast Matα2 type of NLS in which a run of four basic amino acids is interrupted by three hydrophobic amino acids (named after a yeast mating type protein);
- the bi-partite NLS, consisting of two short regions containing basic amino acids (*i.e.* arginine, histidine or lysine) separated by a spacer of up to 10 amino acids (Figure 1.2).

The last of these is by far the commonest in plants. In different proteins, the NLS occurs at different places within the amino acid sequence although many of them are at or near the N-terminus. Presumably what matters is that, in the folded protein, the NLS is 'visible' to the import machinery. The efficacy of a bi-partite NLS is illustrated in Figure 1.3.

Two more things need to be said about nuclear import. Firstly, there are some nuclear proteins that appear to lack completely an NLS (Stiedl *et al.*, 2004) and are thus 'piggy-backed' into the nucleus on an NLS-containing protein (Stiedl *et al.*, 2004; see also Galstyan *et al.*, 2012).

Secondly, there is an increasing list of proteins that appear to have roles in the cytosol and in the nucleus, for example in both glycolysis and in DNA replication or repair. Such dual-function proteins are known as 'moonlighting' proteins. A typical example is phosphoglycerate kinase (PGK), with a primary function in glycolysis but which also enters the nucleus (Anderson *et al.*, 2004; Brice *et al.*, 2004). *In vitro* it stimulates the activity of DNA polymerase-α on poorly primed templates (Burton *et al.*, 1997; Bryant *et al.*, 2000; Bryant, 2008) and it has been suggested that it acts as an accessory protein for the polymerase (Bryant *et al.*, 2000; Bryant, 2008). Concomitant with a

MAT**KRSVGTLKEADLKGKRV**FVR

Figure 1.2 A typical plant bi-partite nuclear localization, situated near the N-terminus. The NLS itself is shown in bold type with the basic residues that are part of the signal sequence underlined (note that in this example there are also two basic amino acids in the intervening sequence between the two halves of the signal).

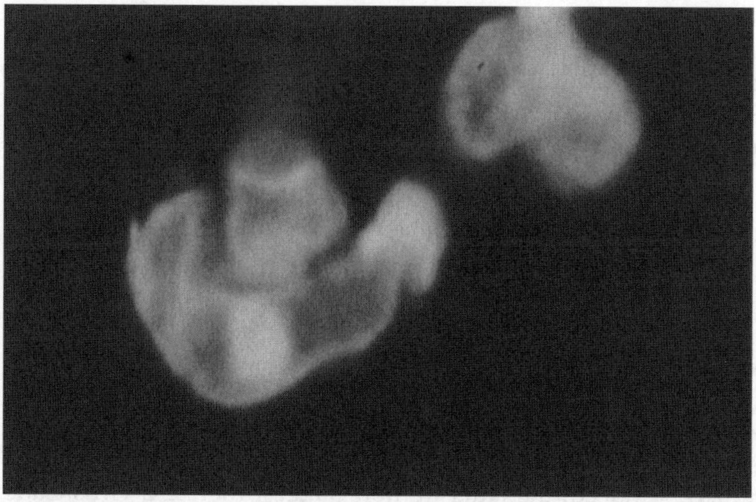

Figure 1.3 A transcriptional fusion was made between the coding sequences for Green Fluorescent Protein (GFP) and the moonlighting protein PGK which has an NLS. The hybrid protein was transported into the nucleus of tobacco cells, as is shown by the bright fluorescence (from Brice *et al.*, 2004).

role in the nucleus, PGK possesses an NLS located at the N-terminus (Brice *et al.*, 2004 and Figure 1.3) and an important question relating to this and, indeed, to all proteins with dual cytosolic/nuclear locations is how they are actually partitioned between the two sites. For PGK, submitting the sequence of the protein to the PSORTII program suggests that *in vivo*, 12% of the protein population will be transported into the nucleus (Bryant, 2008; see also Weis, 2003; Terry *et al.*, 2007; Meier and Somers, 2011). Another question concerns the evolution of moonlighting proteins. Do their dual roles represent an earlier stage in evolution in which a given protein might carry out more than one function? If so, it is interesting that the apparently minor role (as with PGK in the nucleus) has led to the acquisition of an NLS.

1.5 Concluding comments on the evolution of the nucleus

At the beginning of the chapter, reference was made to the mystery surrounding the evolutionary origin of the nucleus. The picture we have of the proto-eukaryote that became the host for the mitochondrial symbiont is certainly of a cell with a nucleus. Further, that picture is becoming clearer and clearer as a result of studies in various aspects of bio-informatics, some of which have been mentioned in this chapter. In analysis of genes and proteins involved in the function of the nucleus, for example, in DNA replication, the Archaea continue to dominate: the archaeal origin of many eukaryotic nuclear enzymes seems very clear (reviewed by Bryant and Aves, 2011). The same picture emerges from analyses based on histones (Pereira and Reeve, 1998), on ribosomal RNA (Xie *et al.*, 2012) and from wider-ranging genomic analyses (Saruhashi *et al.*, 2008). But that still does not tell us how an archaeal cell became a 'proto-eukaryote' with a nucleus. Currently there are three main theories.

Firstly, it has been suggested that an archaeal cell was invaded by an enveloped virus with a linear genome (Bell, 2001, 2006; Villareal and Witzany, 2010). The authors cite the similarity between replicative DNA polymerases encoded by certain DNA viruses and the B-family of replicative DNA polymerases in Archaea and Eukarya (see above). Even so, this may seem far fetched because it is generally held that viruses are a later addition to the rich variety of biology. However, a recent paper in which the ancestry of viruses was investigated through analysis of protein architecture and folding suggests that 'giant' viruses appeared earlier than previously thought (Nasir *et al.*, 2012). Indeed, the latter authors suggest rooting a virus clade alongside the base of the Eukarya. However, we have already seen that linear genomes may arise from circular genomes, so it is not necessary to invoke a virus in order to provide a linear genome. Further, there is the question of the integration of the viral genome into that of the host. Although, based on examples of extant viruses, this could certainly happen, bio-informatic analysis of eukaryotic and archaeal genes (e.g. Lui *et al.*, 2009) gives no indication of a viral ancestry for the proteins involved in the initiation of DNA replication.

Secondly it has been proposed that the nucleus arose through a symbiosis between an archaeal cell and a eubacterial cell, with the eubacterial cell as host and the archaeal cell as invading symbiont (Gupta and Golding, 1996; Ohyanagi *et al.*, 2008). This is clearly different from the view that that the progenitor of the Eukarya was an actual archaean but it does allow for the presence of archaean features. According to this view, the nucleus represents an archaeal symbiont that either took over the genetic function of the host or that contributed to a larger genome derived from both cells. The genomic and rRNA analyses mentioned above are consistent with a dominant archaeal genome in such a symbiosis although analysis of ribosome export factors suggests the possibility of a significant contribution from a eubacterial genome (Ohyanagi *et al.*, 2008).

Finally, some authors propose that there is no need to invoke any sort of invasion or symbiosis to explain the origin of the nucleus. This was firmly stated by Martin (1999); more recent evidence provides some support for this view. The discovery that replication origins in Archaea appear to be attached at the cell's periphery (Gristwood *et al.*, 2012) is consistent with the idea that the archaeal chromatin first became attached to the cell membrane and then was enveloped by invaginations of the membrane, possibly in connection with phagocytosis (Cavalier-Smith, 2009; see also Cavalier-Smith, 2010). Development of a nuclear skeleton is then presumed to have followed. An extension of this view is that surrounding the genetic material with a membrane was a protective measure, evolved in response to the presence of reactive oxygen species that became abundant after the evolution of photosynthesis and the 'great oxidation event.' (Gross and Bhattacharya, 2010). Construction of the nucleus from within an archaeal cell is actually the simplest of the three hypotheses and fits better with the majority of the genomic analyses. It is the hypothesis that is favoured by the present author. Hopefully, time will tell.

References

Anderson, L.E., Bryant, J.A. and Carol, A.A. (2004) Both chloroplastic and cytosolic phosphoglycerate kinase isozymes are present in the pea leaf nucleus. *Protoplasma* **223**,103–110.

Aves, S.J. (2009) DNA replication initiation. *Methods in Molecular Biology* **521**, 3–17.

Bell, P.J.L. (2001) Viral eukaryogenesis: was the ancestor of the nucleus a complex DNA virus? *Journal of Molecular Evolution* **53**, 251–256.

Bell, P.J.L. (2006) Sex and the eukaryotic cell cycle is consistent with a viral ancestry for the eukaryotic nucleus. *Journal of Theoretical Biology* **243**, 54–63.

Bell, S.D, Magill, C.P and Jackson, S.P (2001) Basal and regulated transcription in Archaea. *Biochemical Society Transactions* **29**, 392–395.

Berbenetz, N.M., Nislow, C. and Brown, G.W. (2010) Diversity of eukaryotic DNA replication origins revealed by genome-wide analysis of chromatin structure. *PLOS Genetics* **6**. doi: 10.1371/journal.pgen.1001092.

Borza, T., Redmond, E.K., Laflamme, M. and Lee, R.W. (2009) Mitochondrial DNA in the Oogamochlamys clade (Chlorophyceae): high GC content and unique genome architecture for green algae. *Journal of Phycology* **45**, 1323–1334.

Brice, D.C., Bryant, J.A., Dambrauskas, D. *et al.* (2004) Cloning and expression of cytosolic phosphoglycerate kinase from pea (*Pisum sativum* L.). *Journal of Experimental Botany* **55**, 955–956.

Bryant, J.A. (2008) Copying the template: with a little help from my friends? In: Bryant, J.A. and Francis, D. (eds) *The Eukaryotic Cell Cycle*. Taylor & Francis, Abingdon, pp. 71–80.

Bryant, J.A. (2010) Replication of Nuclear DNA. *Progress in Botany* **71**, 25–60.

Bryant, J.A. and Aves, S.J. (2011) Initiation of DNA replication: functional and evolutionary aspects. *Annals of Botany* **107**, 1119–1126.

Bryant, J.A., Brice, D.C., Fitchett, P.N. and Anderson, L.E. (2000) Novel DNA-binding protein associated with DNA polymerase-α in pea stimulates polymerase activity on infrequently primed templates. *Journal of Experimental Botany* **51**, 1945–1947.

Bryant, J.A., Fitchett, P.N., Hughes, S.G. and Sibson, D.R. (1992) DNA polymerase-α in pea is part of a large multiprotein complex. *Journal of Experimental Botany* **43**, 31–40.

Bryant, J.A. and Francis, D. (2008) Initiation of DNA replication. In: Bryant, J.A. and Francis, D. (eds) *The Eukaryotic Cell Cycle*. Taylor & Francis, Abingdon, pp 29–44.

Bryant, J.A. and Hughes, S.G. (2011) Vicia. In: Cole, C. (ed.) *Wild Crop Relatives: Genomic and Breeding resources. Legume Crops and Forages*, Springer, Berlin, pp 273–289.

Bryant, J.A., Moore, K.A. and Aves, S.J. (2001) Origins and complexes: the initiation of DNA replication. *Journal of Experimental Botany* **52**,193–202.

Burton, S.K., Bryant, J.A. and Van't Hof, J (1997) Novel DNA-binding characteristics of a protein associated with DNA polymerase-α in pea. *Plant Journal* **12**, 357–365.

Cavalier-Smith, T. (2002) The phagotrophic origin of eukaryotes and phylogenetic classification of protozoa. *International Journal of Systematic and Evolutionary Microbiology* **52**, 297–354.

Cavalier-Smith, T. (2009) Predation and eukaryote cell origins: A coevolutionary perspective. *International Journal of Biochemistry and Cell biology* **41**, 307–322.

Cavalier-Smith, T. (2010) Origin of the cell nucleus, mitosis and sex: roles of intracellular coevolution. *Biology Direct* **5**, Article Number: 7 doi: 10.1186/1745-6150-5-7.

Collinge, M.A., Spillane, C., Kohler, C. *et al.* (2004) Genetic interaction of an origin recognition complex subunit and the Polycomb group gene MEDEA during seed development. *Plant Cell* **16**, 1035–1046.

de Duve, C. (2007) The origin of eukaryotes – a reappraisal. *Nature Reviews Genetics* **8**, 395–403.

Devos, D., Dokudovskaya, S., Williams, R. *et al.* (2006). Simple fold composition and modular architecture of the nuclear pore complex. *Proceedings of the National Academy of Sciences, USA.* **103**, 2172–2177.

Dingwall, C. and Laskey, R.A. (1991) Nuclear targeting sequences – a consensus. *Trends in Biochemical Sciences* **16**, 478–481.

Doyle, J.A. (2006) Seed ferns and the origin of angiosperms. *Journal of the Torrey Botanical Society* **133**, 169–209.

Doyle, J.A. (2008) Integrating molecular phylogenetic and paleobotanical evidence on origin of the flower. *International Journal of Plant Sciences* **169**, 816–843.

Evans, D.E., Bryant, J.A. and Hutchison, C.J. (2004) The nuclear envelope: a comparative overview. In Evans, D.E., Hutchison, C.J. and Bryant, J.A. (eds), *The Nuclear Envelope*, BIOS, Oxford, pp. 1–8.

Fiserova, J., Kiseleva, E. and Goldberg, M.W. (2009) Nuclear envelope and nuclear pore complex structure and organization in tobacco BY-2 cells. *Plant Journal* **59**, 243–255.

Galstyan, A., Bou-Torrent, J., Roig-Villanova, I. and Martinez-Garcia, J.F. (2012) A dual mechanism ontrols nuclear localization in the atypical basic-helix-loop-helix protein PAR1 of *Arabidopsis thaliana*. *Molecular Plant* **5**, 669–677.

Gristwood, T., Duggin, I.G., Wagner, M. *et al.* (2012) The sub-cellular localization of *Sulfolobus* DNA replication. *Nucleic Acids Research* **40**, 5487–5496.

Gross, J. and Bhattacharya, D. (2010) Uniting sex and eukaryote origins in an emerging oxygenic world. *Biology Direct* **5** Article Number: 53, doi: 10.1186/1745-6150-5-53.

Gupta, R.S. and Golding, G.B. (1996) The origin of the eukaryotic cell. *Trends in Biochemical Sciences* **21**, 166–171.

Hernández, P., Bjerknes, C.A., Lamm, S.S. and Van't Hof, J. (1988) Proximity of an *ARS* consensus sequence to a replication origin of pea (*Pisum sativum*). *Plant Molecular Biology* **10**, 413–422.

Heslop-Harrison, J.S. and Schwarzacher, T. (2011) Organisation of the plant genome in chromosomes. *Plant Journal* **66**, 18–33.

Huang, C.Y., Ayliffe, M.A. and Timmis, J.N. (2004) Simple and complex nuclear loci created by newly transferred chloroplast DNA in tobacco. *Proceedings of the National Academy of Sciences,USA* **101**, 9710–9715.

Keeling, P.J. (2010) The endosymbiotic origin, diversification and fate of plastids. *Philosophical Transactions of the Royal Society, Biological Sciences* **365**, 729–748.

Lee, T.-L., Pascuzzi, P.E., Settlage, S.B. *et al.* (2010) Arabidopsis thaliana chromosome 4 replicates in two phases that correlate with chromatin state. *PLOS Genetics* **6** Article Number: e1000982 doi: 10.1371/journal.pgen.1000982.

Lin, Y. and Moret, B.M.E. (2011) A new genomic evolutionary model for rearrangements, duplications, and losses that applies across eukaryotes and prokaryotes. *Journal of Computational Biology* **18**, 1055–1064.

Lo, Y.-S., Hsiao, L.-J., Cheng, N. *et al.* (2011) Characterization of the structure and DNA complexity of mung bean mitochondrial nucleoids. *Molecules and Cells* **31**, 217–224.

Lui, Y., Richards,T.A. and Aves, S.J. (2009) Ancient diversification of eukaryotic MCM DNA replication proteins. *BMC Evolutionary Biology* **9** Article Number: 60 doi: 10.1186/1471-2148-9-60.

Lundgren, M. and Bernander, R. (2005) Archaeal cell cycle progress. *Current Opinion in Microbiology* **8**, 662–668.

Majernik, A.I. and Chong, J.P.J. (2008) A conserved mechanism for replication origin recognition and binding in Archaea. *Biochemical Journal* **409**, 511–518.

Margulis, L. (1971a) Symbiosis and evolution. *Scientific American* **225**, 48–57.

Margulis, L. (1971b) Origin of plant and animal cells. *American Scientist* **59**, 230–235.

Margulis, L. (1981) *Symbiosis in Cell Evolution: Life and Its Environment on the Early Earth.* W.H. Freeman, New York.

Margulis, L., Chapman, M.J. and Dolan, M.F. (2007) Eukaryosis: phagocytosis and hydrogenases. *Symbiosis* **43**, 161–163.

Martin, W. (1999) A briefly argued case that mitochondria and plastids are descendants of endosymbionts, but that the nuclear compartment is not. *Proceedings of the Royal Society, Series B* **266**, 1387–1395.

Meier, I. and Somers, D.E. (2011) Regulation of nucleocytoplasmic trafficking in plants. *Current Opinion in Plant Biology* **14**, 538–546.

Minge, M.A., Silberman, J.D., Orr, R.J.S. *et al.* (2009) Evolutionary position of breviate amoebae and the primary eukaryote divergence. *Proceedings of the Royal Society B-Biological Sciences* **276**, 597–604.

Mori, Y., Yamamoto, T., Sakaguchi, N. *et al.* (2005) Characterization of the origin-recognition complex (ORC) from a higher plant, rice (*Oryza sativa* L.) *Gene* **353**, 23–30.

Nagaki, K., Talbert, P.B., Zhong, C.X. *et al.* (2003) Chromatin immunoprecipitation reveals that the 180-bp satellite repeat is the key functional DNA element of *Arabidopsis thaliana* centromeres. *Genetics* **163**, 1221–1225.

Nasir, A., Kim, K.M. and Caetano-Anolles, G. (2012) Giant viruses coexisted with the cellular ancestors and represent a distinct supergroup along with super-kingdoms Archaea, Bacteria and Eukarya. *BMC Evolutionary Biology* **12**, 156. Epub ahead of print PMID:22920653.

Neumann, N., Lundin, D. and Poole, A.M. (2010) Comparative genomic evidence for a complete nuclear pore complex in the Last Eukaryotic Common Ancestor. *PLOS ONE* **5**, Article Number: e13241 doi: 10.1371/journal.pone.0013241.

Ohyanagi, H., Ikeo, K. and Gojobori, T. (2008) The origin of nucleus: Rebuild from the prokaryotic ancestors of ribosome export factors. *Gene* **423**, 149–152.

Pan, M., Santangelo, T.J., Li, Z. *et al.* (2011) *Thermococcus kodakarensis* encodes three MCM homologs but only one is essential. *Nucleic Acids Research* **39**, 9671–9680.

Pereira, S.L. and Reeve, J.N. (1998) Histones and nucleosomes in Archaea and Eukarya: a comparative analysis. *Extremophiles* **2**, 141–148.

Robinson, N.P. and Bell, S.D. (2005) Origins of DNA replication in the three domains of life. *FEBS Journal* **272**, 3757–3766.

Round, E.K., Flowers, S.K. and Richards, E.J. (1997) *Arabidopsis thaliana* centromere regions: Genetic map positions and repetitive DNA structure. *Genome Research* **7**, 1045–1053.

Rousseau-Gueutin, M., Ayliffe, M.A. and Timis, J.N. (2011) Conservation of plastid sequences in the plant nuclear genome for millions of years facilitates endosymbiotic evolution. *Plant Physiology*, **157**, 2181–2193.

Sandman, K., Soares, D. and Reeve, J.N. (2001) Molecular components of the archaeal nucleosome. *Biochimie* **83**, 277–281.

Saruhashi, S., Hamada, K., Miyata, D. *et al.* (2008) Comprehensive analysis of the origin of eukaryotic genomes. *Genes and Genetic Systems* **83**, 285–291.

Sato, T. and Atomi, H. (2011) Novel metabolic pathways in Archaea. *Current Opinion in Microbiology* **14**, 307–314.

Shultz, R.W., Tatineni, V.M., Hanley-Bowdoin, L. and Thompson, W.F. (2007) Genome-wide analysis of the core DNA replication machinery in the higher plants Arabidopsis and rice. *Plant Physiology* **144**, 1697–1714.

Sibson, D.R., Hughes, S.G., Bryant, J.A. and Fitchett, P.N. (1991) Sequence organization of simple, highly repetitive DNA elements in *Brassica* species. *Journal of Experimental Botany* **42**, 243–249.

Stiedl, S., Tuncher, A., Goda, H. *et al.* (2004) A single subunit of a heterotrimeric CCAAT-binding complex carries a nuclear localization signal: Piggy back transport of the pre-assembled complex to the nucleus. *Journal of Molecular Biology* **342**, 515–524.

Tanaka, T., Padavattan, S. and Kumarevel, T. (2012) Crystal structure of Archaeal chromatin protein Alba2-double-stranded DNA complex from *Aeropyrum pernix* K1. *Journal of Biological Bhemistry* **287**, 10394–10402.

Terry, L.J., Shows, E.B. and Wente, S.R. (2007) Crossing the nuclear envelope: hierarchical regulation of nucleocytoplasmic transport. *Science* **318**, 1412–1416.

Valach, M., Farkas, Z., Fricova, D. *et al.* (2011) Evolution of linear chromosomes and multipartite genomes in yeast mitochondria. *Nucleic Acids Research* **39**, 4202–4219.

van der Giezen, M. and Tozar, J. (2005) Degenerate mitochondria. *EMBO Reports* **6**, 525–530.

Van't Hof, J. and Lamm, S.S. (1992) Site of initiation of replication of the ribosomal RNA genes of pea (*Pisum sativum*) detected by 2-dimensional gel electrophoresis. *Plant Molecular Biology* **20**, 377–382.

Villarreal, L.P. and Witzany, G. (2010) Viruses are essential agents within the roots and stem of the tree of life. *Journal of Theoretical Biology* **262**, 698–710.

Volff, J.N. and Altenbuchner, J. (2000) A new beginning with new ends: linearisation of circular chromosomes during bacterial evolution. *FEMS Microbiology Letters* **186**, 143–150.

Wang, D., Lloyd, A.H. and Timmis, J.N. (2012) Environmental stress increases the entry of cytoplasmic organellar DNA into the nucleus in plants. *Proceedings of the National Academy of Sciences, USA* **109**, 2444–2448.

Weis, K. (2003) Regulating access to the genome: nucleocytoplasmic transport throughout the cell cycle. *Cell* **112**, 441–451.

Xie, Q., Wang, Y.H., Lin, J.Z. *et al.* (2012) Potential key bases of ribosomal RNA to kingdom-specific spectra of antibiotic susceptibility and the possible archaeal origin of eukaryotes. *PLOS ONE* **7** Article Number: e29468 doi: 10.1371/journal.pone.0029468.

Yoshimochi, T., Fujikane, R., Kawanami, M., Matsunaga, F. and Ishino, Y. (2008) The GINS complex from *Pyrococcus furiosus* stimulates the MCM helicase activity. *Journal of Biological Chemistry* **283**, 1601–1609.

Annual Plant Reviews (2013) **46**, 19–56 http://onlinelibrary.wiley.com
doi: 10.1002/9781118472507.ch2

Chapter 2

THE NUCLEAR ENVELOPE – STRUCTURE AND PROTEIN INTERACTIONS

Katja Graumann and David E. Evans
Department of Biological and Medical Sciences, Faculty of Health and Life Sciences, Oxford Brookes University, Headington Campus, Oxford OX3 0BP, UK

Abstract: The proteins of the nuclear envelope have been characteriszd in detail in non-plant systems, revealing a complex series of interactions between the inner and outer nuclear envelope, with the proteins of the nuclear pore complex and connecting to the nucleoskeleton and cytoskeleton. Many of the proteins described in animal and fungal systems have not been identified in plants; however, recent advances have resulted in the description of several key plant nuclear envelope proteins including members of the Sad1/UNC84 (SUN) domain family and recently the first Klarsicht/Anc-1/Syne-1 Homology (KASH)-domain protein. This chapter describes the emerging networks of protein interactions at the plant nuclear envelope, together with discussion of their likely functions. In particular, the important role of plant SUN domain proteins as a key component of a linker of nucleoskeleton and cytoskeleton complex is described together with prospects for future study of plant nuclear envelope interaction networks.

Keywords: nuclear envelope; INM; ONM; nucleo-cytoskeletal bridging; protein targeting; cell division; pathogen response

2.1 Introduction

The nuclear envelope (NE) plays a fundamental role in eukaryotic cells. By separating the contents of the nucleus from the cytoplasm and by gatekeeping the traffic into and out of the nucleus, it enables the functioning of a complex and sophisticated genome. In consequence, it has been the subject of considerable interest and a substantial knowledge base has been created

Annual Plant Reviews Volume 46: Plant Nuclear Structure, Genome Architecture and Gene Regulation, First Edition. Edited by David E. Evans, Katja Graumann and John A. Bryant.

for animal and yeast cells. This chapter aims to review the current state of knowledge for plants, where recent discoveries are opening the field for rapid advance.

2.2 Organization and structure of the plant nuclear envelope

Plant nuclei are variable in shape and size and in vegetative tissue are often compressed into a narrow region of cytoplasm by a large central vacuole. However, in mitotic and undifferentiated cells, the nucleus is often rounded or ovoid, central and capable of movement within the cell cytoplasm. As in metazoans, chromatin is organized into regions of hetero- and euchromatin and there are great similarities in the overall organization of the internal structure of the nucleus with heterochromatin located adjacent to the NE. The shape and structure of the plant nucleus as well as its position are likely to be maintained by the interaction of the envelope with chromatin and the nucleoskeleton and with the cytoskeleton, just as in metazoans. The plant NE is also structurally similar to those of other kingdoms, with three identifiable domains: the outer nuclear membrane (ONM), the inner nuclear membrane (INM) and the pore membrane (Figure 2.1; Brandizzi *et al.*, 2004). The ONM is linked to the perinuclear endoplasmic reticulum (PNER), functions in protein synthesis and is coated with ribosomes. The proteomes of the PNER and ONM are not identical and the junction region between the two membranes (diameter 2–30 nm) is believed to limit protein movement (Craig and Staehelin, 1988; Staehelin, 1997). Connections between the ONM and cytoskeleton are an important feature of the nucleus. Unlike higher metazoans, the entire plant ONM serves as a microtubule-organizing centre (MTOC) and plants lack centrosomes (Shimamura *et al.*, 2004).

The second domain is the pore domain or pore membrane. As well as connecting INM and ONM and providing the environment in which the nuclear pore complexes (NPCs) are anchored, it helps to maintain the protein composition of the INM as membrane proteins entering the INM must pass through the periphery of the NPC, many requiring a nuclear targeting signal (King *et al.*, 2006; Lusk *et al.*, 2007). The structure and function of the NPCs and their relationship with the pore domain are dealt with in chapter 3. The third domain, the INM, is closely associated with chromatin via numerous protein connections as well as with the ONM by protein bridges spanning the periplasm. In addition, a nucleoskeletal meshwork, termed the lamina in animals, underlies the INM, thereby supporting the structure and shape of the NE and the entire nucleus. To date, only two proteins have been characterized at the plant INM – Sad1/UNC84 (SUN) domain proteins. Hence the protein components that mediate most of the anchorage and functions at the plant INM remain to be identified (Figure 2.1).

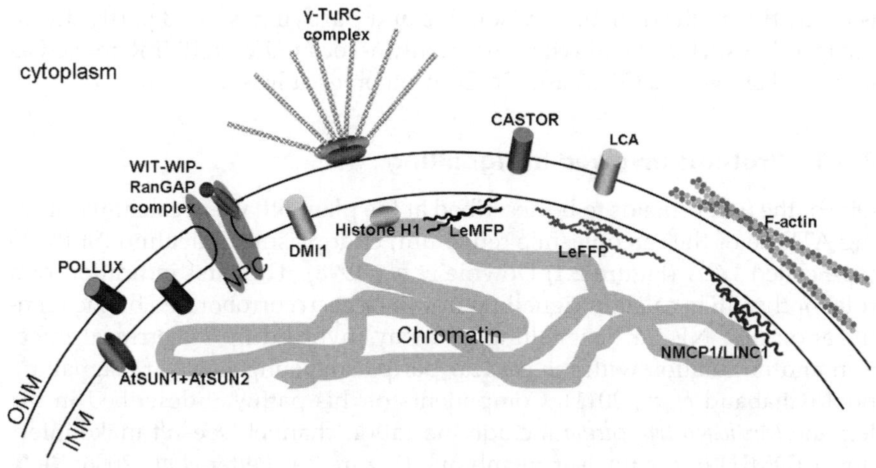

Figure 2.1 Plant NE intrinsic and associated proteins. Membrane-intrinsic plant proteins characterized to date include AtSUN1, AtSUN2, LCA, DMI1, CASTOR, POLLUX as well as WIPs and WITs. Soluble proteins specifically localized at the nuclear periphery include RanGAP, γ-TURC proteins and f-actin on the cytoplasmic face, and NMCP1, LINC1, LeMFP1, LeFFPs and histone H1 at the nucleoplasmic face. It is hypothesized that membrane-intrinsic NE membrane proteins are involved in tethering the soluble proteins to the NE but apart form RanGAP anchoring, the NE proteins and mechanisms involved in this process remain to be elucidated. RanGAP anchorage is mediated via WIP-WIT complexes and also requires the presence of SUN proteins (Zhou *et al.*, 2012; Graumann and Evans, 2010a). This research was originally published in *Biochemical Society Transactions.* Graumann K. and Evans D.E. (2010) The plant nuclear envelope in focus. *Biochemical Society Transactions* **38**, 307–311 ©Portland Press Limited and International Federation for Cell Biology.

Study of the plant NE reveals that while it is structurally similar, many of the key proteins providing these connections are absent. For instance, in metazoans, the lamina consists of filamentous lamin proteins. However, homologues of these are not present in the plant genome (Brandizzi *et al.*, 2004). In addition, many other proteins interlinking the INM and chromatin are absent. These include homologues of animal and yeast NE proteins including members of the LEM1 (lamin–emerin–man1) domain family of proteins and the lamin B receptor (LBR). It may therefore be concluded that the plant NE is homologous to that of metazoans in structure and many of its functions; but is distinct in many of the protein interactions through which these features are achieved.

2.3 Proteins of the plant nuclear envelope

The plant NE proteome has yet to be fully characterized, though recently, significant advances have been made in identifying plant NE proteins. This is partly due to the difficulty in isolating and purifying NE and partly due to the fact that such isolated NE is commonly associated with PNER; indeed as indicated above, the ONM and PNER proteomes at least overlap.

2.3.1 Proteins involved in signalling

One of the first proteins to be described at the plant NE was a calcium pumping ATPase of the sarcoplasmic reticulum/endoplasmic reticulum (SERCA) type called LCA (Figure 2.1; Downie *et al.*, 1998). This first indication of a role for the NE in calcium signalling has since been corroborated by the identification of a NE Ca^{2+} signalling pathway involved in mycorrhizal infection and nodulation with the nuclear periplasm acting as a Ca^{2+} signalling pool (Chabaud *et al.*, 2011). Components of this pathway described in the legume *Medicago truncatula* include the cation channel Doesn't make infection 1 (DMI1) in the nuclear membrane (Figure 2.1; Peiter *et al.*, 2006; Riely *et al.*, 2006); DMI2, a leucine-rich repeat (LRR) receptor-like kinase (Endre *et al.*, 2002) and a calcium- and calmodulin-dependent kinase DMI3 in the nucleus (Mitra *et al.*, 2004; Smit *et al.*, 2005). Nodulation factors produced by the infecting organism trigger Ca^{2+} spikes in the nucleus (Oldroyd and Downie, 2006), detected by DMI3 and resulting in altered transcription required for successful symbiosis (Gleason *et al.*, 2006). Two homologues of DMI1, Castor and Pollux, have also been reported in *Lotus japonicus* and also localize to the NE (Figure 2.1; Charpentiere *et al.*, 2008). In addition, *in silico* studies predict the presence of various other Ca^{2+}-sensitive and voltage-gated cation channels as well as ion pumps in both INM and ONM and suggest the plant NE plays an inherent role in signal transduction events to regulate gene activity (Matzke *et al.*, 2009).

2.3.2 Proteins of the nuclear pore complex

Further significant advances have recently been made in the identification of plant nucleoporins (Tamura *et al.*, 2010). While structure and protein traffic at the NPC is considered in detail elsewhere in this volume, it is useful here to summarize the known components of the NPC. The work of Tamura and co-workers reveals that there are at least 29 nucleoporins present in Arabidopsis, of which 22 had not previously been described in plants (Tamura *et al.*, 2010). Overall, the plant nups show a higher homology with vertebrate than yeast, commensurate with the greater size and complexity of function of the plant NPC (*Nicotiana tabacum* [tobacco] NPC ~ 105 nm diameter, compared with vertebrate 110–120 nm and yeast ~ 95 nm; Kiseleva *et al.*, 2004; Fiserova *et al.*, 2009). Orthologues of six vertebrate nucleoporins are not found in plants (Nup358, Nup188, Nup153, Nup45, Nup37, and Pom121). The major membrane anchor for the NPC found in vertebrates (but not yeast), gp210, is present in plants and is therefore a major component of the pore domain.

Like its animal counterparts, gp210 has a single C-terminal transmembrane domain. While its molecular functions are yet to be studied, it has been identified in a screen of essential Arabidopsis genes to be indispensable for embryo development (Meinke *et al.*, 2008). Two other plant nucleoporins (ALADIN, and Nup43) also have vertebrate, but not yeast, orthologues, which Tamura *et al.* (2010) suggest imply higher eukaryotic functions. Both are WD-repeat proteins likely to be involved in multiprotein complex formation.

Vertebrate NPCs have a cytoplasmic meshwork of filaments of importance in transport through the pores, which is driven by a RanGTP/ RanGDP gradient (Clarke and Zhang, 2008). Development of this gradient is defined by localization of two accessory proteins, Ran GTPase-activating protein (RanGAP) and the nucleotide exchange factor RCC1, which so far has not been identified in plants. In mammals, RanGAP is anchored to the cytoplasmic filaments by binding the small ubiquitin-related modifier protein (SUMO) followed by binding of this SUMOylated domain to Nup358. As the cytoplasmic filaments (Fiserova *et al.*, 2009) and Nup358 are not present in plants (Xu and Meier, 2008; Tamura *et al.*, 2010), the plant NE therefore has an alternative mechanism for localizing RanGAP. Attachment of plant RanGAP to the NE involves a plant-specific targeting domain, termed the WPP (tryptophan proline proline) -domain (Rose and Meier, 2001). This N-terminal domain is found in plant RanGAPs, plant NE-assocaited WPP proteins WPP1 and WPP2 as well as matrix attachment factor 1 (MAF1). The domain is necessary for RanGAP targeting and sufficient to target the heterologous protein GFP to the plant nuclear rim (Rose and Meier, 2001). Plant RanGAP is anchored to the NE by associating with WPP-interacting proteins (WIPs) and WPP interacting tail anchored proteins (WITs), which show predicted NPC localization (Figure 2.1; Xu *et al.*, 2007; Zhao *et al.*, 2008). Plant WIPs and WITs contain a C-terminal TM domain, which anchors the proteins to the NE membranes. A coiled-coil domain at the cytoplasmic side is thought to mediate both dimerization and association with the WPP domain of RanGAP. A triple knockout of WIPs 1–3 abolishes NE anchorage of RanGAP in undifferentiated root tip cells whereas RanGAP anchorage in other tissue seems to be independent of WIP and WIT making this a tissue- and development-specific anchorage process (Xu *et al.*, 2007).

2.3.3 Proteins of the INM

The metazoan INM possesses an array of proteins involved in connecting the INM to the underlying nuclear lamina and to chromatin. Description of these proteins has been driven to a great extent by diseases such as progerias, skeletal defects, muscular dystrophies, lipoatrophy, and epilepsy resulting from INM protein mutations (Stewart *et al.*, 2007). These proteins perform a variety of important tasks, some specific to vertebrates, but others found in all metazoans; their absence from the plant genome raises a number of important questions, given that INM proteins are implicated in nucleic acid metabolism, signal transduction, NPC spacing and the tethering of nuclear matrix and chromatin. Most of these functions also involve the lamina, which is tightly associated with the INM (Gruenbaum *et al.*, 2005). Proteomic studies of the rat liver NE identified over 60 NE transmembrane proteins (Schirmer *et al.*, 2003; Schirmer and Gerace, 2005).

The lamin B receptor (LBR) (Worman *et al.*, 1988) is a key link between the nuclear lamina and INM. It has eight transmembrane domains, forms a multimeric protein complex, shows sterol reductase activity and contains a Tudor domain (involved in histone binding) (Huang *et al.*, 2006). The N-terminus is located in the nucleoplasm and binds to lamin B and chromatin. Both LBR and its binding partners are absent from plants. In a recent review, Olins *et al.* (2010) provide evidence that LBR is conserved in organisms from the vertebrates including avians and amphibians to man, though it has not been detected in lower organisms and (of particular importance for this chapter) plants. Given the range of functions for LBR (reviewed by Olins *et al.*, 2010) in post-mitotic NE reformation, interphase NE growth and heterochromatin positioning in interphase, we suggest that its roles must be undertaken by other plant proteins. The search for INM - chromatin bridges is therefore of high priority. Interestingly, while LBR is not present in plants, a construct derived from the N-terminus of the mammalian LBR including the first transmembrane domain and a nuclear localization signal (NLS) faithfully targets to the plant INM (Irons *et al.*, 2003) and this has been used to investigate INM targeting in plants (Graumann *et al.*, 2007; Graumann and Evans, 2011).

A further major group of proteins not present in plants is the LEM domain family. The LEM domain proteins are named after a conserved domain comprising lamina-associated polypeptide 1 (LAP1) and LAP2 in vertebrates; emerin, identified through its role in X-linked recessive Emery–Dreifuss muscular dystrophy and Man1 (Gruenbaum *et al.*, 2005). The LEM-domain family tether chromatin to the NE (Gruenbaum *et al.*, 2005). They bind directly to lamins and to barrier-to-autointegration factor (BAF), which binds to lamins and chromatin (Gruenbaum *et al.*, 2005). Emerin also links nuclear actin with the INM and MAN1 acts as an antagonist in the transforming growth factor beta (TGFβ) signalling cascade (Gruenbaum *et al.*, 2005; Bengtsson, 2007). Amino acid sensor independent 1-3 (ASI1-3, another LEM family member) are signal transducers that regulate expression of amino-acid permeases (Zargari *et al.*, 2007).

The presence of a nucleoskeletal meshwork underneath the INM and chromatin associations at the NE suggests that the plant INM also contains proteins, which are involved in chromatin and nucleoskeletal anchorage and hence implied in controlling nuclear activities. It is therefore of great importance to identify these components and study their functional relationships with the nucleoskeleton and chromatin. To date two INM intrinsic proteins – the SUN domain proteins AtSUN1 and AtSUN2 have been identified and their functions are discussed in following sections.

2.3.4 Proteins spanning the periplasm and linking the NE membranes

The Linker of Nucleoskeleton and Cytoskeleton (LINC) complex (Crisp *et al.*, 2006) is a multifunctional protein bridge that connects INM and ONM and provides anchorage for a variety of NE associated proteins in the nucleoplasm and cytoplasm supporting, moving and shaping the NE (Figure 2.2). In non-plant systems it is made up of two membrane integral components; SUN domain proteins (in the INM), ubiquitously present in higher and lower eukaryotes (Figure 2.3; Starr, 2009; Graumann *et al.*, 2010) and KASH domain proteins of the ONM. The two families interact through their respective SUN and KASH domains in the periplasmic space separating the two membranes. While SUN-domain proteins show structural and sequence conservation across species (Figure 2.3), KASH-domain proteins are more diverse, permitting interactions with a wide range of cytoskeletal elements (Starr and Fridolfsson, 2010). In yeast, SUN domain proteins Sad1 *(Schizosaccharomyces pombe)* and Mps3 *(Saccharomyces cerevisiae)* are located at spindle pole bodies (SPBs) in addition to the NE (Bupp *et al.*, 2007). Putative plant homologues of SpSad1 were first suggested in Arabidopsis (van Damme *et al.*, 2004), and *Oryza sativa* (rice; Moriguchi *et al.*, 2005) at the phragmoplast and mitotic spindle. The first detailed characterization by the authors (Graumann *et al.*,

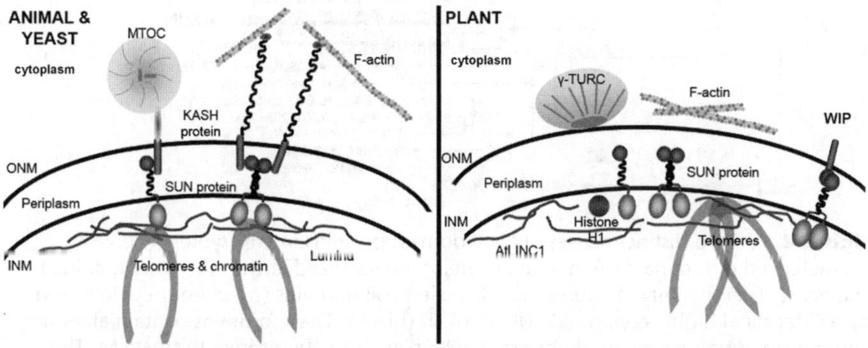

Figure 2.2 Nucleo-cytoskeletal bridging complexes at the NE. The LINC complex in animal and yeast systems connects nuclear and cytoskeletal components across the NE. The outer nuclear membrane (ONM) contains Klarsicht/ANC-1/SYNE-1 homology (KASH)-domain proteins, which link actin, microtubule (MT)-associated elements such as dynamin, kinesin and centrosomes to the NE. The inner nuclear membrane (INM)-associated Sad1/UNC-84 (SUN)-domain proteins anchor chromatin, telomeres and lamins to the nuclear face of the NE. SUN- and KASH-domain proteins interact with each other in the periplasm completing the linkage. In plants, the family of WIP proteins have been found to be functional and structural KASH proteins, which interact with AtSUN1 and AtSUN2. The WIP-SUN protein complexes anchor RanGAP to the NE and play a role in maintaing nuclear shape. However, their functions in cytoskeletal and nucleoskeletal anchorage remain unclear (Zhou *et al.*, 2012; Graumann and Evans, 2010b).

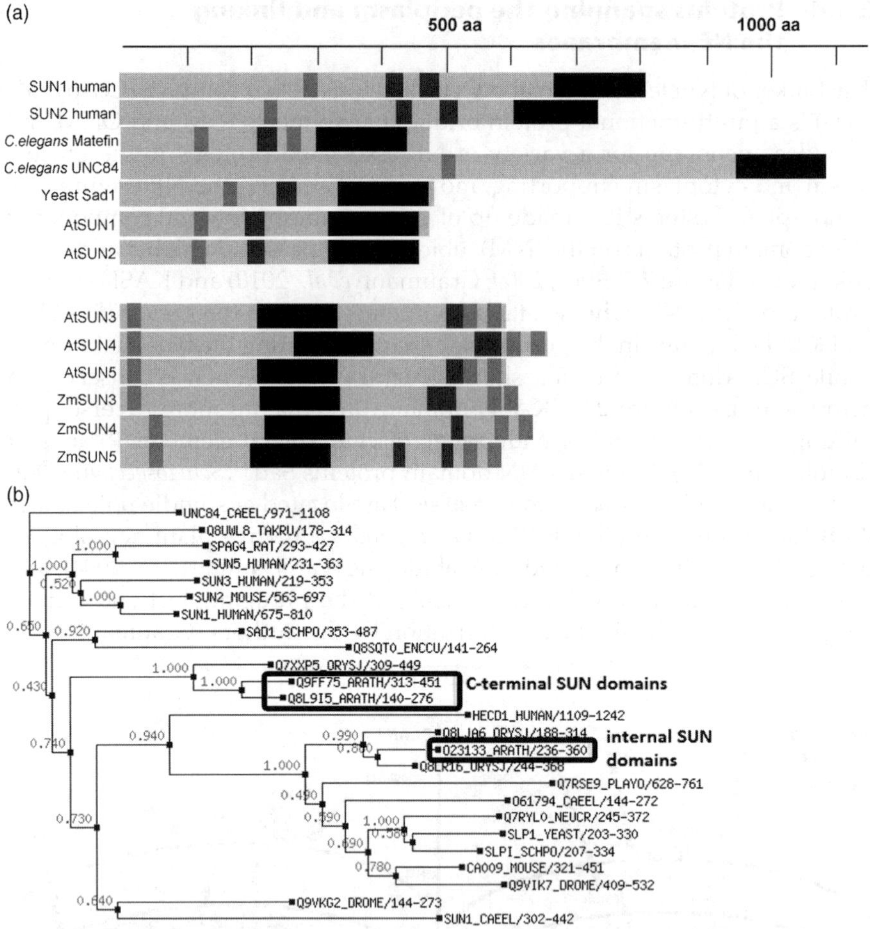

Figure 2.3 Two distinct classes of SUN-domain proteins in eukaryotes. The well characterized c-terminal SUN-domain proteins are localized at the NE in plant, animal and yeast. Their hallmark features include coiled coil domains (medium grey) followed by a C-terminal highly conserved SUN domain (black). These proteins contain at least one transmembrane domain (light grey), which anchors the protein to the INM. The c-terminal SUN domain proteins are part of protein bridges that cross-link nucleoskeletal elements and chromatin to cytoskeletal components. A second class – the mid-SUN domain proteins – have recently been described in maize and *in silico* data suggest the presence of these throughout the plant, animal and fungi kingdoms. These proteins contain three transmembrane domains, at least one coiled coil domain and the highly conserved SUN domain in the centre section of the protein. While the properties and functions of these mid-SUN-domain proteins remain unknown, the maize SUN3 protein was found to be present at the nuclear periphery indicative of NE localization (Murphy *et al.*, 2010; Evans and Tatout, unpublished observations).

2010) revealed them to be classical SUN-domain proteins and showed that AtSUN1 and AtSUN2 are localized to the NE in interphase and provide first evidence of a putative LINC complex in plants (Figure 2.2). The SUN-domain proteins are considered in detail later in this chapter.

The KASH domain is a highly hydrophobic domain comprising a transmembrane region and a C-terminal KASH domain of 6-35 amino acids in the nuclear periplasm, which interacts with the SUN domain (Razafsky and Hodzic, 2009; Starr and Fridolfsson, 2010). Other than a short extreme C-terminal PPPX motif essential for SUN binding, the KASH domain does not show sequence homology sufficient for detection using a conventional bioinformatics approach and this short sequence is not found in plants. The cytoplasmic region of the KASH proteins is highly variable, though commonly containing spectrin repeats or coiled coils. Four KASH domain proteins have been characterized in mammals (nesprins 1–4) and in *Caenorhabditis elegans* (ANC-1, UNC-83, ZYG-12 and KDP), two in *Drosophila melanogaster* (Klarsicht and MSP-300) and in *S. pombe* (KMS1-2) and one in *S. cerevisae* (Cms4) (Wilhelmsen *et al.*, 2006; Roux, *et al.*, 2009; Starr, 2009). Each anchors specific cytoskeletal elements; nesprins 1 and 2 bind actin, nesprin 3 intermediate filaments and nesprin 4 kinesin (McGee *et al.*, 2009; Starr, 2009). *C. elegans*, UNC-83 mediates the movement of the nucleus using kinesin (McGee *et al.*, 2009; Meyerzon *et al.*, 2009) whereas ZYG-12 interacts with dynein and is involved in microtubule organization and nuclear positioning in gonads (Zhou *et al.*, 2009). Nesprins 1 and 2 are involved in nuclear migration during neurogenesis and neuronal migration in mice involving attachment of the nucleus to centrosomes by dynein/dynactin and kinesin (Zhang *et al.*, 2009).

Very recently, the WIP protein family were identified to be structural and functional, but not sequence, homologues of KASH proteins in plants. As mentioned, they also consist of a cytoplasmic coiled-coil domain, are C-terminally anchored to the ONM and have a short, nine amino acid C-terminal tail. While this tail has no sequence homology to animal and yeast KASH tails, it contains a penultimate proline residue embedded in a VVPT motif that is conserved in other plant species (Zhou *et al.*, 2012). In Arabidopsis, the WIP proteins have been shown to interact with SUN proteins and both the SUN domain and the VVPT motif are required for this interaction (Figure 2.2; Zhou *et al.*, 2012). Evidence that WIP proteins are functional KASH homologues comes from observations that both WIP and SUN proteins are required to maintain nuclear shape. Knocking out either protein family results in rounded nuclei in various plant tissues (Zhou *et al.*, 2012). In addition, the plant SUN-WIP complexes are involved in WPP-mediated RanGAP anchorage in undifferentiated root tip cells as knock out of AtSUN1 and AtSUN2 results in loss of RanGAP from the nuclear periphery (Zhou *et al.*, 2012). This implies that LINC complexes can mediate plant-specific functions. Whether the WIP proteins are the only plant KASH proteins and whether they mediate NE-cytoskeleton interactions like their animal and yeast counterparts, remains to be elucidated.

2.3.5 The plant lamina

The presence of a meshwork of proteins, underlying and closely associated with the INM, is a feature of animal (Gruenbaum *et al.*, 2005) and plant (Fiserova *et al.*, 2009) nuclei. While the lamina of animal cells has been well characterized, that of plants is much less well described. The lamina of animal cells consists of lamins, type-5 intermediate filament proteins, together with lamin-associated proteins (reviewed in Wilson and Berk, 2010). Lamins are encoded by three genes, A, B1 and B2, which also generate the alternative splice product lamin C. The lamins are 60–80 kDa proteins and possess a characteristic structure with a long rod domain made up of linked α-helices and a highly conserved, globular N-terminus. A phosphorylation domain acting at the onset of NE breakdown controls association with lamin binding proteins. Plants lack sequence homologs of mammalian lamins (Brandizzi *et al.*, 2004; Meier, 2007; Graumann and Evans, 2010a). However, there is now considerable evidence for a protein meshwork underlying the plant INM and attached to it (Minguez and Diaz de la Espina, 1993; Masuda *et al.*, 1997; Fiserova *et al.*, 2009). Recently, high-resolution field-emission scanning electron microscopy (Fiserova *et al.*, 2009) has provided images revealing a complex filamentous structure at the nucleoplasmic face of the INM. Two types of filaments differing in thickness, 10–13 nm and 5–8 nm, were observed to interconnect to the NPC. The structure was also visible from the cytoplasmic face in detergent extracted nuclei and resembled the lamina meshwork of frog oocytes.

Several approaches have suggested the protein composition of the plant 'lamina'. Initially, immunological studies revealed lamin-like nuclear intermediate filament-like proteins (McNulty and Saunders, 1992; Minguez and Diaz de la Espina, 1993; Masuda *et al.*, 1997). The first of these proteins to be characterized was *Daucus carrota* (carrot) nuclear matrix constituent protein 1 (NMCP1), a 134 kDa NE-associated protein. Although much larger than mammalian lamins, NMCP1 has a central coiled-coil domain and head and tail domains, including a putative NLS (Masuda *et al.*, 1997). There are NMCP proteins present in other dicots including Arabidopsis and *Apium graveolens* (celery) as well as monocots such as rice (Table 2.1; Moriguchi *et al.*, 2005). As well as structurally resembling lamins, NMCP1 also has similar isoelectric point (5.6-5.8) and kinase-recognition domains (Masuda *et al.*, 1997).

Using a reverse genetics approach, Dittmer *et al.*, (2007) were able to characterize the Arabidopsis NMCP homologs they named 'Little Nuclei' or LINC. In addition to the structural characteristics suggested by sequence homology with NMCP1 and 2 (Table 2.1), T-DNA insertional double mutants of *LINC1* and *LINC2* showed significantly smaller nuclei (50% of wild type) as well as reduced plant growth. In addition, the mutant nuclei were more spherical. Expression of LINC1-GFP on its native promoter was high in meristematic tissues, but not differentiated tissues, such as mature root hairs and epidermal cells and the protein was localized to the periphery of the

Table 2.1 Homologues of NMCP1 and little nuclei (LINC) identified by blast searching using NMCP1

Accession	Protein names	Organism	Length	Identity (%)	Score	E value
D2YZU5	Nuclear matrix constituent protein 1	*Apium graveolens*	1171	100.00	5884	0
O04390	Nuclear matrix constituent protein 1	*Daucus carota*	1119	84.00	4688	0
A6BME3	Nuclear matrix constituent protein 1-like	*Petroselinum crispum*	1119	83.00	4673	0
A6BME2	Nuclear matrix constituent protein 1-like	*Foeniculum vulgare*	1119	83.00%	4665	0
A6BME0	Nuclear matrix constituent protein 1-like	*Apium graveolens*	1119	83.00	4659	0
A6BME1	Nuclear matrix constituent protein 1-like	*Coriandrum sativum*	1003	84.00	4190	0
B9N1Z9	Predicted protein	*Populus trichocarpa*	1156	51.00	2698	0
E5GCT1	Nuclear matrix constituent-like protein 1	*Cucumis melo* subsp. melo	1205	48.00	2613	0
A5BQE9	Putative uncharacterized protein	*Vitis vinifera*	1234	46.00	2450	0
F4HRT5	Protein little nuclei1	*Arabidopsis thaliana*	1132	43.00	2240	0
B9SEG9	ATP binding protein, putative	*Ricinus communis*	1172	44.00%	2219	0
Q7XXP7	Os02g0709900 protein (Putative nuclear matrix constituent protein 1)	*Oryza sativa* subsp. *japonica*	1155	32.00	1574	$1.00E^{-172}$
D2YZU8	Nuclear matrix constituent protein 2	*Daucus carota*	927	30.00	1046	$1.00E^{-111}$
D2YZU6	Nuclear matrix constituent protein 2	*Apium graveolens*	925	30.00	1016	$1.00E^{-107}$
B9SX77	Filamin-A-interacting protein, putative	*Ricinus communis*	1052	30.00	992	$1.00E^{-104}$
Q0WQM6	Putative nuclear matrix constituent protein	*Arabidopsis thaliana*	743	33.00	963	$1.00E^{-101}$
F4JXK1	Branched-chain-amino-acid aminotransferase 5	*Arabidopsis thaliana*	1010	29.00	952	$1.00E^{-100}$
Q9FLH0	Putative nuclear matrix constituent protein 1-like protein (NMCP1-like)	*Arabidopsis thaliana*	1042	29.00	940	$1.00E^{-98}$

nucleoplasm (Dittmer *et al.*, 2007). The LINC proteins therefore appear to be strong candidates as components of a plant nucleoskeleton; however, progress still needs to be made to prove that they constitute major components of the fibres observed in the electron micrographs or that they interact with proteins of the NE and with chromatin.

Other potential components of the higher plant nucleoskeleton have also been suggested, although without verification. A family of coiled-coil proteins, the filament-like proteins (FPP), was suggested as a possibility. The *Lycopersicon esculentum* (tomato) homolog, LeFPP, was identified in a yeast-two hybrid screen interacting with LeMAF1, an NE-associated protein involved in RanGAP binding the plant nucleus (Gindullis *et al.*, 1999; Gindullis *et al.*, 2002). In addition, a homolog from *Pisum sativum* (pea) was identified using a lamin B antibody (Blumenthal *et al.*, 2004). There are seven family members of the FPPs in Arabidopsis (FPP1–FPP7; Gindullis *et al.*, 2002) ranging from 615 to 1054 amino acids with four conserved motifs and a long coiled coil core. They remain to be functionally characterized.

Moriguchi *et al.*, 2005, attempted to characterize the potential nucleoskeleton of rice. In addition to identifying a rice homologue of NMCP1 (see above), they also identified a number of other potential nucleoskeleton associated components. Ankyrin is a structural protein usually located at the plasma membrane; however, Nalp1, a 28 kDa rice ankyrin homologue, was located close to the INM, in a region termed the inner nuclear matrix region or chromatin-poor space. AtAnkyrin3, the Arabidopsis counterpart of Nalp1, shows high homology with the ankyrin repeat region of Nalp1 and is of similar size (27 kDa); they are however, much smaller than the human ankyrins and unlikely to perform similar functions (De Ruijter *et al.*, 2000).

As well as proteins of the nucleoskeleton linking to the NE, there is also evidence for a variety of proteins involved in the interaction of the nucleoskeleton with chromatin domains. Plant nuclei contain actin, which forms nucleoplasmic microfilaments and associates with transcription sites (Cruz *et al.*, 2008; Cruz and Moreno Diaz de la Espina, 2009). While the major actin binding proteins of the nucleoskeleton are absent (lamins, lamin associated proteins, nesprins), three actin-binding proteins have been identified (Perez-Muniva and Moreno Diaz de la Espina, 2011); profilin (Kandasamy *et al.*, 2002), nuclear myosin 1 (Cruz *et al.*, 2008; Moreno Diaz de la Espina, 2009) and actin depolymerizing factor (Ruzicka *et al.*, 2007). Their molecular role in maintaining nuclear architecture remains to be determined.

The presence of spectrin-like proteins in pea nuclei was suggested by a study by De Ruijter *et al.* (2000) using antibodies raised to erythrocyte spectrin. Western blots revealed bands of 220–240 Da, similar to erythrocyte spectrin, and a prominent 60 kDa band. In light microscope immunolocalization, the antibodies detected proteins in puncta in the nucleoplasm and with staining increasing as the nuclear matrix was extracted progressively with detergents, DNase I and RNase A, and high salt. The authors suggest the data shows the presence of spectrin-like epitopes in plant nuclei, and postulated a role in stabilizing interchromatin domains. Recently, Pérez-Munive and Moreno Diaz de la Espina (2011), also using an immunocytochemical approach, have provided further evidence for spectrin chain proteins that co-immunoprecipitate and co-localize with nuclear actin after detergent extraction of the nuclear matrix of onion (*Alium cepa*) with nuclease digestion. Using antibodies to chicken spectrin, they observed that spectrin-like proteins and plant intermediate filaments co-localize in the nucleoplasm suggesting a structural role. Spectrin-like proteins were shown to be distributed at or near the NE, associated with chromatin and at the nucleolus. The authors are investigating interactions between the spectrin-like proteins and NMCP family members; however, to date sequence analysis of the proteins do not indicate homology with animal spectrins (Pérez-Munive and Moreno Diaz de la Espina, 2011).

An alternative approach to the identification of putative plant lamins resulted from the isolation of nuclear matrix proteins and use of an anti-intermediate filament monoclonal antibody MAb TIB 131 by several laboratories. In plants, researchers have shown that this antibody and various other anti-lamin antibodies recognize nuclear 60 and 65 kDa proteins (McNulty and Saunders 1992; Minguez and Moreno Diaz de la Espina, 1993). In the most detailed of these studies (Blumenthal *et al.*, 2004) proteins of 54, 60 and 65 kDa were suggested as putative nuclear intermediate filament proteins in pea. Limited sequence data from the 65 and 60 kDa proteins showed homology with intermediate filament proteins and antibodies raised to the 65 kDa protein also cross-reacted with the smaller proteins and with an avian lamin preparation. Immunocytochemistry suggested nucleoplasmic, but not specifically nuclear peripheral, localization. Negatively stained extracts of the fractions show filaments of 6–12 nm diameter.

2.4 The plant nuclear envelope and the nucleoskeleton; attachments at the INM

As many of the key proteins involved in attaching the nucleoskeleton (like members of the LEM domain family) are absent from plants, attention has focussed on plant members of the SUN-domain family as key INM proteins with the potential to act as bridges between envelope and nucleoskeleton.

The SUN-domain proteins occur throughout the plant kingdom. Classical (C-terminal) SUN-domain proteins have been identified in the club moss, *Selaginella meollendorffii*, the moss, *Physcomitrella patens* and in algae (Table 2.2) as well as in a range of both monocot and dicot species. They have been characterized in detail in two species: Arabidopsis (Graumann *et al.*, 2010; Graumann and Evans, 2011; Oda and Fukuda, 2011) and *Zea mays* (maize; Murphy *et al.*, 2010). In each case, two proteins with a C-terminal SUN domain have been described, designated AtSUN1 and AtSUN2 (Graumann *et al.*, 2010) and ZmSUN1 and ZmSUN2 (Murphy *et al.*, 2010). Sequence comparison

Table 2.2 Predicted plant SUN domain proteins included in the pfam phylogentic tree (further information derived from Uniprot annotations or pfam unless otherwise stated)

Uniprot	Species	SUN Domain (pfam)	Annotated Name	Length
Proteins with a predicted C-terminal SUN domains				
Q9FF75_ARATH	*Arabidopsis thaliana*	313-415	AtSUN1	471 aa
Q8L9I5_ARATH	*Arabidopsis thaliana*	140-276		285 aa
Q9SG79_ARATH	*Arabidopsis thaliana*	310-446	AtSUN2	455 aa
A9TD16_PHYPA	*Physcomitrella patens*	145-288		293 aa
A9RVG5_PHYPA	*Physcomitrella patens*	152-295		300 aa
A2WN90_ORYSI	*Oryza sativa indica*	303-444		455 aa
Q5NBL8_ORYSJ	*Oryza sativa japonica*	303-444		455 aa
B6TRH2_MAIZE/ B6TY16_MAIZE	*Zea mays*	289-427	ZmSUN2	439 aa
C5XGK7_SORBI	*Sorghum bicolour*	15-115		126 aa
A2Y2L0_ORYSI	*Oryza sativa indica*	309-449		453 aa
Q7XXP5_ORYSJ	*Oryza sativa japonica*	309-449		453 aa
C5XW88_SORBI	*Sorghum bicolour*	295-435		443 aa
B6THU8_MAIZE	*Zea mays*	314-454	ZmSUN1	462 aa
B9SBM5_RICCO	*Ricinus communis*	322-460		471 aa
A5BJ94_VITVI	*Vitis vinifera*	320-364; 532-629		640 aa
B9HVF3_POPTR/ B9HK83_POPTR	*Populus trichocarpa*	148-286		292 aa
A4RXN9_OSTLU	*Ostreococcus lucimarinus*	1534-1671		1676 aa
C1MM01_MICPC	*Micromonas pusilla*	830-967		683 aa
C1E249_MICPC	*Micromonas pusilla*	381-516		526 aa

of both entire sequences (Table 2.2) and of the SUN domains (Figure 2.3) reveal that they show a greater homology within the monocots and within the dicots than between the two groups; for instance AtSUN1 and AtSUN2 show a higher degree of homology (68% identity, 1.00 E^{-178}) with each other than with either ZmSUN1 or ZmSUN2 (41%, 4.0 E^{-79} and 2.0 E^{-70} respectively). AtSUN1 is located on chromosome 5, AtSUN2 on chromosome 3, while in maize, ZmSUN1 is located on chromosome 5, and ZmSUN2 is located on chromosome 3.

Plant C-terminal SUN-domain proteins are smaller than their metazoan counterparts (Figure 2.3), the plant proteins having between 430 and 480 residues (yeast SAD1 with 524 residues). There is a strong structural resemblance between the plant C-terminal SUN domain proteins (Figure 2.3) and with the mammalian and yeast proteins, with a single transmembrane domain located towards the N-terminus and a coiled-coil domain between the transmembrane domain and SUN domain. In common with the non-plant family members, the N-terminus is located in the nucleoplasm and the coiled-coil domain and SUN domain in the nuclear periplasm (Graumann *et al.*, 2010; Graumann and Evans, 2010b).

Evidence that the plant SUN domain proteins are involved, directly or indirectly, in connections between the envelope and nucleo- or cytoskeleton comes from a number of experimental approaches (Figure 2.2). A number of groups have generated double T-DNA mutant plants for AtSUN1 and AtSUN2 (Armstrong and Osman, unpublished; Zhou *et al.*, 2012); Oda and Fukuda (2011) incorporated a micro (mi) RNA interference construct to improve knockdown efficiency. While they observed no significant major phenotypic differences, for instance in plant growth, development or fertility, they did observe that typical elongated nuclei of root epidermis and root hairs of wild-type plants were rounded in the SUN knockdowns. Movement of the nucleus in developing root hairs was, however, not affected. As interactions between INM proteins and nucleoskeletal elements such as the lamins are known to regulate nuclei shape, it suggests that the roles of AtSUN1 and AtSUN2 in maintaining nuclear shape may also involve interactions with plant nucleoskeletal elements. Further evidence for this comes from a study in which knock down of the lamin-like LINC proteins resulted in similarly spherical nuclei (Dittmer *et al.*, 2007).

Studies of the behaviour of SUN domain proteins in mitosis also strongly points to their association with the nucleoskeleton and cytoskeleton. The NE of higher plants breaks down in what is termed 'open mitosis' and reforms around decondensing chromatin at the end of division. This process involves the breaking of protein bridges between the membrane and its associated proteins, followed by reformation of these links. The SUN-domain proteins in metazoans are among the first proteins to reassociate with chromatin. In tobacco suspension culture cells, AtSUN1 and AtSUN2 migrate to mitotic ER membranes at envelope breakdown and then rapidly accumulate in the reforming NE at the end of division, strongly suggesting direct or indirect association with chromatin. Plant NE reformation is spatially organized as AtSUN1 and AtSUN2 first aggregate at the surface of chromatin facing the spindle pole, then accumulate at the sides and finally facing the cell plate (Figure 2.4; Graumann and Evans, 2011). It has been shown that the lamin-like NE associated NMCP1 and NMCP2 in celery cells behave similarly (Kimura *et al.*, 2010). It is interesting that AtSUN2 accumulates at the NE specifically during prophase and a role in chromatin organization prior to NE breakdown has been suggested (Graumann and Evans, 2011). In a parallel study, Oda and Fukuda (2011) explored the location and involvement of microtubules using a fluorescent tubulin construct with AtSUN1-mRFP. They observed microtubules both in association with the NE during NE breakdown and associated with the mitotic membranes. The SUN-domain proteins also localize to the phragmoplast – an area of high cytoskeleton and membrane activity forming the new cell wall and AtSUN2 may be interacting with other proteins here (Graumann and Evans, 2011; Oda and Fukuda, 2011).

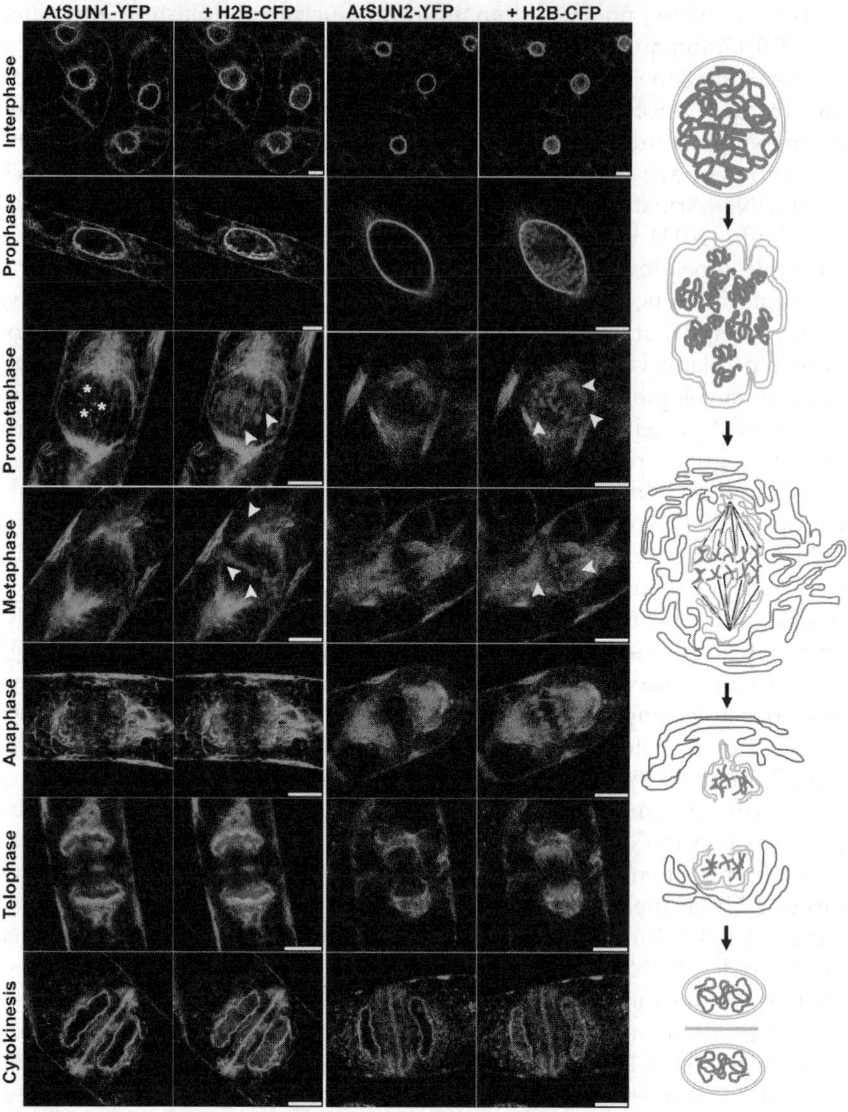

Figure 2.4 NE membrane dynamics during mitosis marked by SUN domain proteins. Stable transformed BY-2 cells co-expressing either AtSUN1-YFP (green in Plate 2.4) and chromatin marker histone H2B-CFP (magenta) or AtSUN2-YFP (green) and H2B-CFP were synchronized and living cells imaged by confocal microscopy. The two SUN proteins are present in the NE around chromatin in interphase and prophase. Upon NEBD, they distribute to mitotic ER and spindle membranes including tubules traversing the division zone. As the sister chromatids are separated, AtSUN1-YFP and AtSUN2-YFP accumulate in the reforming NE around chromatin first facing the spindle pole and finally proximal to the cell plate. In cytokinesis both fusion proteins are present in the expanding NE, phragmoplast and cell plate. AtSUN1-YFP is present in puncta in prometaphase (asterix) and both SUN-domain proteins are present in tubules in close proximity to chromatin (arrow heads). Scale bar = 10 μm (Graumann and Evans, 2011). (*For colour details please see colour plate section.*)

In addition to the classical SUN-domain proteins, recent *in silico* evidence and studies in maize suggest the presence of a second group of SUN-domain proteins, termed the mid-SUN proteins, where the highly conserved SUN domain is localized at the centre of the protein (Figure 2.3). Three of these mid-SUN proteins have been identified in maize, Arabidopsis and *Chlamydomonas reinhardtii* (Murphy *et al.*, 2010; Graumann, Brickley and Tatout, unpublished observations). Immunological evidence suggests that the mid-SUN domain proteins are also localized at the NE. However, their functions and characteristics remain to be studied.

2.5 The plant nuclear envelope and the cytoskeleton; attachments at the ONM

Initial evidence for the attachment of the NE to the cytoskeleton is indirect; in a number of circumstances, in particular in development and in response to stress, plant nuclei move in a manner dependent on the actin cytoskeleton (Chytilova *et al.*, 2000; Ketelaar *et al.*, 2002). Nuclear movement in response to abiotic and biotic stimuli such as bacterial, fungal and viral infections is considered later on in detail. Whether the connection between actin filaments and the NE is direct or mediated through protein complexes is unknown (Figure 2.2). Treatment of cells with latrunculin B or other actin depolymerizing drugs reduces nuclear movement in leaf epidermal cells expressing an NE marker (Graumann *et al.*, 2007). While the full extent of nuclear movement has yet to be described, observations of nuclei in root epidermal cells and root hairs have been made by a number of groups (Chytilova *et al.*, 2000; Ketelaar *et al.*, 2002). Long-distance nuclear migration requires large actin bundles (Iwabuchi *et al.*, 2010) while finer actin filaments position nuclei within fully elongated root hairs (Iwabuchi *et al.*, 2010). The actin-associated motor protein myosin XII has also been found to localize to the plant NE (Avisar *et al.*, 2009), indicating that not only actin bundles per se but also actin-associated proteins and motorproteins may play a role in nuclear movement. The presence of proteins for nucleating tubulin (Seltzer *et al.*, 2007) at the ONM (Figure 2.2) suggests that the microtubule cytoskeleton might additionally play a role in nuclear size or positioning and it has been shown that destabilizing of microtubules increases the rate of movement of the nucleus (Fournier *et al.*, 2008).

Less is known about the protein components by which the cytoskeleton is anchored to the NE. In non-plant systems, the γ-tubulin ring complex (γ-TuRC) forms the central part of microtubule organizing centres (MTOC) and SPB. The γ-TuRC comprises five γ-tubulin complex proteins (GCP) and γ-tubulin itself (Fava *et al.*, 1999; Murphy *et al.*, 2001). Two plant proteins, AtGCP2 and AtGCP3, have been shown to be homologs of Drosophila and yeast GCP2 and 3, and form a soluble complex with γ-tubulin that associates with the plant ONM (Seltzer *et al.*, 2007). Both have NE targeting domains that are thought to target the γ-TuRC to the NE, which is then retained there by associating with an as yet unknown ONM intrinsic protein (Figures 2.1 and 2.2; Seltzer *et al.*, 2007). In metazoan and yeast, MTOCs and SPB are anchored to the NE by KASH proteins. Similarly, actin filaments are associated with the NE by interacting with the actin-binding domain of KASH proteins (Figure 2.2). While WIP proteins have been identified as plant KASH homologs, it is unclear whether they participate in anchoring cytoskeletal elements to the NE. A triple knock-out mutant, which lacks all three WIP proteins results in rounded nuclei indicating these proteins are needed to shape nuclei (Zhou *et al.*, 2012). Whether this process is also dependent on cytoskeletal elements or whether other plant KASH-like proteins mediate this remains to be elucidated.

2.6 Targeting of proteins to the plant NE

The molecular mechanisms of trafficking membrane-intrinsic proteins to the INM have predominantly been studied in metazoan and yeast systems while evidence of these processes in plants remains sparse. The endoplasmic reticulum (ER), ONM, pore membrane and INM are continuous, and therefore all proteins destined to reside in the INM pass through the pore membrane. This was thought to be due to passive diffusion, with the INM proteins retained by binding to nucleoplasmic components (Figure 2.5a; Mattaj, 2004). However, while proteins smaller than 25 kDa may enter the INM by this means, the process is slow and inefficient. It is now clear that INM-intrinsic proteins enter the nucleus using an energy-dependent mechanism (Ohba *et al.*, 2004) similar to that for soluble proteins (see Chapter 3), involving the Ran cycle (King *et al.*, 2006; Lusk *et al.*, 2007) and requiring an NLS. The INM proteins are first sorted at the translocon where association with importin-α occurs (Figure 2.5b). They are differentiated from other membrane proteins by their transmembrane domains (Saksena *et al.*, 2004; Saksena *et al.*, 2006; Braunagel *et al.*, 2007). Many INM proteins have an NLS, which is recognized by the import machinery and is partly essential for correct targeting to the INM (King *et al.*, 2006; Lusk *et al.*, 2007). Entry through the pore (probably the region of the pore membrane) requires changes in NPC structure, provided by flexible, sliding nucleoporins (King *et al.*, 2006; Alber *et al.*, 2007; Lusk *et al.*, 2007). In addition, FG nucleoporins are required for INM protein translocation and some, such as POM121, are associated with, or are in close proximity to, the pore membrane (Alber *et al.*, 2007; Lusk *et al.*, 2007).

Targeting of NE proteins in plants is far less well understood. However, plants show similar mechanisms for the import and export of soluble proteins through NPC (Merkle, 2004; Meier, 2007; Meier and Brkljacic, 2009) with 17 importin-β, eight importin-α and three exportin homologs identified (Brandizzi *et al.*, 2004; Meier, 2007). Most members of the Ran cycle have

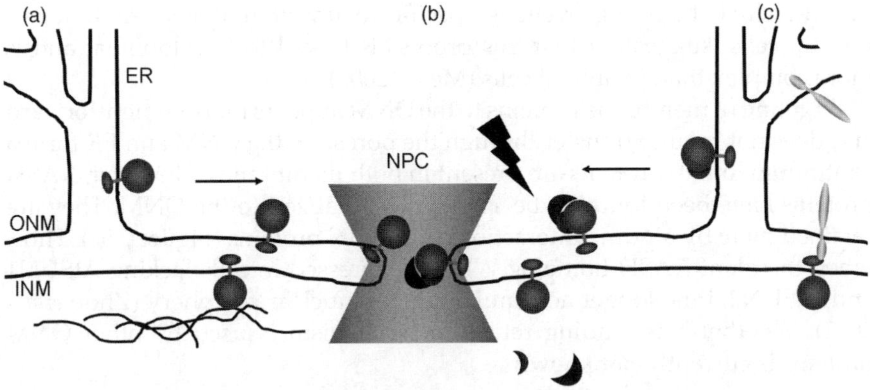

Figure 2.5 Targeting of membrane proteins to the NE. The NE membrane-intrinsic proteins are translocated into the membrane at the site of their synthesis in the ER and ONM. Two models predict how INM intrinsic protein reach their destination. (a) According to the diffusion-retention model, INM proteins freely diffuse through the ER/ONM, the pore membrane and into the INM. There, they are retained by binding interactions, for instance with the nucleoskeleton. (b) A more recent model suggests that INM protein transport is active and regulated. Importin complexes associate with the cytosolic part of the INM protein and are thought to facilitate the transport through the pore. The transport is energy dependent and may require functional and structural involvement of the NPC. On the nucleoplasmic side, the importins are thought to disassemble, similar to soluble protein import complexes. (c) ONM-localized proteins are not trafficked through the pore. For KASH proteins, a class of ONM-specific proteins, interactions with INM-localized SUN proteins are required to anchor them at the ONM. These models are based on animal and yeast systems.

been characterized (Meier and Brkljacic, 2009) and putative nucleoporins with FG repeats have been identified (Tamura *et al.*, 2010). The presence of a classical NLS has also been identiftied in plant NE proteins such as DMI1, AtSUN1 and AtSUN2 and the WIPs (Xu *et al.*, 2007; Charpentier *et al.*, 2008; Graumann *et al.*, 2010). The fact that the N-terminus and first transmembrane domain of human LBR fused to GFP (green fluorescent protein; LBR–GFP) localizes to the INM in tobacco cells, also strongly suggests that mammalian INM targeting signals are recognized by plants (Irons *et al.*, 2003; Graumann *et al.*, 2007). Apart from the seemingly conserved INM targeting mechanisms, plants also have a specific NE targeting mechanism for NE-associated proteins (Meier, 2007). This system is based on the WPP domain, which is present in LeMFP1, MAF1 and its Arabidopsis homologues WPP1 and WPP2 as well as Arabidopsis RanGAP (Meier, 2007; Zhao *et al.*, 2008). Whereas in animals RanGAP is anchored to the ONM by SUMOylation (SUMO is small ubiquitin-related modifier) and binding to RanBP2 (Ran-binding protein 2)/Nup385, in plants the WPP domain is essential for association with the ONM and NPC by interacting with WIPs and WITs (see Section 2.3.2.) (Meier, 2007).

This RanGAP targeting event is specific to undifferentiated Arabidopsis root-tip cells, suggesting that this process is linked to development and is more complex than in animal cells (Meier, 2007).

Targeting of membrane proteins to the ONM appears more straightforward and does not require transfer through the pores. As the ONM and ER form a continuum, many proteins are present in both membranes. However, KASH proteins have been found to be specifically localized to the ONM. They are retained there by binding interactions with SUN proteins (Figure 2.5c). Thus, when the plant KASH homolog WIP1 is expressed in cells lacking AtSUN1 and AtSUN2, it no longer accumulates at the nuclear periphery (Zhou *et al.*, 2012). Whether this binding retention mechanism is used by other ONM proteins is currently not known.

2.7 Nuclear envelope protein dynamics in mitosis

2.7.1 The role of NPC in regulating NE dynamics in cell division

The first major NE event in dividing higher plants and metazoans occurs prior to NE breakdown (NEBD) in G2, when NPCs replicate, the NE enlarges and DNA is duplicated. During this, new NPCs are inserted, either by joining INM and ONM to create a pore into which the NPC proteins insert (requiring protein interaction with the membrane) or nucleoporins are inserted into an intact membrane and cause fusion of the ONM and INM, after which recruitment of the other nucleoporins follows (Hetzer, 2010). Membrane fusion occurs at the luminal face of the membranes to form the pore. Proteins are then added from both the nucleoplasmic and cytoplasmic face (D'Angelo *et al.*, 2006). Drin *et al.* (2007) have shown that Nup133 contains an α- helical domain that senses membrane curvature. If sensing curvature by this domain is a prerequisite for development of the pore, it implies that membrane fusion and pore formation precede the construction of the NPCs (Hetzer, 2010). It is interesting to note that at this stage, as insertion of NPCs has preceded NEBD, the processes parallel those in eukaryotes with closed-cell division like yeast, where pore insertion has been characterized in detail. Here, the pore-domain transmembrane proteins (Pom34, Pom152 and Ndc1) mark the point of insertion of the NPC, possibly by bringing ONM and INM together. Nuclear pore assembly involves Nups59/53 with integral membrane proteins Pom34 and Pom152, to which Nup170 and membrane-integral nucleoporin Ndc1 attach (Onischenko *et al.*, 2009). Depletion of Nup170 and Nup157 results in pores that are mis-located to the INM and cytoplasm rather than creating true pores (Dawson *et al.*, 2009).

The insertion of NPCs into the plant NE in interphase has received little attention. In a study by Fiserova *et al.* (2009), NPCs in three–day old and ten-day old tobacco BY-2 suspension cultures were observed. The NPCs were highly abundant in all cells (40–50 per μm^2). The authors suggest that the three-day old cells were likely to be in S phase. At this stage, NPCs were distributed over the nuclear surface, with around 30% of pores in connected pairs. In older cells, rows of NPC were observed instead. In addition, pores in three day old cells appeared simpler than those in older cells, and the cytoplasmic ring thinner, with structures similar to those observed in *Xenopus leavis* oocytes. However, to date, antibody and molecular probes have not been available to permit detailed sequential analysis of NPC formation in plants and the hypothesis that the processes involved are the same as those in animals or yeast has yet to be tested.

While plants possess the Ran cycle for transport at the pores (see above), the Ran nucleotide exchange factor 'regulator of chromatin condensation 1' (RCC1) is not present and RanGAP is attached to the NPCs by interaction via a C-terminal WPP domain with the WIPs and WITs (Rose and Meier, 2001; Jeong *et al.*, 2005; Xu *et al.*, 2007; Zhao *et al.*, 2008). In addition to this role, Ran is also directly involved in pore assembly. In yeast, high levels of Ran-GTP and decreased importin stimulate pore formation (Harel *et al.*, 2003; Walther *et al.*, 2003). In xenopus oocytes, NE with nuclear pores assemble around beads coated with Ran in the presence of RCC1, forming pseudo-nuclei (Zhang and Clarke, 2000). Importin has also been shown to be essential to the process by which pores assemble on chromatin (Rotem *et al.*, 2009). A role for Ran in NPC insertion in plants has yet to be considered.

Open mitosis requires the detachment of the envelope from its underlying structures (chromatin and the nucleoskeleton) and its breakdown and separation from chromatin. In animals, the NE membranes and some of its protein constituents migrate into the ER network, to form the 'mitotic ER' or 'mitotic membranes'. Before these events, the NPCs are removed, giving exchange between nucleoplasm and cytoplasm. Pore removal is regulated and soluble NPC components migrate to the cytoplasm or become part of the mitotic apparatus, some remaining as protein complexes (Hetzer, 2010).

The first stage in the breakdown of NPCs is the removal of Nup98 (Griffis *et al.*, 2002), followed by other nucleoporins (Dultz *et al.*, 2008). In animal cells, this occurs in prometaphase while in plants it is earlier, in late prophase (Rose, 2007). Linkage of NPC to the lamina is implied for animals as mutation, down-regulation of expression, or introduction of Fab fragments of antibodies to gp210 in *C. elegans* results in failure of the lamin nucleoskeleton to depolymerize. Phosphorylation of the C-terminus of gp210 by cyclin B is essential for NEBD (Galy *et al.*, 2008).

Following NEBD in plants, Tpr, which is associated with the nuclear pore basket, migrates to the mitotic spindle in prometaphase (Xu *et al.*, 2007). Tobacco Rae1 associates with mitotic microtubules including the preprophase band (PPB), spindle and phragmoplast (Lee *et al.*, 2009). Deletion or down regulation of the gene for Rae1 causes defects in the spindle organization, chromatin alignment and segregation as well as decreased levels of cyclin B, cyclin-dependent kinase B1-1 (CDKB1-1) and histones H3 (Lee *et al.*, 2009).

Removal of NPCs has a number of consequences, including the loss of selective permeability of the membrane. Exposure of the nucleoplasm to cytoplasm initiates a series of phosphorylation events. Cyclin dependent kinases (CDKs) and cyclin B1 enter the nucleus and cause dissociation of proteins of the lamina and INM, which are then released into the ER. Breakdown then progresses as NE proteins move to the mitotic ER. Dephosphorylation at the end of division reverses this (Anderson and Hetzer, 2008; Guttinger *et al.*, 2009). Progress of NEBD is regulated by aurora kinases and CDKs. How these kinases regulate plant NEBD and reformation remains to be established; however, aurora kinases are associated with the plant NE in interphase and localize to the mitotic spindle, centromeres and phragmoplast in division. They phosphorylate histone H3 and are involved in chromosome segregation and cytokinesis (Demidov *et al.*, 2005; Kawabe *et al.*, 2005). Plant B1 cyclins also accumulate at the NE (Rose *et al.*, 2004). The down regulation of cyclin B and CDKB1-1 in combination with decreased mitotic activity in NbRae1 mutants suggests their involvement in plant mitosis progression (Lee *et al.*, 2009). In addition to kinases, RanGAP plays a critical role in NEBD (Meier and Brkljacic, 2009).

Loss of physical interactions within the nucleus permits physical forces to tear and open the membrane. Microtubule-dynein interactions draw membrane away from the lamina (Beaudouin *et al.*, 2002; Salina *et al.*, 2002; Muhlhausser and Kutay, 2007; Stewart *et al.*, 2007) although Lenart *et al.* (2003) suggest that NEBD can occur in the absence of a microtubule motor. The process in plants also involves positional information in order to establish the plane of division required in a walled structure, where the formation of a PPB of microtubules predicts the division plane. Tearing of the NE occurs where ONM and PPB are closest to one another and before the PPB disintegrates (Dixit and Cyr, 2002; Brandizzi *et al.*, 2004; Rose *et al.*, 2004; Evans *et al.*, 2009).

2.7.2 NE protein dynamics in division

One of the first studies of the location of a plant NE component in mitosis was in 1998, when Downie *et al.* (1998) used immunostaining to detect NE-specific staining of the Ca^{2+} ATPase LCA in spindle associated membranes of dividing tomato cells. Recently, the behaviour of plant NE components in mitosis has been observed using the chimaeric LBR-GFP probe described earlier (Irons *et al.*, 2003), the plant SUN-domain proteins (Graumann and Evans, 2011; Oda and Fukuda, 2011) and the putative plant lamin-like proteins NMCP1 and NMCP2 (Kimura *et al.*, 2010). As expected, both plant NE membrane markers and the putative plant lamin-like proteins disassemble at the beginning of mitosis (Figure 2.4). As the nucleoskeleton is lost in celery cells, NMCP1 associates with the mitotic spindle while NMCP2 migrates to the mitotic cytoplasm and both join the reforming NE at the end of the process (Kimura *et al.*, 2010). While this study suggested that in fixed celery cells, dye-stained membranes from the NE formed vesicles, studies using live-cell imaging reveal a different behaviour, similar to that observed in metazoans. Human-derived LBR-GFP, in use as a plant NE marker, suggest that NE membranes in tobacco BY-2 cells were distributed to mitotic membranes (Irons *et al.*, 2003; Graumann and Evans, 2011). This was also corroborated by studies using native plant proteins described below.

The native plant INM markers AtSUN1 and AtSUN2 also reveal that membranes of the NE enter the pool of mitotic membranes after NEBD (Graumann and Evans 2011; Oda and Fukuda, 2011). The SUN-domain proteins (like LBR-GFP) localize to spindle membranes and to tubules traversing the division zone in metaphase, while most of the spindle membranes accumulate at the spindle poles (Figure 2.4). In contrast with metazoans, where NE proteins distribute throughout the mitotic ER leaving metaphase chromosomes devoid of membranes, membranes containing NE proteins remain in close proximity to chromatin throughout division in plants (Anderson and Hetzer, 2008; Hetzer, 2010; Graumann and Evans, 2011). In metazoans, NE reformation commences with mitotic ER tubules containing INM proteins contacting decondensing chromatin in anaphase (Webster *et al.*, 2009; Hetzer, 2010) with peripheral and core chromatin areas attracting different INM proteins (Güttinger *et al.*, 2009). Both Graumann and Evans (2011) and Oda and Fukuda (2011) observed that NE reformation is spatially organized as AtSUN1-YFP and AtSUN2-YFP aggregate initially on chromatin facing the spindle pole followed by accumulation around the sides and finally localization of the proteins on chromatin facing the cell plate (Figure 2.4). This has not been observed in metazoans (Ellenberg *et al.*, 1997; Webster *et al.*, 2009; Hetzer, 2010) and may be plant specific, commensurate with a linkage to specific chromatin domains, and suggests a tight spatial regulation of NE reformation. Similar observations were made of NMCP1 and NMCP2. NMCP1 also first assembled on chromatin facing the spindle in late anaphase and at telophase completely surrounded the decondensing chromatin (Kimura *et al.*, 2010). Thus AtSUN1 and AtSUN2 may be recruited to the reforming NE by interactions with chromatin or other specifically targeted nuclear components such as NMCP1.

2.8 The phragmoplast and cell plate and their relationship to the NE

The final process in the production of two daughter cells, accompanying the completion of the formation of the NEs, is the formation of the new dividing cell wall. This is generated along the plane of division predicted by the PPB microtubules and involves considerable secretory vesicle activity. Use of the mammalian-based LBR-GFP expressed in plants shows that it is not only present in the newly forming NEs, but also in the phragmoplast (Irons *et al.*, 2003; Brandizzi *et al.*, 2004; Evans *et al.*, 2011). Other known NE proteins, including the WIPs and WITs (other than WIP3; Zhao *et al.*, 2008) and the SUN-domain proteins (Graumann and Evans, 2011) are also observed in the phragmoplast (Figure 2.4). This may be due to a lack of an effective sorting and retrieval mechanism and to the very high volume of membrane traffic directed to the cell plate. Alternatively, it may suggest a role for these proteins in cell wall formation. Interestingly, RanGAP1 is specifically localized to the position of the PPB and remains in location throughout division during mitosis and cytokinesis (Xu *et al.*, 2007). This suggests that a Ran gradient is formed that directs the vesicle traffic to the phragmoplast. As Ran has also been shown to be involved in the formation of the NE and insertion of NPCs, it seems possible that similar mechanisms are involved in both systems resulting in mis-direction of vesicle traffic. Alternatively, some NE proteins may be caught up in the rapid flow of membrane to the phragmoplast and are subsequently removed by recovery mechanisms. AtSUN2–YFP appears to have a significant immobile fraction at the cell plate indicating interactions and therefore may be functional (Graumann and Evans, 2011). It is interesting to observe that so far only in plants has spatial NE reformation from proximal to the spindle to proximal to the cell plate been observed, and it is intriguing to speculate whether this is associated with the presence of NE proteins in the cell plate. Among other factors such as kinases, phosphatases, Ran and chromatin remodelling components, the amount of available membrane is known to affect NE reformation (Webster *et al.*, 2009) and it is possible that the cell plate and phragmoplast form membrane reservoirs that affect the distribution of membranes in the plant cell. Thus the SUN-domain proteins might be involved in such regulation.

2.9 The plant NE in meiosis

In addition to the events described for mitosis, the NE has a specific function in prophase 1 of meiosis as it anchors and positions telomeres. Anchorage of the telomeres to the NE is mediated by the SUN-domain components of the LINC complex in animal and yeast systems and deletion of these can result in abolition of gametogenesis (Tomita and Cooper, 2006; Ding *et al.*, 2007). In animal, yeast and some plants a chromosome bouquet is formed and anchored to the NE via the telomeres. In animals and yeast this is essential for homologous pairing (Tomita and Cooper, 2006; Roberts *et al.*, 2009). In some plant species, however, this structure is not well defined. For instance, in the model plant Arabidopsis, the telomeres are first clustered around the nucleolus and then move to the nuclear periphery (Roberts *et al.*, 2009). How telomeres are linked to the nuclear periphery in plants remains to be investigated but the presence of plant SUN-domain proteins suggests that similar mechanisms may exist (Graumann and Evans, 2010b).

2.10 Lipid composition of the plant NE and its homeostasis

Eukaryotic endomembrane bilayers consist mainly of phospholipids and glycolipids but also contain smaller amounts of sterols and sphingolipids (Jouhet *et al.*, 2006; Aubert *et al.*, 2011). Instead of being randomly located throughout the membrane, lipid composition of membranes is controlled by protein-protein and protein-lipid interactions and results in specific membrane domains (Jouhet *et al.*, 2006; Aubert *et al.*, 2011). Thus membrane lipids, in addition to membrane imbedded proteins, play a significant role in establishing and maintaining the structure and function of specific membranes. Consequently, the lipid composition of the plant NE is also non-random and highly controlled. Biochemical analysis of onion root and stem tissue found that the NE membranes consist of approximately 60% lecithin and between 20–24% phosphatidyl ethanolamine. Non-polar lipids such as sterols and triglycerides make up 35–45% of the NE lipid content. Indeed, the onion NE has a relatively high sterol content with sitosterol making up 80% of the total sterols. Finally, the study also found that 80% of the fatty acids in the membrane phospholipids are unsaturated, contributing to the fluidity of the membrane (Philipp *et al.*, 1976).

Most of phospholipids, sterols and sphingolipids in endomembranes are synthesized in the ER and from there transported to other internal membranes or the plasma membrane. The physical continuity between ER and NE suggests that they may share a similar lipid composition. In support of this, Philipp *et al.* (1976) found that the rough ER (rER) and NE of onion cells have an identical phospholipid composition. It is therefore reasonable to think that the bulk of NE lipids are transported by lateral diffusion from the ER, through to the ONM, pore membrane and INM (Jouhet *et al.*, 2006). Two other transport modes exist to disperse lipids from the ER – diffusion across the bilayer mediated by flipases and movement of individual lipids outside the bilayer through an aqueous phase (Jouhet *et al.*, 2006). The latter method is mediated through membrane contact sites and lipid transfer proteins. Evidence from yeast and mammalian systems suggests such a transport mode is present at the NE and involved in NE lipid homeostasis (Jouhet *et al.*, 2006; Mijaljica *et al.*, 2010).

2.10.1 Nuclear-vacuolar junctions and lipid homeostasis

Membrane contact sites are areas where two membranes are in very close proximity – around 10 nm. These sites contain lipid transfer proteins, which can shuttle individual lipids through the aqueous phase between the two membranes (Jouhet *et al.*, 2006). The membrane contact site at the NE consists of the ONM and the tonoplast of the lytic vacuole; hence they are referred to as nucleus-vacuole junctions (Kvam and Goldfarb, 2004; Jouhet *et al.*, 2006).

These junctions have been best studied in *S. cerevisiae* but also exist in mammalian cells. Whether they are also present in plants remains to be studied. However, one of the protein components of the nucleus-vacuole junctions has a homolog in Arabidopsis. The two main proteins that maintain this junction through their interaction with each other are the ONM-localized Nvj1p and the tonoplast-localized Vac8p (Kvam and Goldfarb, 2004). Targeting to and specifically interacting with Nvj1p is the soluble oxysterol binding protein Osh1. The Arabidopsis homolog for this protein is At4g12460 (Jouhet *et al.*, 2006). While the functions of this protein still need to be investigated in plants, the yeast homologue of Osh1 is known to bind oxysterol, an oxygenated cholesterol derivative, is linked to post-synthetic regulation of sterols and mediates lipid exchange at membrane contact sites like the nucleus-vacuole junctions (Kvam and Goldfarb, 2004; Jouhet *et al.*, 2006).

The nucleus-vacuole junction is implied in regulating lipid homeostasis of the NE by a process called piecemeal microautophagy of the nucleus (PMN). During PMN, small portions of the nucleus, including membranes and nuclear contents, are pinched off into the vacuolar lumen by the nucleus-vacuole junctions for degradation. Localization of Osh1 at these junctions is important for this process suggesting that oxysterol is here used as a signalling messenger to induce PMN (Kvam and Goldfarb, 2004). This form of autophagy has been suggests to be a 'house-cleaning' mechanism for the nucleus under both normal and stress conditions by clearing out excess protein and membrane as well as domains of the nucleaus that are damaged or no longer active (Mijalijcs *et al.*, 2010). Again, this process has been well studied in yeast and is also known to occur in mammalian systems. Whether it occurs in plants is so far not known. However, microautophagy certainly takes place in plant cells and the need to regulate the composition of the NE and nucleus per se as well as the presence of a plant Osh1 homologue indicate that a process like PMN could also take place in plants.

2.10.2 NE phospholipid regulation by lipins

Apart from trafficking phospholipids from their place of synthesis, the ER, to the NE, to maintain the composition of the NE membranes, metabolism of phospholipids can also occur directly at the NE. Enzymes responsible for this are phosphatases called lipins and they are involved in two different processes (Sinissoglou, 2009). Firstly, they dephosphorylate phosphatidic acid into diacylglycerol and are thus involved in the triacylglycerol synthesis and phospholipid metabolism. Secondly, they are implied in directly regulating expression of genes involved in lipid metabolism. In addition, the yeast lipin Pah1 is also required for maintaining the spherical shape of nuclei (Sinissoglou, 2009). Knock-outs of Pah1 result in irregularly shaped nuclei with long stacks of NPC-containing membranes associated with the NE. As this soluble protein is translocated into the ONM, it is thought that changes in the ONM phosphatidic acid homeostasis in the knock out lead to the NE expansion

(Sinissoglou, 2009). Two lipin homologues have been identified in Arabidopsis and named AtPah1 and AtPah2 (Nakamura *et al.*, 2009). Like their yeast and mammalian counterparts, they contain phosphatase activity and are essential for maintaining lipid metabolism and adaptation to phosphate starvation. During phosphate starvation in plants, membrane lipid remodelling occurs, whereby phospholipids are converted to non-phosphorus galactolipids to free up phosphorus. In *AtPah1/AtPah2* double knock-out plants this remodelling is severely compromised (Nakamura *et al.*, 2009). However, so far it remains unclear whether either of the two proteins is localized in the plant NE and whether a single or double knock-out affects NE structure and function. Irrespective of this, it is becoming clear that coordinating production, transport and remodelling of phospholipids is essential in maintaining the structure and function of eukaryotic membranes including the plant NE.

2.11 The role of plant NE components in stress responses

2.11.1 Nuclei repositioning in response to environmental stimuli

It was already mentioned (Section 2.5) that cytoskeletal associations with the NE are involved in nuclear movement due to abiotic and biotic stress. Indeed, nuclear repositioning is one of the earliest defensive cell responses to external stimulation. Abiotic stimuli studied so far include blue light exposure and mechanical stress. In Arabidopsis leaf cells, nuclei positioning depends on light exposure. In the dark, nuclei are located at the bottom centre of the cell adjacent to the periclinal wall. Upon exposure of strong blue light of 50 μmol $m^{-2}s^{-1}$ and more the nuclei move to the anticlinal wall. This repositioning event is thought to protect from UV-induced DNA damage. Instead of being caused by general cytoplasmic streaming, the light induced nuclei repositioning is specifically regulated by phototropin receptor 2 and involves significant reorganization of the actin cytoskeleton (Iwabuchi *et al.*, 2007, Iwabuchi *et al.*, 2010).

In addition to light sensitivity, nuclear movement has also been observed in response to mechanical stimulation. Plant cells are exposed to repeated and varied strength mechanical stimulation exerted from neighbouring cells as well as from pathogens such as fungal hyphae. It has been shown that nuclei are highly sensitive to such pressures and can react to repeated stimuli of various strength by moving towards the sites of the stimuli without any loss of velocity or sensitivity (Figure 2.6; Qu and Sun, 2007). Pressures exerted from neighbouring cells are due to growth and cell division. In fact, the intracellular position of the nucleus predefines the division plane, hence appropriate nuclear positioning is essential for correct proliferation and morphology of plant tissues and organs (Qu and Sun, 2008).

While is has been shown that elements of the cell wall, plasma membrane and microtubule cytoskeleton are responsible for transduction of mechanical

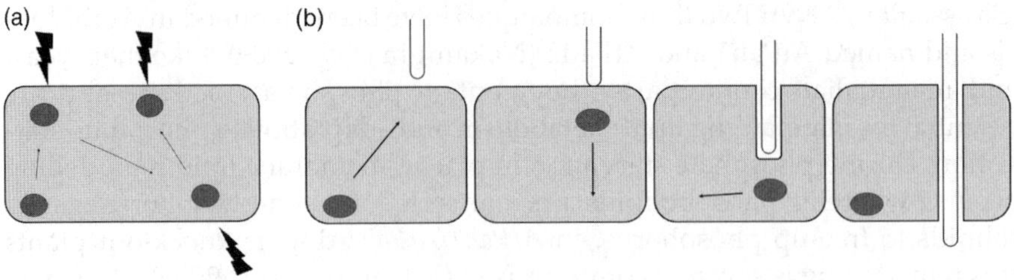

Figure 2.6 Nuclear movements in response to external stimuli. The position of the nucleus (grey) inside the cell is not fixed but changes in response to both abiotic (e.g. light, pressure) and biotic (e.g. fungal infection) stimuli. (a) Qu and Sun (2007) have shown that repeated mechanical stimuli (black) cause nuclear movement (arrows). Specifically, the nucleus moves towards the site where the stimulus has been applied. If several stimuli are applied in sequence, the nucleus moves to the various sites without any loss of speed or reactivity. (b) Genre *et al.* (2005) have shown that nuclear movement in root epidermal cells during fungal infections is a key step in the successful establishment of mycorrhiza. The nucleus first moves close to the site where the invading hyphae contact the cell. This is followed by the nucleus traversing the cell and thereby priming the path the invading hyphae take to traverse the cell.

signals (Qu and Sun, 2008), it is reasonable to speculate that components of the NE are also involved in these signalling events, which may effect NE-cytoskeletal associations, which in turn facilitate the nuclear movement. It will be a significant contribution to study the role of NE components in orchestrating nuclear repositioning events.

More is know about the function of NE components in response to fungal and bacterial pathogens. Plants undergo symbiotic relationships with both fungi and bacteria to effectively take up phosphorous and nitrogen, respectively. *Medicago truncatula* is a model organism to study both of these symbiotic relationships. Gigaspora is an arbuscular mycorrhizal fungus, which infects *M. truncatula* roots. During this infection, fungal hyphae penetrate root epidermal cells to spread towards the inner root cortex. The cell penetration is controlled by both plant and fungus and the plant nucleus plays a crucial role by orchestrating cellular events of this process (Genre *et al.*, 2005). As the hyphae exert pressure onto the epidermal cell, the nucleus moves towards this area (Figure 2.6), called an appressorium. Cytoskeletal elements and ER membranes accumulate in the space between nucleus and appressorium and together with the nucleus form the pre-penetration apparatus (Genre *et al.*, 2005). This assembly step is followed by the nucleus moving away from the appressorium towards the opposite side of the wall at speeds of 15–20 μm/h (Figure 2.6). Following the moving nucleus, endomembranes, actin and microtubule cytoskeleton reorganize to initiate the formation of a cytoplasmic column that stretches between the appressorium and the nucleus. It is inside this cytoplasmic column that the penetrating fungal hyphae traverses the plant cell (Genre *et al.*, 2005). While NE components associated with the cytoskeleton may be hypothesized to function in the nuclear movement, it has been shown that the NE-localized ion channel DMI1 is involved in Ca^{2+} signalling that governs this process (Genre *et al.*, 2005; Chabaud *et al.*, 2011).

This DMI1-dependent Ca^{2+} signalling has also been shown to be required for nodulation upon rhizobial bacteria infection of *M. truncatula* (Riely *et al.*, 2006; Peiter *et al.*, 2007). The bacteria release Nod factors, which trigger calcium spikes associated with the nucleus of the infected root cell. These calcium signals, in turn, activate the expression of genes involved in formation of root nodules. The NE- localized DMI1 (Riely *et al.*, 2006) is required for the generation and control of the calcium spikes but whether it acts as a Ca^{2+} or other cation channel remains unclear (Peiter *et al.*, 2007; Matzke *et al.*, 2009). Rhizobial bacteria enter the root system through infection threads that stretch from the root hair to other root cell layers. Similar to the cytoplasmic tubes that contain the penetrating fungal hyphae these infection threads require the reorganization of cytoskeletal elements and endomembranes. Notably, the growing infection thread in the root hair follows at a set distance to the migrating nucleus, which moves towards the bottom of the cell (Gage, 2004) suggesting that here, too, the nucleus marks the path of infection threads.

2.11.2 Functions of the plant NE during viral infection

Apart from fungal and bacterial infection, the nucleus also moves towards the site of viral infection. This has been studied in *Vigna unguiculata* (cowpea) during infection with cowpea rust virus. Similar to other nuclear movement events, this repositioning is actin dependent (Skalamera and Heath, 1998). During viral infection, NE components may not only play a role in nuclear movement, but the NE itself is utilized by the virus for replication and formation of mature virions. Such events have been studied during barley yellow dwarf virus (BYDV) and Sonchus yellow net virus (SYNV) infections (Dennison *et al.*, 2007; Goodin *et al.*, 2007). Both are RNA viruses whose replication cycles include a nuclear stage. The BYDV has a movement protein that is required for the transport of viral RNA (vRNA) into the nucleus. The exact mechanism remains to be studied, but it was found that this movement protein associates with and causes conformational changes of the NE. The N-terminus of this protein can directly interact with lipids and causes the formation of protrusions that emanate from the NE (Liu *et al.*, 2005). The lipid interaction is probably mediated by an amphiphilic α-helix at the N-terminus (Liu *et al.*, 2005; Dennison *et al.*, 2007). In addition, it was found that the movement protein can alter the physical properties of the NE, specifically its fluidity, and thereby induce surface pressures of as much as 2 mNm^{-1} (Dennison *et al.*, 2007). These alterations of the NE are thought to be involved in integrating the protein into the membrane. The purpose of this remains hypothetical with suggestions that it is either involved in the translocation of vRNA across the NE membranes (Liu *et al.*, 2005) or targets and associates vRNA at the NE (Dennison *et al.*, 2007).

By contrast, during SYNV infection, the ribonucleoprotein core of the virus is trafficked into the nucleus via the NPC (Jackson *et al.*, 2005). SYNV infection also causes NE morphological changes – nuclei are larger and numerous intranuclear membrane protrusions have been observed. Goodin *et al.* (2007) found that these protrusions are extensions of the NE and involved in maturation of virions. The SYN virus consists of a ribonucleoprotein core that is surrounded by a membrane layer. The membrane contains a viral glycoprotein and is thought to be attached to the viral core by a viral matrix protein. Viral replication and assembly of the core occurs inside the nucleus. Using fluorescent fusions to the matrix and glycoprotein, Goodin *et al.* (2007) have shown that the glycoprotein is located in the NE and the matrix protein associated with the nuclear side of the NE. The assembled core virions are thought to bud through the INM at the intranuclear membrane extensions. By budding through the INM, the virions acquire their membrane coat. Once in the periplasm, it is thought that the mature virions move into the ER lumen and traffic throughout the ER to either the cell periphery or plasmodesmata for cell-to-cell infection (Goodin *et al.*, 2007). The formation of the intranuclear membranes and the presence of viral glycoprotein in the plant NE demonstrate that the viral infection interferes with nucleoskeletal and NE membrane components to effect such dramatic changes.

2.12 Concluding remarks

The NE is an important structure as it is essential for a variety of cellular and nuclear processes. We have only just caught a glimpse of the functional capacity of the plant NE with many questions remaining to be answered. What are the protein components of the plant NE and what are their functional roles? Why is the plant NE structurally similar to yeast and metazoan NEs, yet seems to contain plant-specific components? How is the NE involved in chromatin anchorage and organization? What molecular processes define the roles the NE plays in stress and disease responses? The field of plant NE biology should have a bright future with a lot of potential for novel and exciting research.

Acknowledgements

DE and KG Acknowledge the support of the Leverhulme Trust under grant F/00 382/H and through a Leverhulme Early Career Fellowship to KG.

References

Alber, F., Dokudovskaya, S., Veenhoff, L.M. *et al.* (2007) The nolecular architecture of the nuclear pore complex. *Nature* **450**, 695–701.

Anderson, D.J. and Hetzer, M.W. (2008) Shaping the endoplasmic reticulum into the nuclear envelope. *Journal of Cell Science* **121**, 137–142.

Aubert, A., Marion, J., Boulogne, C. *et al.* (2011) Sphingolipids involvement in plant endomembrane differentiation: the BY2 case. *The Plant Journal* doi: 10.1111/j.1365-313X.2011.04481.x.

Avisar, D., Abu-Abied, M., Belausov, E. *et al.* (2009) A comparative study of the involvment of 17 Arabidopsis myosin family members on the motility of Golgi and other organelles. *Plant Physiology* **150**, 700–709.

Beaudouin, J., Gerlich, D., Daigle, N. *et al.* (2002) Nuclear envelope breakdown proceeds by microtubule-induced tearing of the lamina. *Cell* **108**, 83–96.

Bengtsson, L. (2007) What MAN1 does to the Smads TGFb/BMP signaling and the nuclear envelope. *The FEBS Journal* **274**, 1374–1382.

Blumenthal, S.S., Clark, G.B. and Roux, S.J. (2004) Biochemical and immunological characterization of pea nuclear intermediate filament proteins. *Planta* **218**, 965–975.

Brandizzi, F., Irons, S.L. and Evans, D.E. (2004) The plant nuclear envelope: new prospects for a poorly understood structure. *New Phytologist* **163**, 227–246.

Braunagel, S.C., Williamson, S.T., Ding, Q. *et al.* (2007) Early sorting of inner nuclear envelope proteins is conserved. *Proceedings of the National Academy of Sciences USA* **104**, 9307–9312.

Bupp, J.M., Martin, A.E., Stensrud, S.E. and Jaspersen, S.L. (2007) Telomere anchoring at the nuclear periphery requires the budding yest Sad1-UNC-84 domain proteins Mps3. *Journal of Cell Biology* **179**, 845–854.

Chabaud, M., Genre, A., Sieberer, B.J. *et al.* (2011) Arbuscular mycorrhizal hyphopodia and germinated spore exudates trigger Ca^{2+} spiking in the legume and nonlegume root epidermis. *New Phytologist* **189**, 347–355.

Charpentier, M., Bredemeier, R., Wanner, G. *et al.* (2008) Lotus japonicus CASTOR and POLLUX are ion channels essential for perinuclear calcium spiking in legume root endosymbiosis. *Plant Cell* **20**, 3467–3479.

Chytilova, E., Macas, J., Sliwinska, E. *et al.* (2000) Nuclear dynamics in Arabidopsis thaliana. *Molecular Biology of the Cell* **11**, 2733–2741.

Clarke, P.R. and Zhang, C. (2008) Spatial and temporal coordination of mitosis by Ran GTPase. *Nature Review Molecular Cell Biology* **9**, 464–477.

Craig, S. and Staehelin, L. (1988) High pressure freezing of intact plant tissues. Evaluation and characterization of novel features of the endoplasmic reticulum and associated membrane systems. *European Journal of Cell Biology* **46**, 81–93.

Crisp, M., Qian Liu, Q., Roux, K. *et al.* (2006) Coupling of the nucleus and cytoplasm: role of the LINC complex. *The Journal of Cell Biology* **172**, 41–53.

Cruz, J.R., de la Torre, C. and Moreno Diaz de la Espina, S. (2008) Nuclear actin in plants. *Cell Biology International* **32**, 584–587.

Cruz, J.R. and Moreno Diaz de la Espina, S. (2009) Subnuclear compartmentalisation and function of actin and nuclear myosin I in plants. *Chromosoma* **118**, 193–207.

D'Angelo, M.A., Anderson, D.J., Richard, E. and Hetzer, M.W. (2006) Nuclear pores form de novo from both sides of the nuclear envelope. *Science* **312**, 440–443.

Dawson, T.R., Lazarus, M.D., Hetzer, M.W. and Wente, S.R. (2009) ER membrane-bending proteins are necessary for de novo nuclear pore formation. *Journal of Cell Biology* **184**, 659–675.

Demidov, D., Van Damme, D., Geelen, D. *et al.* (2005) Identification and dynamics of two classes of aurora-like kinases in Arabidopsis and other plants. *The Plant Cell* **17**, 836–848.

Dennison, S., Harris, F., Brandenburg, K. and Phoenix, D. (2007) Characterization of the N-terminal segment used by the barley yellow dwarf virus movement protein to promote interaction with the nuclear membrane of host plant cells. *Peptides* **28**, 2091–2097.

De Ruijter, N., Ketelaar, T., Blumenthal, S.S. *et al.* (2000) Spectrin-like proteins in plant nuclei. *Cell Biology International* **24**, 427–438.

Ding, X., Xu, R., Yu, J. *et al.* (2007) SUN1 is required for telomere attachment to nuclear envelope and gametogenesis in mice. *Developmental Cell* **12**, 863–872.

Dittmer, T.A., Stacey, N.J., Sugimoto-Shirasu, K. and Richards, E.J. (2007) LITTLE NUCLEI genes affecting nuclear morphology in Arabidopsis thaliana. *The Plant Cell* **19**, 2793–2803.

Dixit, R. and Cyr, R.J. (2002) Spatio-temporal relationship between nuclear-envelope breakdown and preprophase band disappearance in cultured tobacco cells. *Protoplasma* **219**, 116–121.

Downie, L., Priddle, J., Hawes, C. and Evans, D.E. (1998) A calcium pump at the higher plant nuclear envelope. *Febs Letters* **429**, 44–48.

Drin, G., Casella, J.F., Gautier, R. *et al.* (2007) A general amphipathic alpha-helical motif for sensing membrane curvature. *Nature Sructural and Molecular Biology* **14**, 1380146.

Dultz, E., Zanin, E., Wurzenberger, C. *et al.* (2008) Systematic kinetic analysis of mitotic dis- and reassembly of the nuclear pore in living cells. *The Journal of Cell Biology* **180**, 857–865.

Ellenberg, J., Siggia, E.D., Moreira, J.E. *et al.* (1997) Nuclear membrane dynamics and reassembly in living cells: targeting of an inner nuclear membrane protein in interphase and mitosis. *The Journal of Cell Biology* **138**, 1193–1206.

Endre, G., Kereszt, A., Kevei, Z. *et al.* (2002) A receptor kinase gene regulating symbiotic nodule development. *Nature* **417**, 962–966.

Evans, D.E., Irons, S.L., Graumann, K. and Runions, J. (2009) The Plant Nuclear Envelope. In: Meier, I. (ed.), *Functional Organisation of the Plant Nucleus*, Springer, Berlin, Vol. **14**, p. 20.

Evans, D.E., Shvedunova, M. and Graumann, K. (2011) The nuclear envelope in the plant cell cycle: structure, function and regulation. *Annals of Botany* **107**, 1111–1118.

Fava, F., Raynaud-Messina, B., Leung-Tack, J. *et al.* (1999) Human 76p: a new member of the gamma-tubulin-associated protein family. *The Journal of Cell Biology* **147**, 857–868.

Fiserova, J., Kiseleva, E. and Goldberg, M.W. (2009) Nuclear envelope and nuclear pore complex structure and organisation in tobacco BY-2 cells. *The Plant Journal* **59**, 243–255.

Fournier, J., Timmers, A.C., Sieberer, B.J. *et al.* (2008) Mechanism of infection thread elongation in root hairs of Medicago truncatula and dynamic interplay with assocaited rhizobial colonization. *Plant Physiology* **148**, 1985–1995.

Gage, D. (2004) Infection and invasion of roots by symbiotic, nitrogen-fixing rhizobia during nodulation of temperate legumes. *Microbiol Mol Biol Rev* **68**, 280–300.

Galy, V., Antonin, W., Jaedicke, A. *et al.* (2008) A role for gp210 in mitotic nuclear-envelope breakdown. *Journal of Cell Science* **121**, 317–328.

Genre, A., Chabaud, M., Timmers, T. *et al.* (2005) Arbuscular mycorrhizal fungi elicit a novel intracellular apparatus in Medicago truncatula root epidermal cells before infection. *The Plant Cell* **17**, 3489–3499.

Gindullis, F., Peffer, N.J. and Meier, I. (1999) MAF1, a novel plant protein interacting with matrix attachment region binding protein MFP1, is located at the nuclear envelope. *Plant Cell* **11**, 1755–1768.

Gindullis, F., Rose, A., Patel, S. and Meier, I. (2002) Four signature motifs define the first class of structurally related large coiled-coil proteins in plants. *BMC Genomics* **3**.

Gleason, C., Chaudhurin, C., Yang, T. *et al.* (2006) Nodulation independent of rhizobia induced by a calcium-activated kinase lacking autoinhibition. *Nature* **441**, 1149–1152.

Goodin, M., Chakrabarty, R., Yelton, S. *et al.* (2007) Membrane and protein dynamics in live plant nuclei infected with Sonchus yellow net virus, a plant-adapted rhabdovirus. *Journal of General Virology* **88**, 1810–1820.

Graumann, K. and Evans, D.E. (2010a) The plant nuclear envelope in focus. *Biochemical Society Transactions* **38**, 307–311.

Graumann, K. and Evans, D.E. (2010b) Plant SUN domain proteins: Components of putative plant LINC complexes? *Plant Signalling and Behaviour* **5**, 154–156.

Graumann, K. and Evans, D.E. (2011) Nuclear envelope dynamics during plant cell division suggest common mechanisms between kingdoms. *Biochemical Journal* **435**, 661–667.

Graumann, K., Irons, S.L., Runions, J. and Evans, D.E. (2007) Retention and mobility of the mamalian lamin B receptor in the plant nuclear envelope. *Biology of the Cell* **99**, 553–562.

Graumann, K., Runions, J. and Evans, D.E. (2010) Characterisation of SUN-domain proteins at the higher plant nuclear envelope. *The Plant Journal* **61**, 134–144.

Griffis, E.R., Altan, N., Lippincott-Schwartz, J. and Powers, M.A. (2002) Nup98 is a mobile nucleoporin with transcription-dependent dynamics. *Molecular Biology of the Cell* **13**, 1282–1297.

Gruenbaum, Y., Margalit, A., Goldman, R.D. *et al.* (2005) The nuclear lamina comes of age. *Nature Reviews* **6**, 21–31.

Guttinger, S., Laurell, E. and Kutay, U. (2009) Orchestrating nuclear envelope disassembly and reassembly during mitosis. *Nature Review Molecular Cell Biology* **10**, 178–191.

Harel, A., Orjalo, A.V., Vincent, T. *et al.* (2003) Removal of a single pore subcomplex results in vertebrate nuclei devoid of nuclear pores. *Molecular Cell* **11**, 853–864.

Hetzer, M.W. (2010) The nuclear envelope. *Cold Spring Harbor Perspectives in Biology* **2**.

Huang, Y., Fang, J., Bedford, M.T. *et al.* (2006) Recognition of histone H3 lysine-4 methylation by the double tudor domain of JMJD2A. *Science* **312**, 748–751.

Irons, S.L., Evans, D.E. and Brandizzi, F. (2003) The first 238 amino acids of the human lamin B receptor are targeted to the nuclear envelope in plants. *Journal of Experimental Botany* **54**, 943–950.

Iwabuchi, K., Minamino, R. and Takagi, S. (2010) Actin reorganization underlies phototropin-dependent positioning of nuclei in Arabidopsis leaf cells. *Plant Physiology* **152**, 1309–1319.

Iwabuchi, K., Sakai, T. and Takagi, S. (2007) Blue light-dependent nuclear positioning in Arabidopsis thaliana leaf cells. *Plant and Cell Physiology* **48**, 1291–1298.

Jackson, A., Dietzgen, R., Goodin, M. *et al.* (2005) Biology of plant rhabdoviruses. *Annual Review of Phytopathology* **43**, 623–660.

Jeong, S.Y., Rose, A., Joseph, J. *et al.* (2005) Plant-specific mitotic targeting of RanGAP requires a functional WPP domain. *The Plant Journal* **42**, 270–282.

Jouhet, J., Marechel, E. and Block, M.A. (2006) Glycerolipid transfer for the building of membranes in plant cells. *Progress in Lipid Research* **46**, 37–55.

Kandasamy, M.K., McKinney, E.C., Meagher, R.B. (2002) Plant profilin isovariants are distinctly regulated in vegetative and reproductive tissues. *Cell Motility and the Cytoskeleton* **52** (1), 22–32.

Kawabe, A., Matsunaga, S., Nakagawa, K. *et al.* (2005) Characterization of plant Aurora kinases during mitosis. *Plant and Molecular Biology* **58**, 1–13.

Ketelaar, T., Moskalenko, C.F., Esseling, J.J. *et al.* (2002) Positioning of nuclei in Arabidopsis root hairs: an actin-regulated process of tip growth. *Plant Cell* **14**, 2941–2955.

Kimura, Y., Kuroda, C. and Masuda, K. (2010) Differential nuclear envelope assembly at the end of mitosis in suspension-cultured *Apium graveolens* cells. *Chromosoma* **119**, 195–204.

King, M.C., Lusk, C.P. and Blobel, G. (2006) Karyopherin-mediated import of integral inner nuclear membrane proteins. *Nature* **442**, 1003–1007.

Kiseleva, E., Allen, T.D., Rutherford, S. *et al.* (2004) Yeast nuclear pore complexes have a cytoplasmic ring and internal filaments. *Journal of Structural Biology* **145**, 272–288.

Kvam, E. and Goldfarb, D.S. (2004) Nvj1p is the outer-nuclear-membrane receptor for oxysterol-binding protein homolog Osh1p in Saccharomyces cerevisiae. *Journal of Cell Science* **117**, 4959–4968.

Lee, J.Y., Lee, H.S., Wi, S.J. *et al.* (2009) Dual functions of Nicotiana benthamiana Rae1 in interphase and mitosis. *The Plant Journal* **59**, 278–291.

Lenart, P., Rabut, G., Daigle, N. *et al.* (2003) Nuclear envelope breakdown in starfish oocytes proceeds by partial NPC disassembly followed by a rapidly spreading fenestration of nuclear membranes. *Journal of Cell Biology* **160**, 1055–1068.

Liu, K., Xia, Z., Wang, W. and Brandenburg, K. (2005) Interaction between the movement protein of barley yellow dwarf virus and the cell nuclear envelope: role of a putative amphiphilic alpha-helix at the N-terminus of the movement protein. *Biopolymers* **79**, 86–96.

Lusk, C.P., Blobel, G. and King, M.C. (2007) Highway to the inner nuclear membrane: rules for the road. *Nature Review Molecular Cell Biology* **8**, 414–420.

Masuda, K., Xu, Z.J., Takahashi, S. *et al.* (1997) Peripheral framework of carrot cell nucleus contains a novel protein predicted to exhibit a long a-helical domain. *Experimental Cell Research* **232**, 173 –181.

Mattaj, I.W. (2004) Sorting out the nuclear envelope from the endoplasmic reticulum. *Nature Review Molecular Cell Biology*, **5**, 65–69.

Matzke, M., Weiger, T.M., Papp, I. and Matzke, A.J.M. (2009) Nuclear membrane ion channels mediate root nodule development. *Trends in Plant Science* **14**, 295–298.

McGee, M.D., Stagljar, I. and Starr, D.A. (2009) KDP-1 is a nuclear envelope KASH proteins required for cell-cycle progression. *Journal of Cell Science* **122**, 2895–2905.

McNulty, A.K. and Saunders, M.J. (1992) Purification and immunological detection of pea nuclear intermediate filaments: evidence for plant nuclear lamins. *Journal of Cell Science* **103**, 407–414.

Meier, I. (2007) Composition of the plant nuclear envelope: theme and variations. *Journal of Experimental Botany* **58**, 27–34.

Meier, I. and Brkljacic, J. (2009) The nuclear pore and plant development. *Current Opinion in Plant Biology* **12**, 87–95.

Meinke, D., Muralla, R., Sweeney, C. and Dickerman, A. (2008) Identifying essential genes in Arabidopsis thaliana. *Trends in Plant Science* **13**, 483–491.

Merkle, T. (2004) Nucleo-cytoplasmic partitioning of proteins in plants: implications for the regulation of environmental and developmental signalling. *Current Genetics* **44**, 231–260.

Meyerzon, M., Fridolfsson, H.M., Ly, N. *et al.* (2009) UNC-93 is a nuclear-specific cargo adaptor for kinesin-1-mediated nuclear migration. *Development* **136**, 2725–2733.
Mijalijca, D., Prescott, M. and Devenish, R. (2010) The intricacy of nuclear membrane dynamics during nucleophagy. *Nucleus* **1**, 213–223.
Minguez, A. and Moreno Diaz de la Espina, S. (1993) Immunological characterisation of lamins in the nuclear matrix of onion cells. *Journal of Cell Science* **106**, 431–439.
Mitra, R.M., Gleason, R.A., Edwards, A. *et al.* (2004) A Ca2+/calmodulin-dependent protein kinase required for symbiotic nodule development: Gene identification by transcript-based cloning. *PNAS* **101**, 4701–4705.
Moriguchi, K., Suzuki, T., Ito, Y. *et al.* (2005) Functional isolation of novel nuclear proteins showing a variety of subnuclear localisations. *The Plant Cell* **17**, 389–403.
Muhlhausser, P. and Kutay, U. (2007) An in vitro nuclear disassembly system reveals a role for the RanGTPase system and microtubule-dependent steps in nuclear envelope breakdown. *Journal of Cell Biology* **178**, 595–610.
Murphy, S.M., Preble, A.M., Patel, U.K. *et al.* (2001) GCP5 and GCP6: two new members of the human gamma-tubulin complex. *Molecular Biology of the Cell* **12**, 3340–3352.
Murphy, S.P., Simmons, C.R. and Bass, H.W. (2010) Structure and expression of the maize (Zea mays L.) SUN-domain protein gene family: evidence for the existence of two divergent classes of SUN proteins in plants. *BMC Plant Biology* **10**.
Nakamura, Y., Koizumi, R., Shui, G. *et al.* (2009) Arabidopsis lipins mediate eukaryotic pathway of lipid metabolism and cope critically with phosphate starvation. *PNAS* **106**, 20978–20983.
Oda, Y. and Fukuda, H. (2011) Dynamics of Arabidopsis SUN proteins during mitosis and their involvement in nuclear shaping. *Plant Journal* **66**, 629–641.
Ohba, T., Schirmer, E.C., Nishimoto, T. and Gerace, L. (2004) Energy- and temperature-dependent transport of integral proteins to the inner nuclear membrane via the nuclear pore. *The Journal of Cell Biology* **167**, 1051–1062.
Oldroyd, G.E. and Downie, J.A. (2006) Nuclear calcium changes at the core of symbiosis signalling. *Current Opinion in Plant Biology* **9**, 351–357.
Olins, A.L., Rhodes, G., Welch, D.B. *et al.* (2010) Lamin B receptor: multi-tasking at the nuclear envelope. *Nucleus* **1**, 53–70.
Onischenko, E., Stanton, L.H., Madrid, A.S. *et al.* (2009) Role of the Ndc1 interaction network in yeast nuclear pore complex assembly and maintenance. *Journal of Cell Biology* **185**, 475–491.
Peiter, E., Sun, J., Heckmann, A.B. *et al.* (2007) The Medicago truncatula DMI1 protein modulates cytosolic calcium signaling. *Plant Physiology Preview* **145**, 192–203.
Nuclear spectrin-like proteins are structural actin-binding proteins in plants.
Pérez-Munive, C. and Moreno Diaz de la Espina, S. (2011) Nuclear spectrin-like proteins are structural actin-binding proteins in plants. *Biology of the Cell* **103**, 145–157.
Philipp, E., Franke, W., Keenan, T. *et al.* (1976) Characterisation of nuclear membranes and endoplasmic reticulum isolated from plant tissue. *The Journal of Cell Biology* **68**, 11–29.
Qu, L. and Sun, M. (2007) The plant cell nucleus is constantly alert and highly sensitive to repetitive local mechanical stimulations. *Plant Cell Reports* **26**, 1187–1193.
Qu, L. and Sun, M. (2008) The nucleus as a chief cellular organizer and active defender in response to mechanical stimulation. *Plant Signalling and Behaviour* **3**, 678–680.
Razafsky, D. and Hodzic, D. (2009) Bringing KASH under the SUN: the many faces of nucleo-cytoskeletal connections. *The Journal of Cell Biology* **186**, 461–472.

Riely, B.K., Lougnon, G., Ane, J.M. and Cook, D.R. (2006) The symbiotic ion channel homolog DMI1 is localized in the nuclear membrane of Medicago truncatula roots. *The Plant Journal* **49**, 208–216.

Roberts, N.Y., Osman, K. and Armstrong, S.J. (2009) Telomere distribution and dynamics in somatic and meiotic nuclei of *Arabidopsis thaliana*. *Cytogenetic and Genome Research* **124**, 193–201.

Rose, A. (2007) Open mitosis: nuclear envelope dynamics. In: Verma, D.P.S. and Hong, Z. (eds), *Cell Division Control in Plants*, Springer-Verlag, Heidelberg.

Rose, A. and Meier, I. (2001) A domain unique to plant RanGAP is responsible for its targeting to the plant nuclear rim. *Proceedings of the National Acadent of Science USA* **98**, 15377–15382.

Rose, A., Patel, S. and Meier, I. (2004) The plant nuclear envelope. *Planta* **218**, 327–336.

Rotem, A., Gruber, R., Shorer, H. *et al.* (2009) Importin beta regulates the seeding of chromatin with initiation sites for nuclear pore assembly. *Molecular Biology of the Cell* **20**, 4031–4042.

Roux, K.J., Crisp, M.L., Liu, Q. *et al.* (2009) Nesprin 4 is an outer nuclear membrane protein that can induce kinesin-mediated cell polarisation. *Proceedings of the National Academy of Sciences USA* **106**, 2194–2199.

Ruzicka, D.R., Kandasamy, M.K., McKinney, E.C. *et al.* (2007) The ancient subclasses of Arabidopsis actin depolymerising factor genes exhibit novel and differential expression. *The Plant Journal* **52**, 460–472.

Saksena, S., Shao, Y., Braunagel, S.C. *et al.* (2004) Cotranslational integration and initial sorting at the endoplasmic reticulum translocon of proteins destined for the inner nuclear membrane. *Proceedings of the National Academy of Sciences* **101**, 12537–12542.

Saksena, S., Summers, M.D., Burks, J.K. *et al.* (2006) Importin-alpha-16 is a translocon-associated protein involved in sorting membrane proteins to the nuclear envelope. *Nature, Stuctural and Molecular Biology* **13**, 500–508.

Salina, D., Bodoor, K., Eckley, D.M. *et al.* (2002) Cytoplasmic dynein as a facilitator of nuclear envelope breakdown. *Cell* **108**, 97–107.

Schirmer, E.C., Florens, L., Guan, T. *et al.* (2003) Nuclear membrane proteins with potential disease links found by subtractive proteomics. *Science* **301**, 1380–1382.

Schirmer, E.C. and Gerace, L. (2005) The nuclear membrane proteome: extending the envelope. *Trends in Biochemical Science* **30**, 551–558.

Seltzer, V., Janski, N., Canaday, J. *et al.* (2007) Arabidopsis GCP2 and GCP3 are part of a soluble c-tubulin complex and have nuclear envelope targeting domains. *The Plant Journal* **52**, 322–331.

Shimamura, M., Brown, R.C., Lemmon, B.E. *et al.* (2004) Gamma-tubulin in basal land plants: characterization, localization, and implication in the evolution of acentriolar microtubule organizing centers. *Plant Cell* **16**, 45–59.

Siniossoglou, S. (2009) Lipins, Lipids and Nuclear Envelope Structure. *Traffic* **10**, 1181–1187.

Skalamera, D. and Heath, M. (1998) Changes in the cytoskeleton accompanying infection-induced nuclear movements and the hypersensitive response in plant cells invaded by rust fungi. *The Plant Journal* **16**, 191–200.

Smit, P., Raedts, J., Portyanko, V. *et al.* (2005) NSP1 of the GRAS protein family is essential for rhizobial Nod factor-induced transcription. *Science* **308**, 1789–1791.

Staehelin, L.A. (1997) The plant ER: a dynamic organelle composed of a large number of discrete functional domains. *Plant J* **11**, 1151–1165.

Starr, D.A. (2009) A nuclear envelope bridge positions nuclei and moves chromosomes. *Journal of Cell Science* **122**, 577–586.

Starr, D.A. and Fridolfsson, H.N. (2010) Interactions between nuclei and the cytoskeleton are mediated by SUN-KASH nuclear-envelope bridges. *Annual Review of cell and Developmental Biology* **10**, 421–444.

Stewart, C.L., Roux, K.J. and Burke, B. (2007) Blurring the boundary: The nuclear envelope extends its reach. *Science* **318**, 1408–1412.

Tamura, K., Fukao, Y., Iwamoto, M., Haraguchi, T. and Hara-Nishimura, I. (2010) Identification and characterization of nuclear pore complex components in Arabidopsis thaliana. *Plant Cell* **22**, 4084–4097.

Tomita, K. and Cooper, J.P. (2006) The meiotic chromosomal bouquet: SUN collects flowers. *Cell* **125**, 19–21.

Van Damme, D., Bouget, F.Y., Van Poucke, K. *et al.* (2004) Molecular dissection of plant cytokinesis and phragmoplast structure: a survey of GFP-tagged proteins. *The Plant Journal* **40**, 386–398.

Walther, T.C., Askjaer, P., Gentzel, M. *et al.* (2003) RanGTP mediates nuclear pore complex assembly *Nature* **424**, 689–694.

Webster, M., Witkin, K. and Cohen-Fix, O. (2009) Sizing up the nucleus: nuclear shape, size and nuclear-envelope assembly. *Journal of Cell Science* **122**, 1477–1486.

Wilhelmsen, K., Ketema, M., Truong, H. and Sonnenberg, A. (2006) KASH-domain proteins in nuclear migration, anchorage and other processes. *Journal of Cell Science* **119**, 5021–5029.

Wilson, K.L. and Berk, J.M. (2010) The nuclear envelope at a glance. *Journal of Cell Science* **123**, 1973–1978.

Worman, H.J., Yuan, J., Blobel, G. and Georgatos, S.D. (1988) A lamin B receptor in the nuclear envelope. *Proc Natl Acad Sci USA* **85**, 8531–8534.

Xu, X.M. and Meier, I. (2008) The nuclear pore comes to the fore. *Trends in Plant Science* **13**, 20–27.

Xu, X.M., Meulia, T. and Meier, I. (2007) Anchorage of plant RanGAP to the nuclear envelope involves novel nuclear-pore-associated proteins. *Current Biology* **17**, 1157–1163.

Zargari, A., Boban, M., Heessen, S. *et al.* (2007) Inner nuclear membrane proteins Asi1, Asi2, and Asi3 function in concert to maintain the latent properties of transcription factors Stp1 and Stp2. *Journal of Biological Chemistry* **282**, 594–605.

Zhang, C.M. and Clarke, P.R. (2000) Chromatin-independent nuclear envelope assembly induced by Ran GTPase in Xenopus egg extracts. *Science* **288**, 1429–1432.

Zhang, X., Ley, K., Yuan, X. *et al.* (2009) SUN1/2 and Syne/Nesprin-1/2 complexes connect centrosomes to the nucleus during neurogenesis and neuronal migration in mice. *Neuron* **64**, 173–187.

Zhao, Q., Brkljacic, J. and Meier, I. (2008) Two distinct interacting classes of nuclear envelope-associated coiled-coil proteins are required for the tissue specific nuclear envelope targeting of Arabidopsis RanGAP. *Plant Cell* **20**, 1639–1651.

Zhou, K., Rolls, M.M., Hall, D.H., Malone, C.J. and Hanna-Rose, W. (2009) A Zyg-12-dynein interaction at the nuclear envelope defines cytoskeletal architecture in the C. elegand gonad. *Journal of Cell Biology* **186**, 229–241.

Zhou, X., Graumann, K., Evans, D.E. and Meier, I. (2012) Novel plant KASH-SUN bridges are involved in RanGAP anchoring and nuclear shape determination. *Journal of Cell Biology* **196**, 203–211.

Annual Plant Reviews (2013) **46**, 57–92
doi: 10.1002/9781118472507.ch3 http://onlinelibrary.wiley.com

Annual Plant Reviews

Chapter 3
THE PLANT NUCLEAR PORE COMPLEX – THE NUCLEOCYTOPLASMIC BARRIER AND BEYOND

Xiao Zhou, Joanna Boruc and Iris Meier
Department of Molecular Genetics, The Ohio State University, Columbus, OH 43210

Abstract: The nuclear pore complex (NPC) provides a highly organized pathway for selective transport between the nucleoplasm and the cytoplasm. Recent work undertaken to characterize the mechanisms and regulation of transport through the plant NPC has resulted in the identification of key components showing similarities – and also significant differences – between plants and other organisms. Mutant studies reveal roles for nucleoporins in plant–microbe interactions, hormone response, abiotic-stress tolerance, plant development, and flowering-time regulation. Recently, significant progress has been made in identifying about 30 proteins that constitute the plant NPC as well as several NPC-associated proteins and in characterizing the structure of the plant NPC. A newly discovered connection between nuclear pore-associated proteins and inner nuclear envelope proteins expands our knowledge of plant nuclear envelope architecture. Dynamic patterns of subcellular localization suggest mitotic functions of plant nucleoporins away from the nuclear pore.

Keywords: nuclear pore complex; nucleoporin; nucleo-cytoplasmic transport; Ran cycle; KASH proteins; karyopherins; biotic and abiotic response; mitosis

3.1 Nuclear pore complex structure

The evolution of an intracellular, membrane-enclosed apparatus distinguishes eukaryotes from prokaryotes and enables eukaryotes to adopt a more complex multicellular lifestyle. An important component of this

---

Annual Plant Reviews Volume 46: Plant Nuclear Structure, Genome Architecture and Gene Regulation, First Edition. Edited by David E. Evans, Katja Graumann and John A. Bryant.

endomembrane system is the nucleus, which gives the eukaryotes (Greek for 'true nucleus') their name. A double-membrane structure, the nuclear envelope (NE), confines most of the cell's genetic material inside the nucleus. This cellular innovation created a new intracellular trafficking issue – namely the need to selectively transport macromolecules between the nucleoplasm and the cytoplasm – nucleocytoplasmic transport. The nuclear pore complexes (NPCs) are the sole site of this transport. NPCs are large protein complexes embedded in the NE, connecting the nucleoplasm and the cytoplasm. Here, our knowledge of the plant NPC will be discussed and compared with our more advanced understanding of the yeast and vertebrate NPC.

3.1.1 Structure of the NPC

Electron microscopy revealed an overall NPC appearance that is conserved throughout eukaryotes (reviewed in Wente, 2000; Lim *et al.*, 2008; Brohawn *et al.*, 2009; Wente and Rout, 2010). The NPC consists of a ring of eight units arranged in eightfold radial symmetry. Two rings (as viewed from both the cytoplasmic and nucleoplasmic side) sandwich a core ring containing eight spokes surrounding a central channel. The central channel is sometimes occupied by unstructured material referred to as the 'central plug' or the 'transporter'. This ring-spoke structure makes the NPC resemble an eight-spoke wheel. At the cytoplasmic side, eight filaments extend from this 'wheel' and protrude into the cytoplasm, while at the nucleoplasmic side a basket structure formed by filaments reaches the nucleoplasm. The whole NPC has a diameter of 100–150 nm and a depth of 50–70 nm, depending on the organism.

Having a diameter of ~105 nm, tobacco BY-2 cell NPCs fall between those of *Xenopus* and yeast (Fiserova *et al.*, 2009). Although the basic organization is conserved, NPCs of BY-2 cells can be grouped into four categories based on their conformation (Fiserova *et al.*, 2009). Viewed from the cytoplasmic side, category I has a star-like cytoplasmic ring instead of a circular ring; category II is featured by round cytoplasmic rings with undistinguishable subunits; category III has a round ring with defined subunits and a central plug; category IV is similar to category III but with filaments emanating from the ring subunits. The presence of these categories appears to be age-related. Category I NPCs are very rare and are only observed in rapidly proliferating three-day-old BY-2 cells, which have mainly category II NPCs, while category II and III are predominant in ten-day-old BY-2 cells (the stationary phase).

Another feature that is possibly affected by cell age is the spatial arrangement of NPCs (Fiserova *et al.*, 2009). In three-day-old BY-2 cells, NPCs tends to show up in pairs with an inter-NPC distance of 25 nm. In ten-day-old BY-2 cells, NPCs form lines of 5–30 pores in length, reducing the inter-NPC distance to 6.5 nm. An NPC density variance was also observed in neuron cells of mice during postnatal development (Lodin *et al.*, 1978).

In animals, NPCs are arranged in distinct patterns at the NE due to their attachment to the nucleoskeleton, the lamina. In contrast, yeast NPCs were shown to form highly mobile clusters, correlating with the absence of a lamina in yeast (Belgareh and Doye, 1997, Winey *et al.*, 1997, Bucci and Wente, 1998). Although no genes for *bona fide* lamins have been found in plant genomes, the carrot (*Daucus carota*) coiled-coil protein Nuclear Matrix Constituent Protein1 (NMCP1) and its Arabidopsis homologues LITTLE NUCLEI (LINC) have been discussed as candidates for plant lamin-like genes (Masuda *et al.*, 1997; Dittmer *et al.*, 2007). Moreover, a filamentous network beneath the inner NE of the tobacco BY-2 cells was observed to spread among and be connected to the nucleoplasmic ring of the NPCs (Fiserova *et al.*, 2009). This structure resembles the animal nuclear lamina, though the identity of these filaments still needs to be established. It is thus possible that a plant lamina-like structure (termed the 'plamina' by Fiserova *et al.*) is involved in the spatial arrangements of the BY2 NPCs. Since only BY-2 suspension cell nuclei have so far been thoroughly investigated by electron microscopy, it is unclear whether the cell-age-related NPC features are also present in other plant species and play a role during plant development. However, one early study documented that the NPC density of the vegetative pollen nuclei was highly increased on the surface facing the generative cells (Shi *et al.*, 1991). Thus, the spatial arrangement of NPCs might be developmentally regulated in plants, too.

3.1.2 Molecular composition of the NPC

The molecular composition of the NPC has been well studied in yeast and vertebrates, and a detailed map of the position of each component has been proposed (reviewed in Brohawn *et al.*, 2009; Strambio-De-Castillia *et al.*, 2010; Wente and Rout, 2010; Hoelz *et al.*, 2011). From these studies, the NPC is composed of ~500 individual polypeptides and has a molecular mass of ~50 MDa, which makes it one of the largest nano-machines in a cell. Despite its high molecular mass, the NPC is composed of multiple copies of only ~30 proteins called nucleoporins (Nups). Correlated with the eightfold symmetrical appearance of the NPC, Nups are arranged in eight spokes, each of which contains five subcomplexes. The Nups can be divided into three categories: the transmembrane Nups, the scaffold Nups, and the Phe-Gly (FG) Nups (Figure 3.1). The transmembrane Nups span the nuclear membrane at the NE equator where the NPC is embedded, forming the outmost ring-like subcomplex (also known as the luminal ring, Plate 3.1, yellow frame) that anchors the NPC to the NE. The scaffold Nups constitute the inner ring of the NPC and are organized in two subcomplexes (in *Saccharomyces cerevisiae* the Nup84 and the Nic96 complexes; in vertebrate the Nup107-Nup160 and the Nup35-Nup155 complexes, Plate 3.1, purple and blue frames). The Nup84 or the Nup107-Nup160 complex, which exhibits a Y-shape in electron micrographs, is universally conserved and has been shown to be critical for NPC assembly in different organisms (Boehmer *et al.*, 2003; Galy *et al.*, 2003; Harel

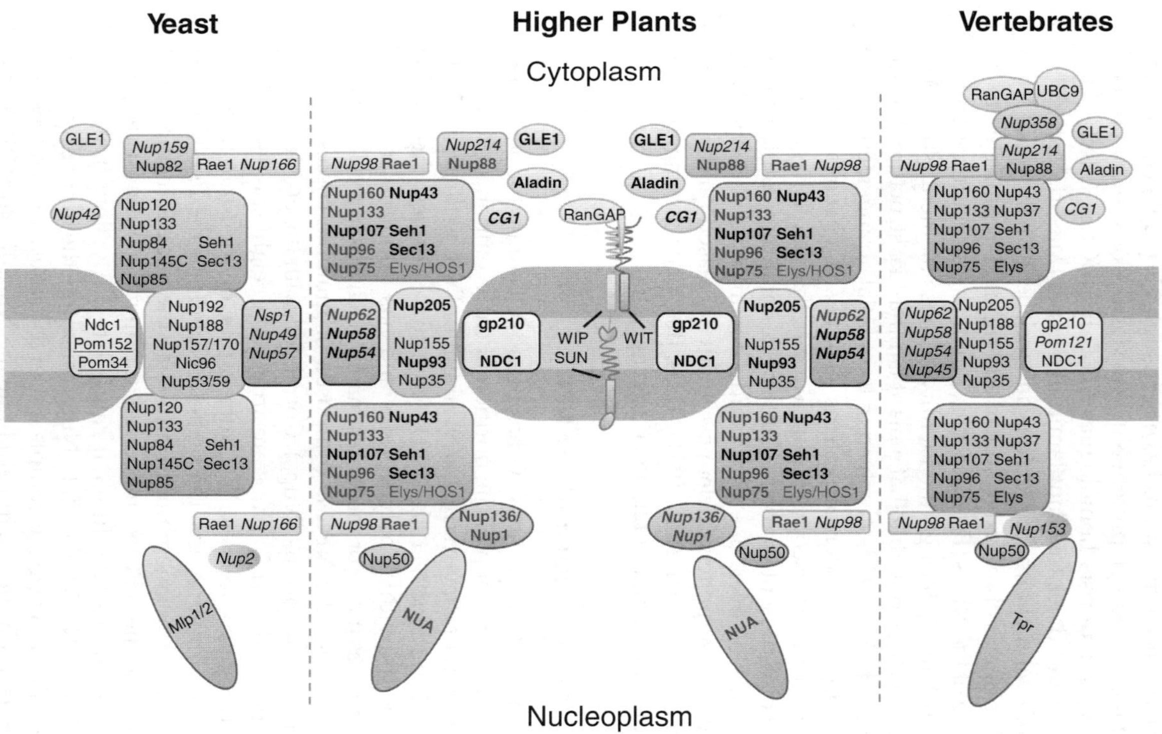

Figure 3.1 Comparison of the NPC components among higher plants, yeast and vertebrates. In Plate 3.1 protein complexes are grouped in single units, and the NPC subcomplexes are indicated in different colours – orange, cytoplasmic filaments; purple, Y-shape subcomplex; blue, Nup35-155/Nic96 subcomplex; grey, central FG subcomplex; and red, nuclear basket subcomplex. Other proteins or protein complexes are shown in green. FG Nups are indicated by italic font. Contacting units indicate confirmed interactions. In the higher plant NPC, bold protein names indicate confirmed NE localization, and mutant phenotypes have been reported for plant Nups indicated in red. In yeast, the underlined Pom152 and Pom34 have no homologues in vertebrates or higher plants. Positioning of plant Nups is based on their vertebrate counterparts. (*For colour details please see colour plate section.*)

et al., 2003, Walther *et al.*, 2003). In most NPC models, the Y-shaped complexes are present at both the cytoplasmic and the nucleoplasmic side, sandwiching the Nup35-Nup155 or Nic96 complex (Brohawn *et al.*, 2009). Attached to the scaffold complexes are the FG Nups, which contain unfolded FG repeats. The FG Nups count for one-third of the total mass of the NPC. Positioned at the center of the NPC, they form the central FG subcomplex that plays a crucial role in the selectivity of the NPC (Plate 3.1, gray frame and see below). FG Nups are also present in the cytoplasmic filament sub-complex (Plate 3.1, orange) and nuclear basket sub-complex (Plate 3.1, red). The cytoplasmic filaments are suggested to facilitate export cargo release (reviewed in Strambio-De-Castillia *et al.*, 2010). In vertebrates, the cytoplasmic filament component Nup358 anchors the Ran GTPase-activating protein (RanGAP), which is a key player in nucleo-cytoplasmic trafficking (see below). The nuclear basket is mainly composed of the long coiled-coil proteins Mlp1/Mlp2 (for Myosin-like protein 1/2) in yeast and Tpr (for translocated promoter region) in vertebrates. Coiled coils often function in protein-protein interactions, and indeed, the nuclear basket and its associated factors are involved in gene expression regulation, chromatin maintenance, and mediating cell division (Strambio-De-Castillia *et al.*, 2010). In vertebrates, NPCs are further connected to the lamina through the lamin-Nup153 interaction at the nuclear basket (not shown in Plate 3.1) (Walther *et al.*, 2001).

The protein composition of the plant NPC was a mystery, since only eight plant Nups had been identified – far less than the predicted number. This situation was changed by a recent Arabidopsis proteomic study using RAE1-GFP (an mRNA export factor) as bait. A combination of immunoprecipitations, mass spectrometry analysis, homology study, and NE localization confirmation identified 22 new plant Nups (Tamura *et al.*, 2010). In addition, a second Arabidopsis transmembrane Nup (NDC1) has been reported (Stavru *et al.*, 2006a; Boruc *et al.*, 2011), adding up to a total of 31 plant Nups, likely to be close to the complete protein complement of the plant NPC. Based on a comparison of protein similarity and composition, the plant NPC is more similar to the NPC of vertebrates than of yeast (Tamura *et al.*, 2010). Although the five sub-complexes and three Nup categories are well conserved, several Nups are missing or replaced by plant-specific proteins in each sub-complex when compared with vertebrates. Pom121, Nup188, Nup37, Nup45, and Nup358 are missing in the luminal ring complex, the Nup35-Nup155 complex, the Nup107-Nup160 complex, the central FG Nup complex, and the cytoplasmic filament complex, respectively (Figure 3.1).

In plants, the role of Nup358 in anchoring RanGAP to the NE is taken over by two plant-specific NE proteins – WPP (for highly conserved tryptophan and proline residues) domain interacting proteins (WIPs) and WPP-domain tail-anchored proteins (WITs) (Xu *et al.*, 2007a; Zhao *et al.*, 2008). Nup358 contains a small ubiquitin-related modifier (SUMO) ligase E3 domain, which interacts with SUMO-conjugating protein 9 (UBC9) and forms a triple complex with SUMO-conjugated RanGAP (Matunis *et al.*, 1996, 1998; Reverter and

Lima, 2005). In plants, RanGAP seems not to be SUMOylated but rather contains a plant-specific WPP domain that interacts with the coiled-coil domain of WIPs and WITs (Xu *et al.*, 2007a; Zhao *et al.*, 2008). Recently, it has been shown that AtWIPs are plant-specific KASH (for Klarsicht/ANC-1/Syne-1 homology) proteins interacting with *Arabidopsis* SUN (for Sad1/UNC-84 homology) proteins (AtSUNs). The interaction is mediated through a highly conserved, yet plant-specific, C-terminal tail of AtWIPs and an extended SUN domain of AtSUNs, which is conserved among land plants. The AtSUN-AtWIP1 interaction is required for the NE localization of AtWIP1 and consequently the NE localization of AtRanGAP1, a function for SUN-KASH complexes not reported in any other organism. AtWIPs and AtSUNs are necessary for maintaining the elongated nuclear shape of *Arabidopsis* epidermal cells. Together, these data identify the first KASH members in the plant kingdom and provide a novel function of SUN-KASH complexes (Zhou *et al.*, 2012; see also Chapter 2 of the current volume).

Nup358 is a multifunctional protein that not only binds RanGAP1 but contains multiple domains that are involved in SUMO E3 activity, Ran-binding, COPI (coatomer protein I) interaction, and karyopherin docking (Wu *et al.*, 1995; Matunis *et al.*, 1998; Saitoh *et al.*, 2006). Interestingly, these domains are absent in both WIPs and WITs. Similarly, plant-specific Nup136 (also known as Nup1) is proposed to be a functional analogue of the vertebrate nuclear basket component Nup153, which also contains multiple domains (Tamura *et al.*, 2010). Although both Nup136 and Nup153 have FG repeats at their C-termini, bind importins, and show similar dynamics upon NE assembly, Nup153 contains additional domains that are missing in Nup136. These are the Nup153 N-terminal domain, which acts in bridging Tpr to the Nup107-Nup160 subcomplex, and the zinc-finger motif which interacts with COPI and Ran (Vasu *et al.*, 2001; Walther *et al.*, 2001; Hase and Cordes, 2003; Boehmer *et al.*, 2003). Interestingly, both Nup385 and Nup153 interact with COPI (Prunuske *et al.*, 2006) and Ran (Higa *et al.*, 2007), and none of their plant functional counterparts contains the corresponding functional domains. The third example for a smaller plant analogue of a more complex vertebrate Nup is HOS1 (for HIGH EXPRESSION OF OSMOTICALLY RESPONSE GENE1), which contains a region homologous to Elys (for Embryonic large molecule derived from yolk sac) but has only half the size of Elys (Tamura *et al.*, 2010). It is plausible that the functions of missing Nups or missing domains are either specific to the vertebrate NPC or fulfilled by other plant Nups (Tamura *et al.*, 2010).

3.1.3 Nucleocytoplasmic trafficking

The main function of NPCs is to control molecular trafficking between the cytoplasm and the nucleoplasm. Small molecules with a molecular weight of less than ~40 kDa or a diameter smaller than ~5 nm can pass NPCs relatively freely (Feldherr and Akin, 1997; Keminer and Peters, 1999). Water, ions and metabolites fall into this category (Figure 3.2). Transport of larger

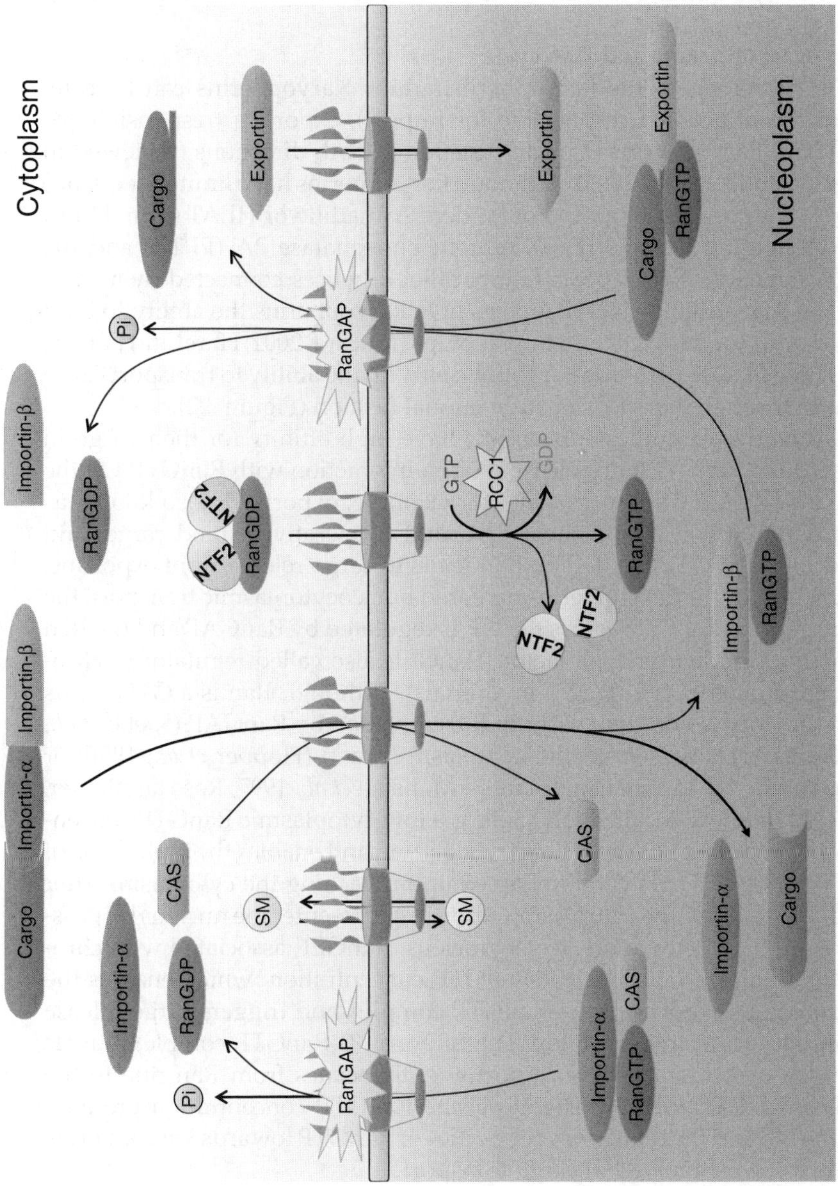

Figure 3.2 The Ran cycle. Small molecules, less than 40 KDa (SM), can travel through the NPC freely, whereas molecules with higher molecular weight need NTF-assistant transport to pass through the NPC. Please refer to the main text for details. The high-to-low RanGTP concentration gradient across the NE is presented by the greyscale change.

molecules, including proteins, ribosomal units, and messenger ribonucleoprotein (mRNPs) requires an active mechanism facilitated by soluble nuclear transport factors (NTFs), which have the abilities of cargo binding, NPC docking, and a cargo-binding-affinity switch to impart translocation directionality.

3.1.3.1 Karyopherins and Ran cycle

Most NTFs belong to the karyopherin family. Karyopherins can be categorized into importins (responsible for import), exportins (responsible for export), and karyopherins that can function in both directions (reviewed in Pemberton and Paschal, 2005). Although karyopherins have limited sequence similarity, they are all composed of tandem Armadillo or HEAT (from Huntingtin, elongation factor 3 (EF3), protein phosphatase 2A (PP2A) and the yeast PI3-kinase TOR1) repeats (antiparallel α-helices connected by a short turn). The Armadillo or HEAT repeats give karyopherins the ability to bind cargo, RanGTP or RanGDP, and the FG Nups (Macara, 2001; Harel and Forbes, 2004). These features provide karyopherins with the ability to transport cargo molecules through the NPCs in a directional fashion (Figure 3.2).

Importins (for example, Importin-β) have high affinity for their cargo in the absence of RanGTP, and release it upon interaction with RanGTP. On the other hand, RanGTP and exportins (for example, exportin 1, also known as Chromosome region maintenance 1, Crm1) cooperatively bind cargo, and the hydrolysis of GTP to GDP by Ran leads to cargo release from exportins. Being critical for the karyopherin-mediated nucleocytoplasmic transport, the concentration of RanGTP across the NE is regulated by RanGAP and the Ran guanine nucleotide exchange factor (RanGEF, also called regulator of chromosome condensation 1, RCC1, in animals). Although Ran is a GTPase, its intrinsic activity is very low without the activation by RanGAP (Klebe *et al.*, 1995). RanGAP is present in the cytoplasm (yeast) (Hopper *et al.*, 1990) or enriched at the NE (animal and plants) (Mahajan *et al.*, 1997; Rose and Meier, 2001). This RanGAP localization leads to a low cytoplasmic RanGTP concentration environment for importin-cargo loading and enables the hydrolysis of RanGTP in the exportin-cargo complex upon reaching the cytoplasm. After cargo dissociation in the cytoplasm, exportin can re-enter the nucleus by passing through the nuclear pore. In the nucleus, RanGEF, associated with chromatin, maintains a high nuclear RanGTP concentration, which enables the formation of an exportin-cargo-RanGTP complex and triggers cargo release from an importin-cargo complex. The Importin-β-RanGTP complex travels back to the cytoplasm, where Importin-β dissociates from Ran due to the hydrolysis of GTP to GDP. Therefore, the RanGTP concentration gradient across the NE and the regulatory properties of RanGTP towards karyopherins facilitate the directionality of cargo transport.

The recognition of cargo by transport receptors is determined by transport signals: the nuclear localization signal (NLS) and the nuclear export signal (NES). For protein cargo, the NLS can be a short basic peptide that allows binding of Importin-α. Despite the similar name, Importin-α belongs to a different protein family and serves as a cargo-adaptor for Importin-β by providing an Importin-β-binding domain for the cargo. Unlike Importin-β, Importin-α cannot penetrate the NPC on its own. Instead, it requires a special exportin, CAS (for Cellular Apoptosis Susceptibility) in order to be exported back to the cytoplasm (Kutay *et al.*, 1997; Haasen and Merkle, 2002). Snurportin is another nuclear import adaptor involved in the trimethylguanosine-cap-dependent nuclear import of uridine-rich small nuclear ribonucleoproteins (Huber *et al.*, 1998). Unlike Importin-α, it is re-exported to the cytoplasm by Exportin 1/Crm1 (Gorlich *et al.*, 1999). Proteins with an NES can form complexes with exportins and RanGTP for nuclear export and there are also proteins that serve as export adaptors. For example, NMD3 and PHAX (for phosphorylated adaptor for RNA export) serve as export adaptors for Crm1-mediated export of the 60S ribosomal subunit and small nuclear RNA, respectively (Ho *et al.*, 2000; Ohno *et al.*, 2000; Gadal *et al.*, 2001; Thomas and Kutay, 2003; Mourao *et al.*, 2010; Sengupta *et al.*, 2010). Furthermore, a protein can have both NLS and NES to shuttle between the cytoplasm and the nucleus. The accessibility of these signals can be regulated, for example by phosphorylation or protein-protein interactions and can depend on external signals, the phase of cell cycle, or the developmental stage (Weis, 2003; Terry *et al.*, 2007; Meier and Somers, 2011).

To date, three plant importins (SAD2 (for Super sensitive to ABA and Drought), Transportin 1, and Importin-β) and four exportins (Exportin 1/Crm1, Exportin 2/CAS, Exportin-T/PAUSED, and Exportin 5/Hasty) have been functionally characterized (reviewed in Merkle, 2011). For adaptors, only Importin-α has been characterized in plants but it has been reported that putative orthologues of Snurportin1, NMD3, and PHAX are encoded in the Arabidopsis genome (Merkle, 2011).

The karyopherin-mediated Ran cycle involves another family of proteins, represented by RanBP1 (for Ran-binding protein 1). RanBP1 family proteins interact with Ran but do not function as NTFs (Merkle, 2011). Instead, through their Ran-binding affinity, they regulate the hydrolysis of RanGTP (RanBP1) (Bischoff *et al.*, 1995), anchor RanGAP to the NE (RanBP2, which does not exist in the plant genome, see above), and serve as a cofactor for the Crm1 export complex (RanBP3) (Lindsay *et al.*, 2001; Mattaj *et al.*, 2001). In Arabidopsis, six proteins with sequence similarity to RanBP1 have been identified (Merkle, 2011). AtRanBP1a has specific binding affinity for RanGTP (Haizel *et al.*, 1997), and AtRanBP1c is a co-activator of RanGAP (Kim and Roux, 2003; Roux and Kim, 2003). However, the functions of the other AtRanBP1 proteins are largely unknown.

3.1.3.2 Non-karyopherin transport

Besides karyopherin family proteins, there are other carriers that function as NTFs. NTF2 is a RanGDP nuclear import factor and has higher affinity for RanGDP than RanGTP (Macara *et al.*, 1998). NTF2-mediated nuclear import of Ran is enforced by nuclear RanGEF, which recharges the imported RanGDP with GTP thereby triggering the release of RanGTP from NTF2 (Macara *et al.*, 1998).

Other non-karyopherin mediated nuclear trafficking includes the export of mRNPs that mainly depend on NXF1 (for Nuclear RNA Export Factor 1) in animals or Mex67 (for Messenger RNA Export factor of 67 kDa) in yeast (Erkmann and Kutay, 2004). NXF1 heterodimerizes with NXT1 (for NTF2-like Export Factor 1), while Mex67 heterodimerizes with the mRNA transport regulator Mtr2 (Cullen, 2003; Izaurralde and Stutz, 2003). NXT1 and Mtr2 have no sequence similarity but they are functional homologues (Conti and Fribourg, 2003). Although NXT1 contains a NTF2-like domain, it does not function like NTF2 (Black *et al.*, 1999). Like karyopherins, Mex67-Mtr2 heterodimers form complexes with their cargo – mRNPs – and bind directly to FG Nups, but it seems that they prefer a different set of FG Nups from those that interact with the karyopherins (Wente and Terry, 2007; Wente and Carmody, 2009). Upon arrival at the cytoplasmic side of the NPC, Mex67-Mtr2 interacts with Dbp5 (DEAD box protein 5), an ATP-dependent RNA helicase, to release mRNPs by altering their protein composition (Chang *et al.*, 1998; Cole *et al.*, 1998). Dbp5 shuttles between the cytoplasm and the nucleus but it is also enriched on the cytoplasmic side of the NPC (Daneholt *et al.*, 2002). Activation of Dbp5 requires Gle1 and the small molecule inositol hexakisphosphate (York *et al.*, 1999; Alcazar-Roman *et al.*, 2006; Weirich *et al.*, 2006). The dissociation of the mRNP complex from Mex67-Mtr2 by Dbp5 finalizes the mRNP export and ensures its directionality.

In plants, genes encoding NXF1 or Mex67 are missing and the mechanism of plant mRNA nuclear export is poorly understood (Merkle, 2011). The Arabidopsis NTL (NTF2-like) protein, which could not functionally replace yeast NTF2, might be a homologue of NXT1 (Merkle, 2011). The Arabidopsis genome does encode a DEAD-box RNA helicase, LOS4 (for low expression of osmotically responsive genes 4), which is enriched at the nuclear rim. In *los4* mutants, mRNA accumulates in the nucleus, suggesting that LOS4 might be a homolog of Dbp5 (Gong *et al.*, 2005; Merkle, 2011).

3.1.3.3 Models explaining 'virtual gating' of the NPC

The selectivity of the NPC is based on the central meshwork of FG repeat-domains. This meshwork creates an entropic barrier that excludes macromolecules. However, NTFs overcome this barrier by cancelling out the entropic potential through their interactions with FG Nups. This concept is termed 'virtual gating' (Rout *et al.*, 2003) and has been experimentally supported by creating an FG Nup-coated nanopore, which replicated many features of the NPC (Chait *et al.*, 2009). However, the detailed mechanism of this gating is still unclear and several models have been suggested (reviewed in Strambio-De-Castillia *et al.*, 2010, Wente and Rout, 2010). In one model, FG repeats collapse and open their way when NTFs bind and pass through the NPC. In another 'saturated' model, FG repeats are cross-linked through the F residues forming a dense gel, and the binding of NTFs to FG repeats dissolves the cross-link. In the 'reduction in dimensionality' model, on the other hand, binding to FG repeats is assumed to be the entrance for NTFs, followed by a rapid search of the FG surface by a two-dimensional random walk for the channel exit. In this model, nonbinding molecules can only pass through the central narrow channel where the network of peptide chains is loose. Another model suggests that the selectivity is enhanced by the competition of NTFs with nonspecific macromolecules. While these models are not mutually exclusive, different NTFs seem to prefer different sets of FG Nups, suggesting the existence of multiple pathways through the FG Nups (Strawn *et al.*, 2004; Wente and Terry, 2007). None of the models has been tested in plants; however, this process might be assumed to function very similarly across kingdoms.

3.2 Physiological and developmental roles of plant nuclear pore components

Several putative plant Nups, as well as components of the nuclear import and export pathways have been identified by forward and reverse genetic approaches. Forward genetic approaches included screens for nodulation, innate immunity, hormone responses and different abiotic stress responses. Some genes were identified in more than one screen, suggesting that each mutant screen revealed only one out of several functions of the protein. Table 3.1 summarizes the plant nuclear pore and nuclear transport-related proteins with functional roles in plant–microbe interactions, hormone response, abiotic stress response, as well as plant growth and development.

3.2.1 Plant–microbe interactions

While most genes were first identified in Arabidopsis, other plant species used to dissect specific pathways are also included. An example is nodulation and mycorrhiza formation, investigated in *Lotus japonicus*. Two loss-of-function alleles of putative Nup133 and Nup85 were identified that have an effect on bacterial symbiosis (Kanamori *et al.*, 2006; Saito *et al.*, 2007). In the mutants, nucleus-associated calcium spiking during symbiotic signal transduction is affected, which suggests that the nuclear pore might play a role in calcium signalling.

R proteins are intracellular immune sensors, typically of the NB (nucleotide binding)-LRR (leucine-rich repeat) type, which act as regulatory signal transduction switches upon sensing isolate-specific pathogen effectors. *Snc1* is a gain-of-function mutation of the Arabidopsis *SNC1* R gene, which expresses a constitutive defence response. A suppressor screen using this gain-of-function mutant has identified a surprising number of nuclear pore and nuclear transport-related genes. Some are shared with the auxin pathway, as discussed below. Modifier of *snc1* (*MOS*) genes include MOS3 (Nup96), which

Table 3.1 Plant nuclear pore and nuclear transport-related proteins with roles in plant stress response, hormone physiology, or growth and development

Protein	Predicted Function	Molecular/Cellular Features	Mutant Phenotype	Reference
Plant–microbe interactions				
MOS2	Predicted RNA-binding protein	Not known	Modifier of snc1	(Zhang *et al.*, 2005)
MOS3	Nup96, scaffold nucleoporin	Nuclear mRNA accumulation, reduced nuclear accumulation of IAA17-GUS	Suppressor of axr1 and of snc1, early flowering, stunted growth	(Zhang and Li, 2005)
MOS7	Nup88	Reduced nuclear accumulation of NPR1-GFP, reduced total and nuclear accumulation of EDS1, increased ratio of cytoplasmic to nuclear snc1-GFP Proposed mechanism: attenuation of Xpo1 mediated nuclear export	Modifier of snc1, defects in basal and systemic acquired resistance	(Wiermer *et al.*, 2007; Cheng *et al.*, 2009; Wiermer *et al.*, 2010).
MOS6	Importin-$\alpha 3$, predicted nuclear import adaptor	Not known	Modifier of snc1, enhanced disease susceptibility	(Palma *et al.*, 2005)
MOS11	Predicted homolog of CIP29, human RNA-binding protein	Nuclear mRNA accumulation	Modifier of snc1	(Germain *et al.*, 2010)
LjNup85	Nucleoporin	Calcium spiking	Deficient in nodulation, mycorrhiza formation, impaired seed production	(Kanamori *et al.*, 2006; Saito *et al.*, 2007)
LjNup133	Nucleoporin	Calcium spiking	Deficient in nodulation, mycorrhiza formation, impaired seed production	(Kanamori *et al.*, 2006)
NbRanGAP2	Ran GTPase activating protein	Binds the coiled-coil domains of two Rx-like proteins, Rx2 and Gpa2	Compromises Rx-mediated resistance to Potato Virus X and against the nematode Globodera pallida	(Rairdan *et al.*, 2008; Sacco *et al.*, 2007, 2009)

Hormone responses				
SAD2	Importin-β, predicted nuclear import receptor	Increased expression of stress- and ABA-induced expression of RD29A-LUC reporter, impaired nuclear import of MYB4, constitutive expression of MYB4-repressed MYB4 and C4H genes	Increased drought sensitivity, ABA sensitivity, increased UV tolerance	(Verslues *et al.*, 2006, Zhao *et al.*, 2007)
SAR3	Nup96, scaffold nucleoporin	Nuclear mRNA accumulation, reduced nuclear accumulation of IAA17-GUS	Suppressor of axr1 and of snc1, early flowering, stunted growth	(Parry *et al.*, 2006)
NUA	Tpr homologue, nuclear basket nucleoporin	Nuclear mRNA accumulation, decrease in miRNA abundance, increased abundance of nuclear sumoylated proteins, downregulation of FLC and MAF4, upregulation of SOC1, LFY, MYB33 and MYB65	Early flowering, phyllotaxy defects, reduced stamen length, fertility, stunted growth, supressor of axr1	(Jacob *et al.*, 2007; Xu *et al.*, 2007a).
TaRAN1	Small GTP binding protein involved in nucleocytoplasmic trafficking	Not known	Overexpression leads to alterations in cell cycle, root and shoot meristem development, auxin hypersensitivity	(Wang *et al.*, 2006)
RanBP1	Ran binding protein 1, stimulates Ran GTPase activity together with RanGAP	Not known	Overexpression leads to auxin hypersensitivity	(Kim *et al.*, 2001)
OsRAN2	Small GTP binding protein involved in nucleocytoplasmic trafficking	Not known	Overexpression leads to hypersensitivity to ABA, salinity and osmotic stress	(Zang *et al.*, 2010)

(continued)

Table 3.1 (*Continued*)

Protein	Predicted Function	Molecular/Cellular Features	Mutant Phenotype	Reference
Abiotic stress responses				
SAD2	Importin-β, predicted nuclear import receptor	Increased expression of stress- and ABA-induced RD29A-LUC reporter, impaired nuclear import of MYB4, constitutive expression of MYB4-repressed *MYB4* and *C4H* genes	Increased drought sensitivity, ABA sensitivity, increased UV tolerance	(Verslues *et al.*, 2006; Zhao *et al.*, 2007; Gao *et al.*, 2008)
SAR1/Nup160	Nup160, scaffold nucleoporin	Nuclear mRNA accumulation, increased abundance of sumoylated proteins	Suppressor of axr1, early flowering, stunted growth, chilling sensitive, impaired acquired freezing tolerance	(Parry *et al.*, 2006; Muthuswamy and Meier, 2011)
LOS4	RNA helicase	Nuclear mRNA accumulation	Early flowering, cold stress response, heat tolerance, stunted growth	(Gong *et al.*, 2005)
TaRAN1	Small GTP binding protein involved in nucleocytoplasmic trafficking	Not known	Overexpression leads to alterations in cell cycle, root and shoot meristem development, auxin hypersensitivity	(Wang *et al.*, 2006)
OsRAN2	Small GTP binding protein involved in nucleocytoplasmic trafficking	Not known	Overexpression leads to hypersensitivity to ABA, salinity and osmotic stress	(Zang *et al.*, 2010)

Growth and development				
SAR3	Nup96, scaffold nucleoporin	Nuclear mRNA accumulation, reduced nuclear accumulation of IAA17-GUS	Suppressor of axr1 and of snc1, early flowering, stunted growth	(Parry *et al.*, 2006).
NUA	Tpr homologue, nuclear basket nucleoporin	Nuclear mRNA accumulation, decrease in miRNA abundance, increased abundance of nuclear sumoylated proteins, downregulation of FLC and MAF4, upregulation of SOC1, LFY, MYB33 and MYB65	Early flowering, phyllotaxy defects, reduced stamen length, fertility, stunted growth, suppressor of axr1	(Jacob *et al.*, 2007; Xu *et al.*, 2007b; Muthuswamy and Meier, 2011).
ESD4	SUMO isopeptidase	Increased abundance of sumoylated nuclear proteins, nuclear mRNA accumulation	Early flowering, stunted growth, stamen development, fertility	(Reeves *et al.*, 2002; Murtas *et al.*, 2003)
RanGAP1/2	Ran GTPase activating protein	Not known	Cytokinesis, female gametophyte development	(Xu *et al.*, 2007a; Rodrigo-Peiris *et al.*, 2011)
TaRAN1	Small GTP binding protein involved in nucleocytoplasmic trafficking	Not known	Overexpression leads to alterations in cell cycle, root and shoot meristem development, auxin hypersensitivity	(Wang *et al.*, 2006)
Nup1/Nup136	Nucleoporin	Altered nuclear morphology	Early flowering, short siliques, reduced fertility	(Lu *et al.*, 2010, Tamura *et al.*, 2010)
SAD2	Importin-β, prediced nuclear import receptor	Reduced expression levels of GL1, MYB23, GL2 and TTG1, but increased expression level of GL3 and EGL3	Disrupted trichome initiation and reduced trichome number	(Gao *et al.*, 2008)

is identical to SAR3 (Zhang and Li, 2005), MOS7, which is a putative Nup88 (Wiermer *et al.*, 2007), MOS6, a member of the Arabidopsis Importin-α family, and two predicted RNA-binding proteins with possible functions in mRNA export (Palma *et al.*, 2005; Zhang *et al.*, 2005; Germain *et al.*, 2010). The relationship between these molecular roles and the mutant phenotypes is not fully understood, but it is likely that impaired nucleo-cytoplasmic transport of proteins or RNAs involved in innate immunity is responsible for the suppression phenotypes. Protein candidates would include the innate-immunity-related factors EDS1 (Enhanced Disease Susceptibility 1), PAD4 (Phytoalexin Deficient 4), SAG101 (Senescence Associated Gene 101), the transcription factor bZIP10, NPR1 and SNC1 itself (Wiermer *et al.*, 2007). Indeed, it was shown that in the partial loss-of function *mos7-1* mutant, the nuclear level of SNC1 and the overall and nuclear levels of EDS1 and NPR1 were reduced (Cheng *et al.*, 2009).

Another connection between the nuclear pore and plant immunity was revealed by finding that RanGAP2 interacts with the coiled-coil domain of the CC-NB-LRR protein Rx, the R protein that confers resistance to potato virus X. (Bendahmane *et al.*, 1999). RanGAP2 also binds the coiled-coil domains of two Rx-like proteins, Rx2 and Gpa2, the latter of which is involved in resistance against the nematode *Globodera pallid* (Sacco *et al.*, 2007, 2009; Tameling and Baulcombe, 2007). Reducing the expression of RanGAP2 leads to reduced Rx-mediated, as well as Gpa2-mediated resistance to their corresponding pathogens (Sacco *et al.*, 2007, 2009; Tameling and Baulcombe, 2007). While the role of RanGAP2 in these responses is not clear, it has been shown that its GTPase activating function for Ran is not required. Equally, the association of RanGAP2 with the NE plays no role (Sacco *et al.*, 2007). Both the coiled-coil domain of the Rx-like protein and the LRR domain are involved in recognition specificity (Rairdan *et al.*, 2008; Sacco *et al.*, 2009). Interestingly, this function is unique to RanGAP2, while for all other aspects investigated, RanGAP gene family members act redundantly.

3.2.2 Hormone responses

3.2.2.1 Abscisic acid signalling

The plant hormone abscisic acid (ABA) is involved in responses to different abiotic stresses such as cold, osmotic and draught stress. A screen was developed to identify mutants with altered expression of a reporter gene induced under these conditions. SAD2 (Sensitive to ABA and drought 2) was identified as a member of the Arabidopsis Importin-β family (Verslues *et al.*, 2006). The *sad2-1* mutant is hypersensitive to ABA in inhibiting seed germination and seedling growth. Interestingly, this phenotype appears to be specific to SAD2 because a mutation in the Importin-β gene most closely related to *SAD2* did not show ABA hypersensitivity. This would be consistent with the hypothesis that some of the 17 Importin-β homologs in Arabidopsis might have different cargo specificities and preferences (Merkle, 2003). *sad2-1* is not specific for ABA; the mutant is also more tolerant to UV-B radiation. In this case, the phenotype was correlated to reduced nuclear import of the transcription factor MYB4, which leads to the upregulation of cinnamate 4-hydroxylase (*C4H*) and the accumulation of more UV-absorbing pigment (Zhao *et al.*, 2007). In addition, other mutations in SAD2 also disrupt trichome initiation, suggesting that this Importin-β isoform is involved in the transport of a number of different cargoes (Gao *et al.*, 2008). One complicating aspect of predicting cargo specificity for Importin-β isoforms is the role of the Importin-α transport adaptors. The prediction would be that different import receptors have different affinities for members of the Importin-α family, which in turn have distinguishing affinity differences for subsets of NLS-containing cargo proteins. A karyopherin-adaptor-cargo interactome study in Arabidopsis would greatly assist in testing this hypothesis.

In addition, overexpression of rice OsRAN2 was found to cause ABA hypersensitivity in transgenic Arabidopsis seedlings (Zang *et al.*, 2010). As discussed for SAD2, reduced nuclear accumulation of important ABA regulators might be the reason for this observation.

3.2.2.2 Auxin signalling

Putative Arabidopsis Nup160 and Nup96 were identified as suppressors of *auxin-resistant 1* (*axr1*), and named SUPRESSOR OF AXR1 1 and 3 (SAR1 and SAR 3) (Parry *et al.*, 2006). SAR3 is identical to MOS3 (Modifier Of SNC1). In *sar1* and *sar3*, the morphological and molecular phenotypes of *axr1* are partially restored. Nuclear import of the repressor IAA17 was affected in *sar1* and *sar3*, leading to the hypothesis that reduced nuclear import of transcriptional Aux/IAA repressors might counteract their over-accumulation based on reduced proteolysis in the *axr1* background. In addition, a mutation in the nuclear basket-associated protein NUA (NUCLEAR PORE ANCHOR)/AtTpr was also able to suppress the auxin resistant phenotype of *axr1* (Jacob *et al.*, 2007).

Another connection between the nuclear pore and auxin comes from the observation that antisense suppression of Arabidopsis *RanBP1* causes auxin hypersensitivity (Kim *et al.*, 2001). Again, this might be caused by impaired protein import of Aux/IAA transcriptional repressors. An auxin hypersensitivity phenotype was also detected in Arabidopsis overexpressing wheat *TaRAN1* (Wang *et al.*, 2006). Interestingly, the overexpression of *TaRAN1* and *OsRAN2* renders Arabidopsis hypersensitive to auxin and ABA, respectively (Wang *et al.*, 2006; Zang *et al.*, 2010), suggesting that these two signalling pathways might be particularly sensitive to perturbing components of nucleocytoplasmic trafficking.

3.2.3 Abiotic stress responses

3.2.3.1 Temperature stress

CRYOPHYTE/LOS4 is a NE-localized DEAD (for Aspartic acid-Glutamic acid-Alanine-Aspartic acid)-box RNA helicase and its mutation causes sensitivity to heat stress. Interestingly, the same mutation also has an effect on the increased tolerance to chilling and freezing (Gong *et al.*, 2005). The *sar1-1* mutation in Nup160 described above sensitizes plants to chilling stress and disrupts acquired freezing tolerance (Dong *et al.*, 2006). A common feature between *sar1-1* and *los4-2* is the increased accumulation of poly(A)RNA in the nucleus, suggestive of impaired mRNA export. If this is the primary molecular phenotype, it remains to be explained why *sar1-1* and *los4-2* have opposite effects on cold tolerance, while both show similar mRNA export defects.

3.2.3.2 Salt and osmotic stress

In addition to the phenotypes described above, overexpression of OsRAN2 in rice and Arabidopsis also leads to hypersensitivity to salinity and osmotic stress (Zang *et al.*, 2010). When the maize transcription factor Lc fused with YFP was used as a nuclear import marker, co-expression of OsRAN2 caused partial retention of Lc-YFP in the cytoplasm, suggesting that overexpression of Ran indeed renders nuclear import less efficient.

3.2.4 Growth and development

Overexpression of TaRan1 in rice and Arabidopsis was shown to cause meristem changes and alterations in mitotic progress (Wang *et al.*, 2006). Induced RNAi of RanGAP1 and RanGAP2 in Arabidopsis leads to problems in growth and to cytokinesis defects (Xu *et al.*, 2008). A double knock-out mutant in Arabidopsis RanGAP1 and RanGAP2 is female gametophyte lethal. The female gametophytes are arrested at interphase, predominantly after the first mitotic division following meiosis. Mutant pollen developed and functioned normally, suggesting a specific effect on female gametophyte development despite the ubiquitous expression of the two RanGAP genes in Arabidopsis. The nuclear division arrest during a mitotic stage indicates that RanGAP might play a role in mitotic cell cycle progression during female gametophyte development (Rodrigo-Peiris *et al.*, 2011).

Many of the mutants in Nups and nuclear-pore associated proteins also have an early-flowering phenotype as well as similar, diverse developmental defects. These include reduced growth, defects in stamen development, a changed juvenile/adult transition and altered phyllotaxy. The genes include SAR1/Nup160, SAR3/MOS3/Nup96, NUA/AtTPR, Nup1/Nup136, Nup62, LOS4 and ESD4 (Table 3.1 and references therein). Nup1/Nup136 has been added to the list recently and appears to be a plant-specific Nup, possibly a functional orthologue of vertebrate Nup153. Knock-out mutants are less impaired in overall growth and development than the other Nup mutants, but share the early-flowering phenotype and impaired fertility (Tamura *et al.*, 2010; Tamura and Hara-Nishimura, 2011). Overexpression-based co-suppression of Nup62 also leads to similar problems with growth and development as described for the other Nup and NPC-associated protein mutants (Zhao and Meier, 2011).

Several of these mutants have common defects in mRNA export, as indicated by a strong increase of intranuclear signal in *in situ* hybridizations with an oligo-dT probe. These include Nup160, Nup96, NUA, ESD4, and LOS4. Others that share the developmental phenotypes have not been tested for nuclear mRNA retention. The open question remains how to explain the pleiotrophic but relatively specific phenotypes if the primary function of the genes is a role in generic "housekeeping" nuclear pore organization and/or nucleocytoplasmic transport of proteins and RNAs.

A possible approach to address this question is the observation that in *nua/tpr* mutants, in addition to the mRNA nuclear retention phenotype, the levels of miR159, miR165, and miR393 are significantly reduced (Jacob *et al.*, 2007). miR393 targets, for example, TIR1, the auxin-receptor that is part of the SCF complex, which triggers Aux/IAA degradation (Sunkar and Zhu, 2004; Si-Ammour *et al.*, 2011). miR393 is highly expressed in inflorescences, and is upregulated by cold, dehydration, NaCl, ABA, and bacterial elicitors, and might thus have additional functions in these pathways (Sunkar and Zhu, 2004; Navarro *et al.*, 2006). The rice homologue of miR393 is salinity- and alkaline-stress regulated and its overexpression in rice and Arabidopsis leads to transgenic plants that were more sensitive to salt and alkali treatment (Gao *et al.*, 2011). miR393b(∗) is involved in regulating innate immunity through targeting a gene for a Golgi-localized SNARE (Zhang *et al.*, 2011). Overexpression of miR165 changes the expression of several auxin signalling-related genes (Zhou *et al.*, 2007). In addition, miR159 targets the two MYB transcription factors MYB33 and MYB65, which were proposed to be involved in GA-induced activation of the flowering regulator LEAFY. If miR159 is overexpressed, flowering time under short-day conditions is delayed, LEAFY transcript levels are reduced, and other defects in development are observed (Achard *et al.*, 2004). miR159 loss of function leads to stunted growth, curly leaves and reduced fertility, reminiscent of several of the mutants discussed here (Achard *et al.*, 2004). Through regulation of the anaphase-promoting complex, miR159 affects pollen development in Arabidopsis (Zheng *et al.*, 2011). Together, a number of these pathways are affected in the nuclear pore mutants. Thus, it is worth investigating whether the major developmental defects of the different mutants discussed here are a consequence of altered miRNA homeostasis, which might be more sensitive to small quantitative changes than the overall mRNA pools.

3.3 The Dynamics of the Nuclear Pore Complex

3.3.1 Types of mitosis

Although basic mitotic events are conserved, there are mitotic variants among eukaryotes. Most unicellular eukaryotes, such as *S. cerevisiae*, undergo closed mitosis during which the NE and the NPCs stay intact. In this process, an intranuclear spindle forms inside the nucleus and separates the chromosomes before karyokinesis occurs. The two daughter nuclei are then separated due to the fission of the NE with a single pinch directly following chromosome separation. The NE of the filamentous ascomycete *Aspergillus nidulans*, on the other hand, undergoes a so-called semi-open mitosis in which the NPCs partially disassemble, allowing the NE – although intact – to become permeable (De Souza *et al.*, 2004; De Souza and Osmani, 2009). In this scenario, the cytoplasm and the nucleoplasm mix and mitotic regulators access the nucleus through passive diffusion to promote the spindle assembly. This semi-open mitosis is thus an intermediate between the closed mitosis of budding yeast and the open mitosis of vertebrates and plants. Similarly, the NE does not completely break down in *Drosophila* early embryos that undergo syncytial mitosis and in the basidiomycete *Ustilago maydis* (Kiseleva *et al.*, 2001; Theisen *et al.*, 2008). During open mitosis NPCs disassemble and the nuclear envelope completely breaks down, allowing the spindle microtubules to attach to kinetochores in the cytoplasm.

3.3.2 NPC disassembly and dynamics of animal NPC components

At the onset of metazoan open mitosis both lamina and NPC subunits disassemble in a stepwise manner in response to their phosphorylation (Gerace and Blobel, 1980; Heald and Mckeon, 1990; Macaulay *et al.*, 1995). Upon the NPC disassembly, Nups localize to various mitotic structures such as kinetochores, spindle pole bodies and chromatin (Galy *et al.*, 2006; Liu *et al.*, 2009; Osmani *et al.*, 2006; Rasala *et al.*, 2006). There is a substantial body of evidence that these Nups exhibit new roles at these mitotic structures, indicating that organisms with open and semi-open mitosis have evolved towards mitotic functions of Nups away from the nuclear pore (Osmani *et al.*, 2006; De Souza and Osmani, 2007, 2009; Guttinger *et al.*, 2009).

For instance, RanBP2 (Nup358) was shown to be required for chromosome segregation in *C. elegans* embryos and mammalians cells (Askjaer *et al.*, 2002; Salina *et al.*, 2003). The siRNA-mediated depletion of human Nup358 revealed that it is essential for kinetochore function. In the absence of Nup358, chromosome congression and segregation are severely perturbed and the assembly of other kinetochore components is strongly inhibited, leading to aberrant kinetochore structure (Salina *et al.*, 2003). In *Xenopus* egg extracts and HeLa cells, RanBP2, which is a component of the short filaments that extend from the cytoplasmic face of the NPC during interphase, relocates to both spindle microtubules and kinetochores (Joseph *et al.*, 2002). This relocation occurs in association with Ran GTPase activating protein 1 (RanGAP1), a protein with which Nup358 also interacts during interphase. In human cells and *Xenopus* egg extracts RanGAP1–SUMO-1 remains associated with RanBP2 in mitosis and RanBP2 is targeted to the same sites on mitotic spindles as RanGAP1, as well as to additional distinct sites (Saitoh *et al.*, 1997; Joseph *et al.*, 2002). Once on kinetochores, the RanBP2-RanGAP1 complex acts in the interaction of kinetochores with k-fibres of the mitotic spindle (Arnaoutov and Dasso, 2005).

A fraction of mammalian RanGAP1 is targeted to the nuclear pore during interphase (Matunis *et al.*, 1996; Mahajan *et al.*, 1997). During cell division, RanGAP1 is associated with the mitotic spindle and kinetochores in a SUMO-1-conjugation-dependent manner (similar to its nuclear pore association) and in a microtubule-dependent manner in HeLa cells (Matunis *et al.*, 1996; Joseph *et al.*, 2002). The interphase and mitotic targeting of metazoan RanGAP1 relies thus on the same protein modification. Gradients of RanGTP could help to orient the mitotic spindle in metazoans by locally stabilizing MTs in the vicinity of chromosomes, based upon the localization of Ran's nucleotide exchange factor, RCC1 and Ran's GTPase-activating protein, RanGAP1 (for reviews see Dasso, 2001; Kahana and Cleveland, 2001). The restriction of SUMO-1 conjugation of RanGAP1 to metazoans, in contrast to yeast and plant cells, may reflect its particular role in the localization of RanGAP1 during spindle assembly in the absence of an intact nuclear envelope and a tight relationship between the nuclear pore and the spindle.

MEL-28, a large protein essential for the assembly of a functional NE in *C. elegans* embryos, localizes to NPCs during interphase, to kinetochores in early to middle mitosis, and is then widely distributed on chromatin late in mitosis. At this stage MEL-28 spreads over the entire chromatin surface and then rapidly relocalizes to the NE. This distribution would be consistent with MEL-28's role in the interactions that occur between chromatin and the assembling NE and NPCs late in mitosis, as well as its involvement in chromosome segregation (Galy *et al.*, 2006). Another Nup, Nup107, is present on kinetochores in mammalian cells (Belgareh *et al.*, 2001), and *C. elegans* Nup107, Nup155 and Pom121 accumulate along the condensed chromosomes in mitosis (Franz *et al.*, 2005). These and many other nucleoporins are recruited to chromosomes at a similar time, presumably reflecting NPC assembly and docking membranes to chromatin very early in nuclear assembly.

Another NPC-associated protein, Rae1, becomes preferentially associated with unattached kinetochores of prometaphase chromosomes (Wang *et al.*, 2001; Babu *et al.*, 2003). Rae1 binds also to MTs and controls MT dynamics in a RanGTP/Importin-β-regulated manner (Blower *et al.*, 2005; Kraemer *et al.*, 2001). It has been shown that interaction of Rae1 with NuMA (nuclear mitotic apparatus protein) during mitosis is critical for proper mitotic spindle formation (Wong *et al.*, 2006). The human mRNA export factor Rae1 (also called Gle2 or mrnp41) binds the nuclear pore complex protein hNUP98 via a short NUP98 motif called GLEBS (for GLE2p-binding sequence) and the mitotic checkpoint proteins hBUB1 (human Budding Uninhibited by Benzimidazoles 1) and hBUBR1 (human BUB1 Budding Uninhibited by Benzimidazoles 1 homologue beta) (Wang *et al.*, 2001). Both Rae1 and Nup98 are not only involved in spindle assembly (Cross and Powers, 2011), but also have a role in proper timing of securin degradation during mitotic exit. Analogously to Mad2, the Rae1/Nup98 NPC subcomplex binds and inhibits an anaphase-promoting complex (APC) activator, Cdh1 (Jeganathan *et al.*, 2005). Therefore, reduced levels of Rae1 and Nup98 lead to premature sister chromatid separation, which results in aneuploidy, revealing a close link between Rae1 and the anaphase promoting complex/cyclosome (APC/C) activity (Guttinger *et al.*, 2009). Additionally, the mitotic checkpoint protein Bub3 shares extensive sequence homology with Rae1 (Taylor *et al.*, 1998; Martinez-Exposito *et al.*, 1999). The loss of a single Rae1 allele in mouse cells causes a mitotic checkpoint defect and chromosome missegregation, and Bub3 haploinsufficient cells exhibit a strikingly similar mitotic phenotype, suggesting that Rae1 may act in an analogous way to Bub3 (Babu *et al.*, 2003). Furthermore, the *Drosophila* and human nucleoporin, Tpr, has been shown to orchestrate the spindle checkpoint control via interaction with Mad1 and Mad2 during cell division (Lee *et al.*, 2008; Lince-Faria *et al.*, 2009; Nakano *et al.*, 2010). Conversely, certain mitotic checkpoint proteins, such as Mad1 and Mad2, which are kinetochore associated during mitosis, are found at the nuclear face of NPCs during interphase (Campbell *et al.*, 2001). The Nup107/160 subcomplex of *X. laevis* egg extract also localizes to spindles and is involved in spindle formation (Orjalo *et al.*, 2006). Interestingly, a fraction of the complex also localizes to kinetochores in both *C. elegans* embryos (Galy *et al.*, 2006) and mammalian cells (Belgareh *et al.*, 2001; Harel *et al.*, 2003; Loiodice *et al.*, 2004). Depletion of this complex leads to a prolonged prometaphase, chromosome missegregation and delayed onset of anaphase (Zuccolo *et al.*, 2007). Some Nups are even involved in the final stages of cell division, such as cytokinesis. For instance, the peripheral mammalian Nup, Nup153, has been shown to play dual roles in both early mitotic progression and the resolution of membrane abscission during cytokinesis in HeLa cells (Mackay *et al.*, 2009). Together, these data provide a molecular basis for a potential interplay between nucleocytoplasmic transport and mitotic machinery in mammalian cells.

3.3.3 Dynamics of fungal NPC components

In filamentous fungi, like *A. nidulans*, the NPC is partially disassembled during mitosis, allowing the mitotic apparatus to enter the nucleus (De Souza *et al.*, 2004). In *A. nidulans*, Nup2 translocates from the NPCs to chromatin during mitosis, similar to its mammalian orthologue Nup50 (Osmani *et al.*, 2006; Dultz *et al.*, 2008). Nup2 is an essential gene in *A. nidulans* and its deletion leads to mitotic defects in spindle assembly and stability (Osmani *et al.*, 2006). A proteomic approach, in which the An-Nup84-120 complex was used as bait, identified two new fungal Nups, An-Nup37 and An-ELYS, previously thought to be vertebrate-specific. During mitosis, the An-Nup84-120 complex is located at the NE and spindle pole bodies but, unlike in vertebrate cells, does not concentrate at kinetochores (Liu *et al.*, 2009). Ndc1, one of transmembrane Nups, is essential in yeasts and is required for the duplication and NE insertion of spindle pole bodies (Winey *et al.*, 1993, West *et al.*, 1998). However, Ndc1 is not essential in either *C. elegans* (Stavru *et al.*, 2006a) or *A. nidulans* (Osmani *et al.*, 2006).

The spindle assembly checkpoint (SAC) proteins Mad1 and Mad2 also locate to the NPCs in *S. cerevisiae* (Iouk *et al.*, 2002) and *A. nidulans* (De Souza *et al.*, 2009), suggesting a functional relationship between the SAC and the nuclear pore. During SAC activation, An-Mlp1 remains associated with kinetochores in a manner similar to An-Mad1 and An-Mad2. Interestingly, in *A. nidulans,* Mlp1 acts as a scaffold to locate Mad1 and Mad2 near kinetochores and the telophase spindle, indicating that it provides a spatial and temporal regulation for these SAC components throughout the cell cycle (De Souza *et al.*, 2009). By maintaining SAC proteins near the mitotic apparatus, An-Mlp1 may help monitor mitotic progression and coordinate efficient mitotic exit. Namely, An-Mad1 and An-Mlp1 relocate from the telophase matrix and associate with segregated kinetochores when mitotic exit is prevented by expression of nondegradable cyclin B (De Souza *et al.*, 2009). In the yeast *S. cerevisiae,* Mad1p is bound to a Nup53p-containing complex and thereby sequesters Mad2p at the NPC until its release by activation of the spindle checkpoint and its accumulation at kinetochores (Iouk *et al.*, 2002). Remarkably, yeast strains deficient in Mad1p exhibit a reduced rate of nuclear protein import, as well as decreased stability of the Nup53p complex (Iouk *et al.*, 2002). In contrast to *A. nidulans*, the mitotic modifications of the NPC of *S. cerevisiae* are more subtle and do not affect global regulation of nuclear transport (Makhnevych *et al.*, 2003).

3.3.4 Dynamics of plant NPC components

Despite the conserved overall structure and function of the NPCs, there is little known about the actual nucleocytoplasmic transport barrier in plants and about the dynamics of its components.

At the onset of mitosis, the plant cortical MTs depolymerize and rearrange into the pre-prophase band (PPB) surrounding the nucleus. The proper formation of the PPB is crucial for the fate of a dividing cell, as this transient MT array demarcates the future cortical division zone where a cell will separate into two daughter cells (Van Damme and Geelen, 2008; Muller *et al.*, 2009). RanGAP1 is anchored at the NPC and just prior to mitosis it is delivered to the PPB in an MT-dependent manner and it remains associated with the cortical division site (CDS) during mitosis and cytokinesis, constituting a continuous positive marker of the plant division plane (Xu *et al.*, 2008). The silencing of RanGAP in Arabidopsis roots leads to mispositioned cell walls similar to other mutants with division plane defects. Therefore, RanGAP is a molecular landmark left behind by the PPB, which probably later guides the phragmoplast and the forming cell plate (Smith, 2001; Xu *et al.*, 2008). Thus, the plant Ran cycle might play roles in mitotic microtubule assembly and vesicle fusion in analogy to its functions in animal cells. Arabidopsis RanGAP1 is localized to kinetochores and the spindle, indicating that it might play a similar role in plant cells (Joseph *et al.*, 2002, 2004; Xu *et al.*, 2008). While mammalian RanGAP1 is targeted to kinetochores in a SUMO-dependent manner (Joseph *et al.*, 2004), it remains enigmatic how Arabidopsis RanGAP1, which lacks the SUMOylation domain, is targeted to kinetochores. As human RanGAP1 is found only on the attached sister chromatids, the exact timing of kinetochore association and the function of plant RanGAP1 at this cellular location awaits verification. As in the animal system, Arabidopsis RanGAP1 might be involved in the release of NLS-containing cargo proteins, such as TPX2, Rae1 and NuMA (Dasso, 2001; Weis, 2003).

Tobacco (*Nicotiana benthamiana*) Rae1 (NbRae1), a homolog of Rae1/mrnp41 in metazoans, Gle2p in *S. cerevisiae*, and Rae1 in *S. pombe*, exhibits a mitotic function, besides its role as an mRNA export factor associated with the NPC (Whalen *et al.*, 1997; Pritchard *et al.*, 1999; Griffis *et al.*, 2004; Lee *et al.*, 2008, 2009). NbRae1 associates with the spindle and was shown to function in the proper spindle organization and chromosome segregation, similar to its mammalian orthologue (Whalen *et al.*, 1997; Babu *et al.*, 2003; Jeganathan *et al.*, 2005; Lee *et al.*, 2009). Interestingly, NbRae1 is targeted to the cell plate and to other mitotic MT arrays, namely the PPB and the phragmoplast, throughout cell division, suggesting a tight linkage between the NPC components and the cytoskeleton during mitosis (Lee *et al.*, 2009). The PPB localization of NbRae1 might reflect partial involvement of the PPB in spindle assembly and bipolarity, because the RNAi inhibition of NbRae1 in BY-2 cells led to formation of disorganized or multipolar spindles and to defects in chromosome segregation. NbRae1 silencing resulted in delayed progression of mitosis, which led to plant-growth arrest, reduced cell division activities in the shoot apex and the vascular cambium, and increased ploidy levels in mature leaves. Together, these results suggest a conserved function of Rae1 in spindle organization among eukaryotes, which is distinct from its role at the interphase NE.

Apart from Rae1, other NPC proteins are localized at the cell plate as well. For instance, the Arabidopsis outer NE proteins that anchor RanGAP1 at the NPC, WIP1, WIP2, WIT1, and WIT2, are redistributed to the cell plate during cytokinesis (Xu *et al.*, 2007a; Zhao, Brkljacic and Meier, 2008). Both WITs and WIPs are required for RanGAP1 anchoring to the NE in the root meristem, but only one of the protein families, either WIPs or WITs, is sufficient to target RanGAP1 to the NE in differentiated cells (Zhao *et al.*, 2008). The cell plate localization of RanGAP1 (as well as its PPB and CDS association), on the other hand, is independent of both WIPs and WITs, suggesting that interphase and mitotic targeting of RanGAP1 require different mechanisms. Therefore, identification of molecular players involved in RanGAP1 localization and function(s) during plant cell division would be of great importance. Similarly, the discovery of other WIP and WIT interactors at the nuclear rim, as well as their mitotic partners, will shed more light on plant-specific functions of NE components in the proper orchestration of mitosis.

In the light of recent discoveries and the isolation of multiple plant NPC proteins, a plethora of data might be on the way depicting, among other things, their dynamics and putative mitotic functions (Zhang and Li, 2005; Dong *et al.*, 2006; Kanamori *et al.*, 2006; Jacob *et al.*, 2007; Saito *et al.*, 2007; Wiermer *et al.*, 2007; Xu *et al.*, 2007b; Tamura *et al.*, 2010; Zhao and Meier, 2011).

3.4 Conclusions

The nuclear pore complex is one of the vital elements of the eukaryotic cell, and its structure and function is conserved across kingdoms. The field has justly received much interest, since aberrations in NPC function often result in severe developmental defects. Even though there has been some delay, plant NPC research is catching up in this regard. Unravelling the structure and the molecular composition of the plant NPC suggests that the overall architecture of NPCs is conserved in eukaryotes. Based on the molecular composition, plant NPCs are close to those of vertebrates, but several plant-unique features also suggest a different evolutionary path. These features include the plant-unique Nups, the plant-unique RanGAP anchoring mechanism, and the plant-specific functions of RanGAP1 and Rae1 during mitosis. Several Nups in vertebrates lack homologues in the plant genome and whether their functional counterparts are simply difficult to identify or if they are indeed missing in plants has not yet been resolved. Recent forward and reverse genetic approaches, as well as proteomic studies, have significantly advanced our understanding of the plant NPC and its roles. One puzzling aspect is the diversity of phenotypes that plant NPC mutants exhibit. The molecular basis of these phenotypes remains to be elucidated and constitutes one of the major challenges in the field. The notion that developmental defects do not result solely from the sensitivity of cells to perturbations in nucleocytoplasmic transport is supported by recent lines of evidence indicating that the NPC acts as a key regulator of events that occur on either side of the NE and at different developmental and cellular stages. Dissection of these NPC functions will shed light on the interplay between the nucleocytoplasmic transport and other pathways, such as mitotic cell division in plants.

References

Achard, P., Herr, A., Baulcombe, D.C. and Harberd, N.P. (2004) Modulation of floral development by a gibberellin-regulated microRNA. *Development* **131**, 3357–3365.

Alcazar-Roman, A.R., Tran, E.J., Guo, S. and Wente, S.R. (2006) Inositol hexakisphosphate and Gle1 activate the DEAD-box protein Dbp5 for nuclear mRNA export. *Nature Cell Biology* **8**, 711–716.

Arnaoutov, A. and Dasso, M. (2005) Ran-GTP regulates kinetochore attachment in somatic cells. *Cell Cycle* **4**, 1161–1165.

Askjaer, P., Galy, V., Hannak, E. and Mattaj, I.W. (2002) Ran GTPase cycle and importins alpha and beta are essential for spindle formation and nuclear envelope assembly in living Caenorhabditis elegans embryos. *Molecular Biology of the Cell* **13**, 4355–4370.

Babu, J.R., Jeganathan, K.B., Baker, D.J. *et al.* (2003) Rae1 is an essential mitotic checkpoint regulator that cooperates with Bub3 to prevent chromosome missegregation. *Journal of Cell Biology* **160**, 341–353.

Belgareh, N. and Doye, V. (1997) Dynamics of nuclear pore distribution in nucleoporin mutant yeast cells. *Journal of Cell Biology* **136**, 747–759.

Belgareh, N., Rabut, G., Bai, S.W. *et al.* (2001) An evolutionarily conserved NPC subcomplex, which redistributes in part to kinetochores in mammalian cells. *Journal of Cell Biology* **154**, 1147–1160.

Bendahmane, A., Kanyuka, K. and Baulcombe, D.C. (1999) The Rx gene from potato controls separate virus resistance and cell death responses. *Plant Cell* **11**, 781–792.

Bischoff, F.R., Krebber, H., Smirnova, E. *et al.* (1995) Co-activation of RanGTPase and inhibition of GTP dissociation by Ran-GTP binding protein RanBP1. *EMBO Journal* **14**, 705–715.

Black, B.E., Levesque, L., Holaska, J.M. *et al.* (1999) Identification of an NTF2-related factor that binds Ran-GTP and regulates nuclear protein export. *Molecular and Cellular Biology* **19**, 8616–8624.

Blower, M.D., Nachury, M., Heald, R. and Weis, K. (2005) A Rae1-containing ribonucleoprotein complex is required for mitotic spindle assembly. *Cell* **121**, 223–234.

Boehmer, T., Enninga, J., Dales, S. *et al.* (2003) Depletion of a single nucleoporin, Nup107, prevents the assembly of a subset of nucleoporins into the nuclear pore complex. *Proceedings of the National Academy of Sciences of the United States of America* **100**, 981–985.

Boruc, J., Zhou, X. and Meier, I. (2011) Dynamics of the plant nuclear envelope and nuclear pore. *Plant Physiology*.

Brohawn, S.G., Partridge, J.R., Whittle, J.R.R. and Schwartz, T.U. (2009) The nuclear pore complex has entered the atomic age. *Structure* **17**, 1156–1168.

Bucci, M. and Wente, S.R. (1998) A novel fluorescence-based genetic strategy identifies mutants of Saccharomyces cerevisiae defective for nuclear pore complex assembly. *Molecular Biology of the Cell* **9**, 2439–2461.

Campbell, M.S., Chan, G.K. and Yen, T.J. (2001) Mitotic checkpoint proteins HsMAD1 and HsMAD2 are associated with nuclear pore complexes in interphase. *Journal of Cell Science* **114**, 953–963.

Chait, B.T., Jovanovic-Talisman, T., Tetenbaum-Novatt, J. *et al.* (2009) Artificial nanopores that mimic the transport selectivity of the nuclear pore complex. *Nature* **457**, 1023–1027.

Chang, T.H., Tseng, S.S.L., Weaver, P.L. *et al.* (1998) Dbp5p, a cytosolic RNA helicase, is required for poly(A)(+) RNA export. *EMBO Journal* **17**, 2651–2662.

Cheng, Y.T., Germain, H., Wiermer, M. *et al.* (2009) Nuclear pore complex component MOS7/Nup88 is required for innate immunity and nuclear accumulation of defense regulators in Arabidopsis. *Plant Cell* **21**, 2503–2516.

Cole, C.N., Snay-Hodge, C.A., Colot, H.V. and Goldstein, A.L. (1998) Dbp5p/Rat8p is a yeast nuclear pore-associated DEAD-box protein essential for RNA export. *EMBO Journal* **17**, 2663–2676.

Conti, E. and Fribourg, S. (2003) Structural similarity in the absence of sequence homology of the messenger RNA export factors Mtr2 and p15. *EMBO Reports* **4**, 699–703.

Cross, M.K. and Powers, M.A. (2011) Nup98 regulates bipolar spindle assembly through association with microtubules and opposition of MCAK. *Molecular Biology of the Cell* **22**, 661–672.

Cullen, B.R. (2003) Nuclear RNA export. *Journal of Cell Science* **116**, 587–597.

Daneholt, B., Zhao, J., Jin, S.B. *et al.* (2002) The mRNA export factor Dbp5 is associated with Balbiani ring mRNP from gene to cytoplasm. *EMBO Journal* **21**, 1177–1187.

Dasso, M. (2001) Running on Ran: nuclear transport and the mitotic spindle. *Cell* **104**, 321–324.

De Souza, C.P., Hashmi, S.B., Nayak, T. *et al.* (2009) Mlp1 Acts as a mitotic scaffold to spatially regulate spindle assembly checkpoint proteins in Aspergillus nidulans. *Molecular Biology of the Cell* **20**, 2146–2159.

De Souza, C.P., Osmani, A.H., Hashmi, S.B. and Osmani, S.A. (2004) Partial nuclear pore complex disassembly during closed mitosis in Aspergillus nidulans. *Current Biology* **14**, 1973–1984.

De Souza, C.P. and Osmani, S.A. (2007) Mitosis, not just open or closed. *Eukaryot Cell* **6**, 1521–1527.

De Souza, C.P. and Osmani, S.A. (2009) Double duty for nuclear proteins – the price of more open forms of mitosis. *Trends in Genetics* **25**, 545–554.

Dittmer, T.A., Stacey, N.J., Sugimoto-Shirasu, K. and Richards, E.J. (2007) *LITTLE NUCLEI* genes affecting nuclear morphology in *Arabidopsis thaliana*. *Plant Cell* **19**, 2793–2803.

Dong, C.H., Hu, X., Tang, W. *et al.* (2006) A putative Arabidopsis nucleoporin, AtNUP160, is critical for RNA export and required for plant tolerance to cold stress. *Molecular and Cellular Biology* **26**, 9533–9543.

Dultz, E., Zanin, E., Wurzenberger, C. *et al.* (2008) Systematic kinetic analysis of mitotic dis- and reassembly of the nuclear pore in living cells. *Journal of Cell Biology* **180**, 857–865.

Erkmann, J.A. and Kutay, U. (2004) Nuclear export of mRNA: from the site of transcription to the cytoplasm. *Experimental Cell Research* **296**, 12–20.

Feldherr, C.M. and Akin, D. (1997) The location of the transport gate in the nuclear pore complex. *Journal of Cell Science* **110** (24), 3065–3070.

Fiserova, J., Kiseleva, E. and Goldberg, M.W. (2009) Nuclear envelope and nuclear pore complex structure and organization in tobacco BY-2 cells. *Plant Journal* **59**, 243–255.

Franz, C., Askjaer, P., Antonin, W. *et al.* (2005) Nup155 regulates nuclear envelope and nuclear pore complex formation in nematodes and vertebrates. *EMBO Journal* **24**, 3519–3531.

Gadal, O., Strauss, D., Kessl, J. *et al.* (2001) Nuclear export of 60s ribosomal subunits depends on Xpo1p and requires a nuclear export sequence-containing factor, Nmd3p, that associates with the large subunit protein Rpl10p. *Molecular and Cellular Biology* **21**, 3405–3415.

Galy, V., Askjaer, P., Franz, C. *et al.* (2006) MEL-28, a novel nuclear-envelope and kinetochore protein essential for zygotic nuclear-envelope assembly in C. elegans. *Current Biology* **16**, 1748–1756.

Galy, V., Mattaj, I.W. and Askjaer, P. (2003) Caenorhabditis elegans nucleoporins Nup93 and Nup205 determine the limit of nuclear pore complex size exclusion in vivo. *Molecular Biology of the Cell* **14**, 5104–5115.

Gao, P., Bai, X., Yang, L., *et al.* (2011) osa-MIR393: a salinity- and alkaline stress-related microRNA gene. *Molecular Biology Reports* **38**, 237–242.

Gao, Y., Gong, X., Cao, W. *et al.* (2008) SAD2 in Arabidopsis functions in trichome initiation through mediating GL3 function and regulating GL1, TTG1 and GL2 expression. *Journal of Integrative Plant Biology* **50**, 906–917.

Gerace, L. and Blobel, G. (1980) The nuclear envelope lamina is reversibly depolymerized during mitosis. *Cell* **19**, 277–287.

Germain, H., Qu, N., Cheng, Y.T. *et al.* (2010) MOS11: a new component in the mRNA export pathway. *PLoS Genetics* **6**, e1001250.

Gong, Z., Dong, C.H., Lee, H. *et al.* (2005) A DEAD box RNA helicase is essential for mRNA export and important for development and stress responses in Arabidopsis. *Plant Cell* **17**, 256–67.

Gorlich, D., Paraskeva, E., Izaurralde, E. *et al.* (1999) CRM1-mediated recycling of snurportin 1 to the cytoplasm. *Journal of Cell Biology* **145**, 255–264.

Griffis, E.R., Craige, B., Dimaano, C. *et al.* (2004) Distinct functional domains within nucleoporins Nup153 and Nup98 mediate transcription-dependent mobility. *Molecular Biology of the Cell* **15**, 1991–2002.

Guttinger, S., Laurell, E. and Kutay, U. (2009) Orchestrating nuclear envelope disassembly and reassembly during mitosis. *Nature Reviews Molecular Cell Biology* **10**, 178–191.

Haasen, D. and Merkle, T. (2002) Characterization of an Arabidopsis thaliana homologue of the nuclear export receptor CAS by its interaction with Importin alpha. *Plant Biology* **4**, 432–439.

Haizel, T., Merkle, T., Pay, A. *et al.* (1997) Characterization of proteins that interact with the GTP-bound form of the regulatory GTPase ran in Arabidopsis. *Plant Journal* **11**, 93–103.

Harel, A. and Forbes, D.J. (2004) Importin beta: conducting a much larger cellular symphony. *Molecular Cell* **16**, 319–330.

Harel, A., Orjalo, A.V., Vincent, T. *et al.* (2003) Removal of a single pore subcomplex results in vertebrate nuclei devoid of nuclear pores. *Molecular Cell* **11**, 853–864.

Hase, M.E. and Cordes, V.C. (2003) Direct interaction with nup153 mediates binding of Tpr to the periphery of the nuclear pore complex. *Molecular Biology of the Cell* **14**, 1923–1940.

Heald, R. and Mckeon, F. (1990) Mutations of phosphorylation sites in lamin A that prevent nuclear lamina disassembly in mitosis. *Cell* **61**, 579–589.

Higa, M.M., Alam, S.L., Sundquist, W.I. and Ullman, K.S. (2007) Molecular characterization of the Ran-binding zinc finger domain of Nup153. *Journal of Biological Chemistry* **282**, 17090–17100.

Ho, J.H., Kallstrom, G. and Johnson, A.W. (2000) Nmd3p is a Crm1p-dependent adapter protein for nuclear export of the large ribosomal subunit. *Journal of Cell Biology* **151**, 1057–1066.

Hoelz, A., Debler, E.W. and Blobel, G. (2011) The structure of the nuclear pore complex. In: R.D. Kornberg, C.R.H. Raetz, J.E. Rothman and J.W. Thorner (eds) *Annual Review of Biochemistry, Vol 80*, Annual Reviews, Palo Alto, pp. 613–643.

Hopper, A.K., Traglia, H.M. and Dunst, R.W. (1990) The yeast RNA1 gene product necessary for RNA processing is located in the cytosol and apparently excluded from the nucleus. *Journal of Cell Biology* **111**, 309–321.

Huber, J., Cronshagen, U., Kadokura, M. *et al.* (1998) Snurportin1, an m3G-cap-specific nuclear import receptor with a novel domain structure. *EMBO Journal* **17**, 4114–4126.

Iouk, T., Kerscher, O., Scott, R.J. *et al.* (2002) The yeast nuclear pore complex functionally interacts with components of the spindle assembly checkpoint. *Journal of Cell Biology* **159**, 807–819.

Izaurralde, E. and Stutz, F. (2003) The interplay of nuclear mRNP assembly, mRNA surveillance and export. *Trends in Cell Biology* **13**, 319–327.

Jacob, Y., Mongkolsiriwatana, C., Veley, K.M. *et al.* (2007) The nuclear pore protein AtTPR is required for RNA homeostasis, flowering time, and auxin signaling. *Plant Physiology* **144**, 1383–1390.

Jeganathan, K.B., Malureanu, L. and Van Deursen, J.M. (2005) The Rae1-Nup98 complex prevents aneuploidy by inhibiting securin degradation. *Nature* **438**, 1036–1039.

Joseph, J., Liu, S.T., Jablonski, S.A. *et al.* (2004) The RanGAP1-RanBP2 complex is essential for microtubule-kinetochore interactions in vivo. *Current Biology* **14**, 611–617.

Joseph, J., Tan, S.H., Karpova, T.S. *et al.* (2002) SUMO-1 targets RanGAP1 to kinetochores and mitotic spindles. *Journal of Cell Biology* **156**, 595–602.

Kahana, J.A. and Cleveland, D.W. (2001) Cell cycle. Some importin news about spindle assembly. *Science* **291**, 1718–1719.

Kanamori, N., Madsen, L.H., Radutoiu, S. *et al.* (2006) A nucleoporin is required for induction of Ca2+ spiking in legume nodule development and essential for rhizobial and fungal symbiosis. *Proceedings of the National Academy of Sciences of the United States of America* **103**, 359–64.

Keminer, O. and Peters, R. (1999) Permeability of single nuclear pores. *Biophysical Journal* **77**, 217–228.

Kim, S.H., Arnold, D., Lloyd, A. (2001) Antisense expression of an Arabidopsis ran binding protein renders transgenic roots hypersensitive to auxin and alters auxin-induced root growth and development by arresting mitotic progress. *Plant Cell* **13**, 2619–2630.

Kim, S.H. and Roux, S.J. (2003) An Arabidopsis Ran-binding protein, AtRanBP1c, is a co-activator of Ran GTPase-activating protein and requires the C-terminus for its cytoplasmic localization. *Planta* **216** (6), 1047–1052.

Kiseleva, E., Rutherford, S., Cotter, L.M. *et al.* (2001) Steps of nuclear pore complex disassembly and reassembly during mitosis in early Drosophila embryos. *Journal of Cell Science* **114**, 3607–3618.

Klebe, C., Bischoff, F.R., Ponstingl, H. and Wittinghofer, A. (1995) Interaction of the nuclear GTP-binding protein Ran with its regulatory proteins RCC1 and RanGAP1. *Biochemistry* **34**, 639–647.

Kraemer, D., Dresbach, T. and Drenckhahn, D. (2001) Mrnp41 (Rae 1p) associates with microtubules in HeLa cells and in neurons. *European Journal of Cell Biology* **80**, 733–740.

Kutay, U., Bischoff, F.R., Kostka, S. *et al.* (1997) Export of importin alpha from the nucleus is mediated by a specific nuclear transport factor. *Cell* **90**, 1061–1071.

Lee, J.Y., Lee, H.S., Wi, S.J. *et al.* (2009) Dual functions of Nicotiana benthamiana Rae1 in interphase and mitosis. *Plant Journal* **59**, 278–291.

Lee, S.H., Sterling, H., Burlingame, A. and Mccormick, F. (2008) Tpr directly binds to Mad1 and Mad2 and is important for the Mad1-Mad2-mediated mitotic spindle checkpoint. *Genes and Development* **22**, 2926–2931.

Lim, R.Y., Ullman, K.S. and Fahrenkrog, B. (2008) Biology and biophysics of the nuclear pore complex and its components. *Int Rev Cell Mol Biol* **267**, 299–342.

Lince-Faria, M., Maffini, S., Orr, B. *et al.* (2009) Spatiotemporal control of mitosis by the conserved spindle matrix protein Megator. *Journal of Cell Biology* **184**, 647–657.

Lindsay, M.E., Holaska, J.M., Welch, K. *et al.* (2001) Ran-binding protein 3 is a cofactor for Crm1-mediated nuclear protein export. *Journal of Cell Biology* **153**, 1391–1402.

Liu, H.L., De Souza, C.P., Osmani, A.H. and Osmani, S.A. (2009) The three fungal transmembrane nuclear pore complex proteins of Aspergillus nidulans are dispensable in the presence of an intact An-Nup84-120 complex. *Molecular Biology of the Cell* **20**, 616–630.

Lodin, Z., Blumajer, J. and Mares, V. (1978) Nuclear pore complexes in cells of the developing mouse cerebral cortex. *Acta Histochemica* **63**, 74–79.

Loiodice, I., Alves, A., Rabut, G. *et al.* (2004) The entire Nup107-160 complex, including three new members, is targeted as one entity to kinetochores in mitosis. *Molecular Biology of the Cell* **15**, 3333–3344.

Lu, Q., Tang, X., Tian, G. *et al.* (2010) Arabidopsis homolog of the yeast TREX-2 mRNA export complex: components and anchoring nucleoporin. *Plant Journal* **61**, 259–270.

Macara, I.G. (2001) Transport into and out of the nucleus. *Microbiology and Molecular Biology Reviews* **65**, 570–594.

Macara, I.G., Smith, A. and Brownawell, A. (1998) Nuclear import of Ran is mediated by the transport factor NTF2. *Current Biology* **8**, 1403–1406.

Macaulay, C., Meier, E. and Forbes, D.J. (1995) Differential mitotic phosphorylation of proteins of the nuclear pore complex. *Journal of Biological Chemistry* **270**, 254–262.

Mackay, D.R., Elgort, S.W. and Ullman, K.S. (2009) The nucleoporin Nup153 has separable roles in both early mitotic progression and the resolution of mitosis. *Molecular Biology of the Cell* **20**, 1652–1660.

Mahajan, R., Delphin, C., Guan, T. *et al.* (1997) A small ubiquitin-related polypeptide involved in targeting RanGAP1 to nuclear pore complex protein RanBP2. *Cell* **88**, 97–107.

Makhnevych, T., Lusk, C.P., Anderson, A.M. *et al.* (2003) Cell cycle regulated transport controlled by alterations in the nuclear pore complex. *Cell* **115**, 813–823.

Martinez-Exposito, M.J., Kaplan, K.B., Copeland, J. and Sorger, P.K. (1999) Retention of the BUB3 checkpoint protein on lagging chromosomes. *Proceedings of the National Academy of Sciences of the United States of America* **96**, 8493–8498.

Masuda, K., Xu, Z.J., Takahashi, S., Ito, A. *et al.* (1997) Peripheral framework of carrot cell nucleus contains a novel protein predicted to exhibit a long alpha-helical domain. *Experimental Cell Research* **232**, 173–181.

Mattaj, I.W., Englmeier, L., Fornerod, M. *et al.* (2001) RanBP3 influences interactions between CRM1 and its nuclear protein export substrates. *EMBO Reports* **2**, 926–932.

Matunis, M.J., Coutavas, E. and Blobel, G. (1996) A novel ubiquitin-like modification modulates the partitioning of the Ran-GTPase-activating protein RanGAP1 between the cytosol and the nuclear pore complex. *Journal of Cell Biology* **135**, 1457–1470.

Matunis, M.J., Wu, J. and Blobel, G. (1998) SUMO-1 modification and its role in targeting the Ran GTPase-activating protein, RanGAP1, to the nuclear pore complex. *Journal of Cell Biology* **140**, 499–509.

Meier, I. and Somers, D.E. (2011) Regulation of nucleocytoplasmic trafficking in plants. *Current Opinion in Plant Biology* **14**, 538–546.

Merkle, T. (2003) Nucleo-cytoplasmic partitioning of proteins in plants: implications for the regulation of environmental and developmental signalling. *Current Genetics* **44**, 231–260.

Merkle, T. (2011) Nucleo-cytoplasmic transport of proteins and RNA in plants. *Plant Cell Reports* **30**, 153–176.

Mourao, A., Varrot, A., Mackereth, C.D. *et al.* (2010) Structure and RNA recognition by the snRNA and snoRNA transport factor PHAX. *RNA* **16**, 1205–1216.

Muller, S., Wright, A.J. and Smith, L.G. (2009) Division plane control in plants: new players in the band. *Trends in Cell Biology* **19**, 180–188.

Murtas, G., Reeves, P.H., Fu, Y.F. *et al.* (2003) A nuclear protease required for flowering-time regulation in Arabidopsis reduces the abundance of SMALL UBIQUITIN-RELATED MODIFIER conjugates. *Plant Cell* **15**, 2308–2319.

Muthuswamy, S. and Meier, I. (2011) Genetic and environmental changes in SUMO homeostasis lead to nuclear mRNA retention in plants. *Planta* **233**, 201–208.

Nakano, H., Funasaka, T., Hashizume, C. and Wong, R.W. (2010) Nucleoporin translocated promoter region (Tpr) associates with dynein complex, preventing chromosome lagging formation during mitosis. *Journal of Biological Chemistry* **285**, 10841–10849.

Navarro, L., Dunoyer, P., Jay, F. *et al.* (2006) A plant miRNA contributes to antibacterial resistance by repressing auxin signaling. *Science* **312**, 436–439.

Ohno, M., Segref, A., Bachi, A. *et al.* (2000) PHAX, a mediator of U snRNA nuclear export whose activity is regulated by phosphorylation. *Cell* **101**, 187–198.

Orjalo, A.V., Arnaoutov, A., Shen, Z. *et al.* (2006) The Nup107-160 nucleoporin complex is required for correct bipolar spindle assembly. *Molecular Biology of the Cell* **17**, 3806–3818.

Osmani, A.H., Davies, J., Liu, H.L. *et al.* (2006) Systematic deletion and mitotic localization of the nuclear pore complex proteins of Aspergillus nidulans. *Molecular Biology of the Cell* **17**, 4946–4961.

Palma, K., Zhang, Y. and Li, X. (2005) An importin alpha homolog, MOS6, plays an important role in plant innate immunity. *Current Biology* **15**, 1129–1135.

Parry, G., Ward, S., Cernac, A. *et al.* (2006) The Arabidopsis SUPPRESSOR OF AUXIN RESISTANCE proteins are nucleoporins with an important role in hormone signaling and development. *Plant Cell* **18**, 1590–1603.

Pemberton, L.F. and Paschal, B.M. (2005) Mechanisms of receptor-mediated nuclear import and nuclear export. *Traffic* **6**, 187–198.

Pritchard, C.E., Fornerod, M., Kasper, L.H. and Van Deursen, J.M. (1999) RAE1 is a shuttling mRNA export factor that binds to a GLEBS-like NUP98 motif at the nuclear pore complex through multiple domains. *Journal of Cell Biology* **145**, 237–254.

Prunuske, A.J., Liu, J., Elgort, S. *et al.* (2006) Nuclear envelope breakdown is coordinated by both Nup358/RanBP2 and Nup153, two nucleoporins with zinc finger modules. *Molecular Biology of the Cell* **17**, 760–769.

Rairdan, G.J., Collier, S.M., Sacco, M.A. *et al.* (2008) The coiled-coil and nucleotide binding domains of the Potato Rx disease resistance protein function in pathogen recognition and signaling. *Plant Cell* **20**, 739–751.

Rasala, B.A., Orjalo, A.V., Shen, Z. *et al.* (2006) ELYS is a dual nucleoporin/kinetochore protein required for nuclear pore assembly and proper cell division. *Proceedings of the National Academy of Sciences of the United States of America* **103**, 17801–17806.

Reeves, P.H., Murtas, G., Dash, S. and Coupland, G. (2002) Early in short days 4, a mutation in Arabidopsis that causes early flowering and reduces the mRNA abundance of the floral repressor FLC. *Development* **129**, 5349–5361.

Reverter, D. and Lima, C.D. (2005) Insights into E3 ligase activity revealed by a SUMO-RanGAP1-Ubc9-Nup358 complex. *Nature* **435**, 687–692.

Rodrigo-Peiris, T., Xu, X.M., Zhao, Q. *et al.* (2011) RanGAP is required for post-meiotic mitosis in female gametophyte development in Arabidopsis thaliana. *Journal of Experimental Botany* **62**, 2705–2714.

Rose, A. and Meier, I. (2001) A domain unique to plant RanGAP is responsible for its targeting to the plant nuclear rim. *Proceedings of the National Academy of Sciences of the United States of America* **98**, 15377–15382.

Rout, M.P., Aitchison, J.D., Magnasco, M.O. and Chait, B.T. (2003) Virtual gating and nuclear transport: the hole picture. *Trends in Cell Biology* **13**, 622–628.

Roux, S.J. and Kim, S.H. (2003) An Arabidopsis Ran-binding protein, AtRanBP1c, is a co-activator of Ran GTPase-activating protein and requires the C-terminus for its cytoplasmic localization. *Planta* **216**, 1047–1051.

Sacco, M.A., Koropacka, K., Grenier, E. *et al.* (2009) The cyst nematode SPRYSEC protein RBP-1 elicits Gpa2- and RanGAP2-dependent plant cell death. *PLoS Pathogens* **5**, e1000564.

Sacco, M.A., Mansoor, S. and Moffett, P. (2007) A RanGAP protein physically interacts with the NB-LRR protein Rx, and is required for Rx-mediated viral resistance. *Plant Journal* **52**, 82–93.

Saito, K., Yoshikawa, M., Yano, K. *et al.* (2007) NUCLEOPORIN85 is required for calcium spiking, fungal and bacterial symbioses, and seed production in Lotus japonicus. *Plant Cell* **19**, 610–624.

Saitoh, H., Pu, R., Cavenagh, M. and Dasso, M. (1997) RanBP2 associates with Ubc9p and a modified form of RanGAP1. *Proceedings of the National Academy of Sciences of the United States of America* **94**, 3736–3741.

Saitoh, N., Uchimura, Y., Tachibana, T., Sugahara, S. *et al.* (2006) In situ SUMOylation analysis reveals a modulatory role of RanBP2 in the nuclear rim and PML bodies. *Experimental Cell Research* **312**, 1418–1430.

Salina, D., Enarson, P., Rattner, J.B. and Burke, B. (2003) Nup358 integrates nuclear envelope breakdown with kinetochore assembly. *Journal of Cell Biology* **162**, 991–1001.

Sengupta, J., Bussiere, C., Pallesen, J. *et al.* (2010) Characterization of the nuclear export adaptor protein Nmd3 in association with the 60S ribosomal subunit. *Journal of Cell Biology* **189**, 1079–1086.

Shi, L., Mogensen, H.L., Zhu, T. and Smith, S.E. (1991) Dynamics of nuclear pore density and distribution patterns within developing pollen: implications for a functional relationship between the vegetative nucleus and the generative cell. *Journal of Cell Science* **99**, 115–120.

Si-Ammour, A., Windels, D., Arn-Bouldoires, E. *et al.* (2011) MiR393 and secondary siRNAs regulate expression of the TIR1/AFB2 auxin receptor clade and auxin-related development of Arabidopsis leaves. *Plant Physiology* **157**, 683–691.

Smith, L.G. (2001) Plant cell division: building walls in the right places. *Nature Reviews Molecular Cell Biology* **2**, 33–9.

Stavru, F., Hülsmann, B.B., Spang, A. *et al.* (2006a) NDC1: a crucial membrane-integral nucleoporin of metazoan nuclear pore complexes. *Journal of Cell Biology* **173**, 509–519.

Strambio-De-Castillia, C., Niepel, M. and Rout, M.P. (2010) The nuclear pore complex: bridging nuclear transport and gene regulation. *Nature Reviews Molecular Cell Biology* **11**, 490–501.

Strawn, L.A., Shen, T., Shulga, N. *et al.* (2004) Minimal nuclear pore complexes define FG repeat domains essential for transport. *Nature Cell Biology* **6**, 197–206.

Sunkar, R. and Zhu, J.K. (2004) Novel and stress-regulated microRNAs and other small RNAs from Arabidopsis. *Plant Cell* **16**, 2001–2019.

Tameling, W.I. and Baulcombe, D.C. (2007) Physical association of the NB-LRR resistance protein Rx with a Ran GTPase-activating protein is required for extreme resistance to Potato virus X. *Plant Cell* **19**, 1682–1694.

Tamura, K., Fukao, Y., Iwamoto, M. *et al.* (2010) Identification and characterization of nuclear pore complex components in Arabidopsis thaliana. *Plant Cell* **22**, 4084–4097.

Tamura, K. and Hara-Nishimura, I. (2011) Involvement of the nuclear pore complex in morphology of the plant nucleus. *Nucleus* **2**, 168–172.

Taylor, S.S., Ha, E. and Mckeon, F. (1998) The human homologue of Bub3 is required for kinetochore localization of Bub1 and a Mad3/Bub1-related protein kinase. *Journal of Cell Biology* **142**, 1–11.

Terry, L.J., Shows, E.B. and Wente, S.R. (2007) Crossing the nuclear envelope: hierarchical regulation of nucleocytoplasmic transport. *Science* **318**, 1412–1416.

Theisen, U., Straube, A. and Steinberg, G. (2008) Dynamic rearrangement of nucleoporins during fungal 'open' mitosis. *Molecular Biology of the Cell* **19**, 1230–1240.

Thomas, F. and Kutay, U. (2003) Biogenesis and nuclear export of ribosomal subunits in higher eukaryotes depend on the CRM1 export pathway. *Journal of Cell Science* **116**, 2409–2419.

Van Damme, D. and Geelen, D. (2008) Demarcation of the cortical division zone in dividing plant cells. *Cell Biology International* **32**, 178–187.

Vasu, S., Shah, S., Orjalo, A. *et al.* (2001) Novel vertebrate nucleoporins Nup133 and Nup160 play a role in mRNA export. *Journal of Cell Biology* **155**, 339–354.

Verslues, P.E., Guo, Y., Dong, C.H. *et al.* (2006) Mutation of SAD2, an importin beta-domain protein in Arabidopsis, alters abscisic acid sensitivity. *Plant Journal* **47**, 776–787.

Walther, T.C., Alves, A., Pickersgill, H., *et al.* (2003) The conserved Nup107-160 complex is critical for nuclear pore complex assembly. *Cell* **113**, 195–206.

Walther, T.C., Fornerod, M., Pickersgill, H. *et al.* (2001) The nucleoporin Nup153 is required for nuclear pore basket formation, nuclear pore complex anchoring and import of a subset of nuclear proteins. *EMBO Journal* **20**, 5703–5714.

Wang, X., Babu, J.R., Harden, J.M. *et al.* (2001) The mitotic checkpoint protein hBUB3 and the mRNA export factor hRAE1 interact with GLE2p-binding sequence (GLEBS)-containing proteins. *Journal of Biological Chemistry* **276**, 26559–26567.

Wang, X., Xu, Y., Han, Y. *et al.* (2006) Overexpression of RAN1 in rice and Arabidopsis alters primordial meristem, mitotic progress, and sensitivity to auxin. *Plant Physiology* **140**, 91–101.

Weirich, C.S., Erzberger, J.P., Flick, J.S. *et al.* (2006) Activation of the DExD/H-box protein Dbp5 by the nuclear-pore protein Gle1 and its coactivator InsP6 is required for mRNA export. *Nature Cell Biology* **8**, 668–676.

Weis, K. (2003) Regulating access to the genome: nucleocytoplasmic transport throughout the cell cycle. *Cell* **112**, 441–451.

Wente, S.R. (2000) Gatekeepers of the nucleus. *Science* **288**, 1374–1377.

Wente, S.R. and Carmody, S.R. (2009) mRNA nuclear export at a glance. *Journal of Cell Science* **122**, 1933–1937.

Wente, S.R. and Rout, M.P. (2010) The nuclear pore complex and nuclear transport. *Cold Spring Harbor Perspectives in Biology* **2**.

Wente, S.R. and Terry, L.J. (2007) Nuclear mRNA export requires specific FG nucleoporins for translocation through the nuclear pore complex. *Journal of Cell Biology* **178**, 1121–1132.

West, R.R., Vaisberg, E.V., Ding, R. *et al.* (1998) cut11(+): A gene required for cell cycle-dependent spindle pole body anchoring in the nuclear envelope and bipolar spindle formation in Schizosaccharomyces pombe. *Molecular Biology of the Cell* **9**, 2839–2855.

Whalen, W.A., Bharathi, A., Danielewicz, D. and Dhar, R. (1997) Advancement through mitosis requires rae1 gene function in fission yeast. *Yeast* **13**, 1167–1179.

Wiermer, M., Germain, H., Cheng, Y.T. *et al.* (2010) Nucleoporin MOS7/Nup88 contributes to plant immunity and nuclear accumulation of defense regulators. *Nucleus* **1**, 332–336.

Wiermer, M., Palma, K., Zhang, Y. and Li, X. (2007) Should I stay or should I go? Nucleocytoplasmic trafficking in plant innate immunity. *Cell Microbiol* **9**, 1880–1890.

Winey, M., Hoyt, M.A., Chan, C. *et al.* (1993) NDC1: a nuclear periphery component required for yeast spindle pole body duplication. *Journal of Cell Biology* **122**, 743–751.

Winey, M., Yarar, D., Giddings, T.H., Jr. and Mastronarde, D.N. (1997) Nuclear pore complex number and distribution throughout the Saccharomyces cerevisiae cell cycle by three-dimensional reconstruction from electron micrographs of nuclear envelopes. *Molecular Biology of the Cell* **8**, 2119–2132.

Wong, R.W., Blobel, G. and Coutavas, E. (2006) Rae1 interaction with NuMA is required for bipolar spindle formation. *Proceedings of the National Academy of Sciences of the United States of America* **103**, 19783–19787.

Wu, J., Matunis, M.J., Kraemer, D. *et al.* (1995) Nup358, a cytoplasmically exposed nucleoporin with peptide repeats, Ran-GTP binding sites, zinc fingers, a cyclophilin A homologous domain, and a leucine-rich region. *Journal of Biological Chemistry* **270**, 14209–14213.

Xu, X.M., Meulia, T. and Meier, I. (2007a) Anchorage of plant RanGAP to the nuclear envelope involves novel nuclear-pore-associated proteins. *Current Biology* **17**, 1157–1163.

Xu, X.M., Rose, A., Muthuswamy, S. *et al.* (2007b) NUCLEAR PORE ANCHOR, the Arabidopsis homolog of Tpr/Mlp1/Mlp2/megator, is involved in mRNA export and SUMO homeostasis and affects diverse aspects of plant development. *Plant Cell* **19**, 1537–1548.

Xu, X.M., Zhao, Q., Rodrigo-Peiris, T. *et al.* (2008) RanGAP1 is a continuous marker of the Arabidopsis cell division plane. *Proceedings of the National Academy of Sciences of the United States of America* **105**, 18637–18642.

York, J.D., Odom, A.R., Murphy, R. *et al.* (1999) A phospholipase C-dependent inositol polyphosphate kinase pathway required for efficient messenger RNA export. *Science* **285**, 96–100.

Zang, A., Xu, X., Neill, S. and Cai, W. (2010) Overexpression of OsRAN2 in rice and Arabidopsis renders transgenic plants hypersensitive to salinity and osmotic stress. *Journal of Expimental Botany* **61**, 777–789.

Zhang, X., Zhao, H., Gao, S. *et al.* (2011) Arabidopsis Argonaute 2 regulates innate immunity via miRNA393(∗)-mediated silencing of a Golgi-localized SNARE gene, MEMB12. *Molecular Cell* **42**, 356–366.

Zhang, Y., Cheng, Y.T., Bi, D. *et al.* (2005) MOS2, a protein containing G-patch and KOW motifs, is essential for innate immunity in Arabidopsis thaliana. *Current Biology* **15**, 1936–1942.

Zhang, Y. and Li, X. (2005) A putative nucleoporin 96 Is required for both basal defense and constitutive resistance responses mediated by suppressor of npr1-1,constitutive 1. *Plant Cell* **17**, 1306–1316.

Zhao, J., Zhang, W., Zhao, Y. *et al.* (2007) SAD2, an importin -like protein, is required for UV-B response in Arabidopsis by mediating MYB4 nuclear trafficking. *Plant Cell* **19**, 3805–3818.

Zhao, Q., Brkljacic, J. and Meier, I. (2008) Two distinct interacting classes of nuclear envelope-associated coiled-coil proteins are required for the tissue-specific nuclear envelope targeting of Arabidopsis RanGAP. *Plant Cell* **20**, 1639–1651.

Zhao, Q. and Meier, I. (2011) Identification and characterization of the Arabidopsis FG-repeat nucleoporin Nup62. *7Plant signaling & behavior* **6**, 330–334.

Zheng, B., Chen, X. and Mccormick, S. (2011) The anaphase-promoting complex is a dual integrator that regulates both MicroRNA-mediated transcriptional regulation of cyclin B1 and degradation of Cyclin B1 during Arabidopsis male gametophyte development. *Plant Cell* **23**, 1033–1046.

Zhou, G.K., Kubo, M., Zhong, R. *et al.* (2007) Overexpression of miR165 affects apical meristem formation, organ polarity establishment and vascular development in Arabidopsis. *Plant and Cell Physiology* **48**, 391–404.

Zhou, X., Graumann, K., Evans, D.E. and Meier, I. (2012) Novel plant SUN–KASH bridges are involved in RanGAP anchoring and nuclear shape determination. *The Journal of Cell Biology* **196**, 203–211.

Zuccolo, M., Alves, A., Galy, V. *et al.* (2007) The human Nup107-160 nuclear pore subcomplex contributes to proper kinetochore functions. *EMBO Journal* **26**, 1853–1864.

Annual Plant Reviews (2013) **46**, 93–122 http://onlinelibrary.wiley.com
doi: 10.1002/9781118472507.ch4

Chapter 4

NUCLEOSKELETON IN PLANTS: THE FUNCTIONAL ORGANIZATION OF FILAMENTS IN THE NUCLEUS

Martin W. Goldberg
Department of Biological and Biomedical Sciences, University of Durham, Durham, DH1 3LE, UK

Abstract: The eukaryotic cytoplasm is a complex, organized and highly dynamic environment, whose dynamic organization is dependent on a network of filaments and associated proteins. These networks are established, accepted and well studied. The nucleus is no less complex, organized or dynamic, but the existence of an equivalent nuclear network in is not universally accepted and has proved difficult to study. Theoretically there seems an almost overwhelming requirement for a nucleoskeleton, which is needed to provide a structural framework to organize the genome, as well as other subnuclear components, into functionally distinct regions. Such organization must be, and is, highly dynamic so that it can change during development and in response to changing requirements of the cell. Unfortunately, filamentous structures in the nucleus are difficult to detect amongst all the other fibrous material (the chromatin), which has to be removed in order to visualize the putative nucleoskeleton. Therefore, although such structures can be prepared from both plants and animals, their *in vivo* relevance has remained contentious. Nucleoskeletal filaments have the appearance of intermediate filaments and in fact animal cells have a clearly defined intermediate filament network at the nuclear periphery: the nuclear lamina. Plant cells, however, have no proteins that are clear equivalents of the lamins, or indeed any other intermediate filament protein. In this review we discuss the evidence for nuclear intermediate filament-like proteins and other potential nucleoskeleton components in plants and discuss their possible roles in plant nuclear organization.

Keywords: nucleoskeleton; lamina; plamina; intermediate filament; coiled coil; electron microscope; matrix attachment regions; NMCP/LINC

Annual Plant Reviews Volume 46: Plant Nuclear Structure, Genome Architecture and Gene Regulation, First Edition. Edited by David E. Evans, Katja Graumann and John A. Bryant.

4.1 Introduction

The nucleus is a complex, dynamic and highly organized machine. It is finely tuned to respond to the changing demands of cells. Such cells may be terminally differentiated, dividing or developing. Understanding how this machine is constructed and how it adapts, globally or to rapidly changing signals, is a longstanding but active and important area of research. Even simple machines must have a structural framework, a system to deliver energy, remove waste and send and receive information (e.g. start/stop, faster/slower). In this review, the evidence for a structural framework within the plant cell nucleus will be considered. Alas, the nucleus is not a simple machine. There are many thousands of genes, each with unique requirements for control. There are many processes in addition to gene expression, such as DNA replication, recombination, repair, RNA processing and ribosomal subunit assembly. During interphase, individual chromosomes are held in defined 'territories' (Cremer and Cremer, 2010). In mammalian cells, DNA replication is organized into clusters that persist from one cell cycle to the next (Nakamura *et al.*, 1986; Jackson and Pombo, 1998), although we do not know if this is the case in plants. There is evidence that transcription, too, is organized into 'factories' (Cook, 2010), with genes being sequestered and then released from apparently immobile clusters of RNA polymerase and transcription factors. Nuclear speckles are organized sites of RNA processing (Spector and Lamond, 2011) and are associated with active genes (Brown *et al.*, 2008). Nucleoli are specialized domains containing highly active ribosomal genes as well as ribosomal subunit assembly factories (Pederson, 2011). Nucleoli are ordered regions of the nucleus, rather than clearly bounded compartments, but they can be isolated (Figure 4.1) and in plants as well as animals they can be analysed (Pendle *et al.*, 2005). This indicates that they have a distinct integrity and composition, suggesting an underlying structure. Indeed, evidence from all these examples suggests that there must be some underlying framework to organize these territories, clusters and factories.

The maize genome is about 10^9bp in length, which, if fully stretched out, would extend to about 1 m. Crudely, this has to be squeezed into a nucleus of about 10 μm in diameter. More importantly, inside this incredibly crowded space, transcription, DNA replication, recombination, DNA repair and other processes involving DNA must be organized and controlled precisely in time and space. We do not understand this. We do know that DNA is organized by histones into chromatin, chromatin has a higher order, but variable and controlled structure (Joffe *et al.*, 2010) and chromatin is further organized, probably by a protein scaffold that permeates the nucleus. DNA sequences have been identified that appear to tether the chromatin to the putative scaffold. These are known as Matrix (or Scaffold) Attachment Regions or MARs (or SARs) (Wang *et al.*, 2010).

The nuclear envelope is a special part of the nuclear framework, making up the outer shell. It is constituted from a complex organization of

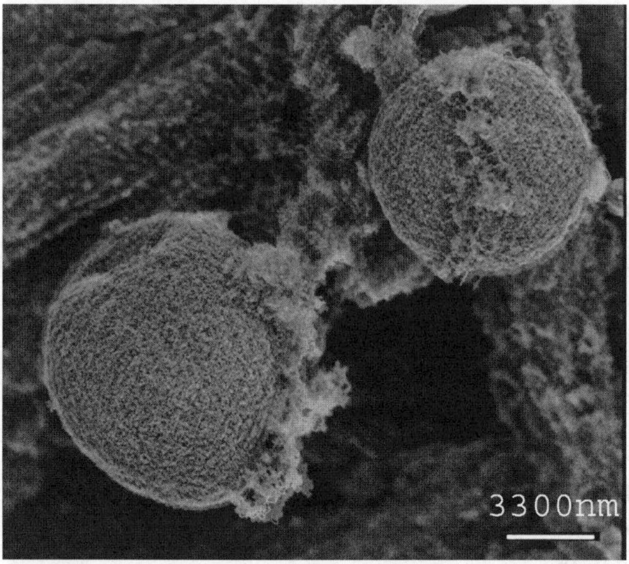

Figure 4.1 Stereo pair (use red/green glasses to view in Plate 4.1) of nucleoli from *Xenopus laevis* oocyte. (*For colour details please see colour plate section.*)

membranes, structural proteins, transmembrane channels (e.g. ion channels) and trans-envelope channels (nuclear pore complexes or NPCs). Together, these physically and functionally link the nuclear and cytoplasmic compartments. In animals, a network of intermediate filaments consisting of lamins, together with associated proteins, lines the inner nuclear membrane. The nuclear lamina not only gives enhanced structural properties to the nucleus, but also has complex roles in cellular functions. It also provides a rigid interface between the cytoskeleton and structural components of the nucleus (Dechat *et al.*, 2008). Even though intermediate filaments in general, and lamins in particular, are not thought to exist in plants, there is increasing evidence for equivalent structures (most recently in Fiserova *et al.*, 2009) and some candidate proteins (Dittmer *et al.*, 2007). There is clear evidence for the existence of other nuclear envelope proteins, such as SUN (Sad1/UNC-84 homology) domain proteins (Graumann *et al.*, 2010a), which are thought to be involved in linking the cytoskeleton to the nucleoskeleton in animals (Schneider *et al.*, 2011). The NPCs have a clearly defined role in the transport of soluble proteins and other molecules to and from the nucleus (Wente and Rout, 2010). They also have a rather less well understood role in the movement of integral membrane proteins and lipids between the outer and inner nuclear membranes, and the endoplasmic reticulum (ER) (King *et al.*, 2006; Lusk *et al.*, 2007; Zuleger *et al.*, 2011). On the other hand, NPCs are also part of the nucleoskeletal preparations, but in contrast to other components of the nuclear envelope, which appear to have roles in genetic repression, NPCs appear to be associated with active genes, at least in some organisms (Blobel, 1985; Dieppois and Stutz, 2010).

What we call the 'scaffold' depends on how it was prepared and its aliases include the 'nuclear matrix', 'nuclear scaffold' or 'nucleoskeleton' (Martelli *et al.*, 2002). We will try to refer to it here as the 'nucleoskeleton' on the grounds that whatever organizes the nuclear interior, it does, comparable to the cytoskeleton, appear to consist of a combination of relatively stable filamentous networks (such as the peripheral nuclear lamina in animals) and more dynamic entities, as well as nuclear specific structures. The terms 'scaffold' and 'matrix' imply a rather rigid and stable framework, which can be no more than a partial description of this organizing network. These terms, however, reflect the origin of discovery from static electron micrographs of highly extracted preparations (Berezney and Coffey, 1974). Moreover, the nucleoskeleton does appear to be linked to the cytoskeleton via the nuclear envelope and therefore could be considered as a specialized 'nucleo-' domain of the cytoskeleton. However, much of the evidence comes from what is operationally described as nuclear matrix preparations, so it is often more appropriate to use this term.

4.2 Intermediate filaments and the nucleoskeleton

Within the cytoplasm of animal cells there are three distinct cytoskeletal components: microtubules, the actin filament network and intermediate filaments. The first two of these also have well established roles in plant cells (Petrásek and Schwarzerová, 2009). F-Actin and microtubules are both highly dynamic filaments constructed from globular monomers, which are in a dynamic flux at either end, depending on their nucleotide bound state and on other interacting proteins that can stabilize or destabilize the filament (Piette *et al.*, 2009). They are primarily used for cell movement (Ridley, 2011) and intra-cellular transport processes, such as vesicle trafficking (Neumann *et al.*, 2003) and chromosome separation (Zhang and Dawe, 2011). Transport involves a myriad of ATP-hydrolysing motor proteins, which link to cargoes and travel along the filaments (Reddy, 2001; Cai and Cresti, 2010).

All three of the cytoskeletal networks interact with each other (Petrásek and Schwarzerová, 2010), to differing degrees in different cell types. There is also clear evidence that the cytoskeleton physically associates with the intranuclear network via the so-called LINC (**LI**nker of **N**ucleo- and **C**ytoskeleton) complex (Mellad *et al.*, 2011).

In contrast to microfilaments and microtubules, intermediate filaments have not been clearly identified in either plants or fungi. This could be where this review ends, because one of the clearest components of the nucleoskeleton is a class of the intermediate filament proteins, the lamins. It has also been suggested that the nucleoskeletal network within the nuclear interior consists of intermediate filament-type proteins (Jackson and Cook, 1988), which may also include A-type lamins, but not B-type lamins (Hozák *et al.*, 1995). Moreover, intermediate filament proteins form relatively stable filaments and are easy to isolate from animal cells as they resist harsh extraction procedures that remove other cytoskeletal components as well as most of the rest of the cell (Berezney and Coffey, 1974). On the other hand, although there are isoforms of actin as well as α-tubulin and β-tubulin, the variations are small (Perrin and Ervasti, 2010; Jost *et al.*, 2004) and essentially a microtubule is a microtubule, and F-actin is F-actin, although their activities, dynamics and functions are highly dependent on many accessory protein (Hamada, 2007; Ketelaar *et al.*, 2010; van der Honing *et al.*, 2007).

Intermediate filaments, on the other hand, represent a large family of diverse proteins (at least 70 genes and isoforms (Hesse *et al.*, 2001), and it could be that structural and functional equivalence is difficult to determine between organisms that have long since diverged such as plants and animals. What have been defined as intermediate filaments, however, have clearly defined features, much of which can be deduced from primary sequences, and these primary structural features have not been identified in plants.

Bone fide intermediate filaments have a long central coiled coil domain, flanked by a variable N-terminal head domain and C-terminal tail domain. The coiled coil is determined by the heptad repeat pattern of hydrophobic amino acids (Herrmann and Aebi, 2004), which results in the assembly of parallel dimers. Recent structural work has modified the description of this coiled-coil rod domain (Nicolet *et al.*, 2010). The rod consists of a structurally conserved C-terminal half, called Coil 2, which is a 142 amino acid, 21nm long rigid rod preceded, at the N-terminal end, by a parallel α-helical bundle, then a flexible linker (called L12). This links to Coil 1B, which like the other coiled coil domains is conserved in length between the different intermediate filament proteins (101 amino acids), with the notable exception of nuclear lamins, where it is 42 amino acids longer. There is then a second flexible linker (L1) connecting to the short Coil 1A (35 amino acids) (Herrmann *et al.*, 2009). The N- and C-terminal domains are highly variable and presumably determine the cell-specific functions.

In order to create filaments, that may be micrometers long, or form part of an extensive network, such as the nuclear lamina, the 40–50 nm long dimers must assemble into higher order structures. The first level of organization, deduced from *in vitro* assembly studies (Herrmann and Aebi, 2004), is the formation of overlapping head-to-head anti-parallel tetramers, which then undergo lateral associations to form ~10 nm filaments. In the case of B-type lamins the filaments appear to be cross-linked into two-dimensional arrays (Goldberg *et al.*, 2008; see also Figure 4.2). The anti-parallel nature of intermediate filament building blocks is a crucial difference from microtubule and actin networks, which use their polarized organization to direct the movement of motor proteins and their cargoes in specific directions. There is no evidence that intermediate filaments facilitate transport in any way. Vimentin, however, could have a role in modulating microtubule-mediated

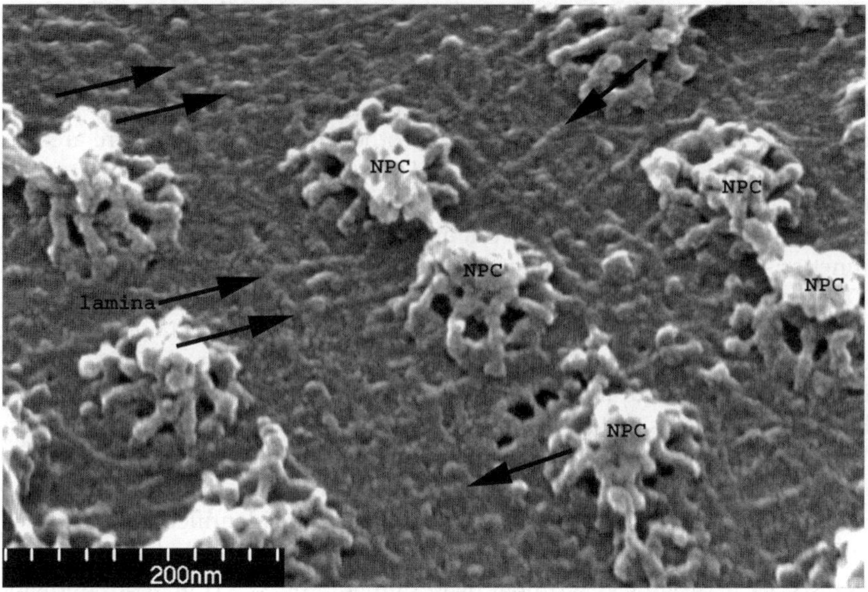

Figure 4.2 Nucleoplasmic face of nuclear envelope isolated from *Xenopus* oocyte imaged using field emission scanning electron microscopy, showing nuclear pore complexes interconnected by nuclear lamin filaments.

movement of melanosome granules by forming a surrounding cage (Chang *et al.*, 2009).

It must be noted however, that intermediate filaments are not the dull, static purely structural filaments that they were once thought to be. Fluorescence recovery after photo-bleaching (FRAP) studies showed that, like in microtubules and F-actin, subunits are continually exchanged (Yoon *et al.*, 2001), although, unlike the other cytoskeletal networks, exchange is not restricted to the ends, but is evenly distributed along the filament, and is controlled by phosphorylation (Goto *et al.*, 2002) rather than nucleotide exchange/hydrolysis. Intermediate filaments also move and change shape, although this movement is dependent on the microtubule and actin networks (reviewed in Eriksson *et al.*, 2009).

Recent studies on human diseases, that can be traced to mutations or perturbations of intermediate filament proteins, have also suggested interesting and diverse roles of intermediate filaments, some of which cannot be explained by a purely structural role (reviewed in Eriksson *et al.*, 2009). The nuclear lamins provide the most striking example of this. Most notably, A-type lamins can be mutated in numerous positions leading to a bewildering array of mostly degenerative diseases (Shimi *et al.*, 2010). Although we are a long way from understanding these diseases, studies have suggested that lamins and their associated proteins have direct roles in maintaining specific chromatin organizations (Puckelwartz *et al.*, 2011), in controlling the activity of transcription factors (Markiewicz *et al.*, 2005; Emerson *et al.*, 2009) and in signalling pathways (Maraldi *et al.*, 2011). It must not be forgotten, however, that they *do* have important mechanical roles in cells and tissues. α-keratin, which forms nails and hair, is the archetypal structural protein (Crick, 1952) and mutations in certain keratins cause skin blistering diseases (Chamcheu *et al.*, 2011). Expression of A-type lamins in *Xenopus laevis* oocytes (which naturally only have B-type lamins), significantly increases the stiffness of the nuclear envelope (Schäpe *et al.*, 2009) and lamin A/C mutations that cause muscular dystrophy result in fragile and damaged nuclei (Shimi *et al.*, 2010).

4.3 Plants do not have intermediate filaments but they may have functional equivalents

So intermediate filaments are important and interesting and, at least as far as lamins are concerned, they are an important component of the nucleoskeleton. However, when we look at the genome of plants such as *Arabidopsis thaliana*, we see no evidence for any protein that could fulfil the criteria for being a *bona fide* intermediate filament protein. On the other hand, it seems very likely that all, or most of, the properties that they impart upon the animal cell should also be important in plants.

To summarize the functional properties of intermediate filaments, they:

- constitute a diverse family of structural proteins;
- are long coiled-coil proteins;
- have a common domain organization (head-rod-tail);
- have highly variable head and tail domains;
- form apolar filaments;
- are dynamic via subunit exchange along the length of the filament;
- are chemically resistant;
- impart enhanced mechanical properties by forming anchored networks;
- provide structural anchors;
- provide biochemical anchors (e.g. for controlling the location of transcription factors).

Although sequence comparisons have not proved successful in identifying plant intermediate filaments, we now know enough about intermediate filaments to try to identify distantly related or unrelated proteins that could possess these properties. This has not yet been achieved but the list of properties above could be a 'check-list' for candidate proteins.

One way to identify candidates is to use antibodies to known metazoan intermediate filament proteins to identify plant proteins. Several earlier studies used this approach to identify putative lamin equivalents, but firm conclusions were difficult to draw from mainly immuno-histological data. More recently, two Pea (*Pisum sativum L.*) proteins, p65 and p60, were recognized by a general anti-intermediate filament antibody in nuclear matrix preparations (Blumenthal *et al.*, 2004). These were shown to have some sequence similarity to lamins and keratin, respectively and crude preparations of these and associated proteins had an intermediate filament-like ultrastructure. However, detailed biochemical and structural studies are yet to be done and immuno-fluorescence of fixed pea plumule cells did not give a clear localization or structural organization (Blumenthal *et al.*, 2004). A careful and extensive genomic study (Rose *et al.*, 2004, 2005) used prediction algorithms to search for genes encoding long coiled coil proteins in *Arabidopsis thaliana* and compared them to other organisms. Although long coiled-coil proteins were identified, no intermediate filament-related proteins could be recognized. On the other hand, spectrin repeat proteins were also not found, despite immuno-histological evidence for such proteins in algae (*Chara globularis* Thuill) (Braun, 2001) and onion (*Allium cepa*) (Pérez-Munive *et al.*, 2011) using antibodies specific for spectrin repeats. It is therefore plausible that intermediate filaments and other structural proteins like the spectrin repeat proteins, such as the nuclear envelope nesprins, do have structural and functional equivalents in plants that cannot be identified by traditional algorithms.

4.4 Plants can evolve different solutions to the same problem

Another possibility is that equivalent protein structures have not evolved via the same route. An interesting example of this can be found in the nucleo-cytoplasmic transport system (for review of plant nucleo-cytoplasmic transport see Merkle, 2011). The direction of transport is controlled by the small GTPase Ran, which is GTP-bound in the nucleus due to the location of the Ran **G**uanine nucleotide **E**xchange **F**actor (RanGEF/RCC1) and becomes GDP-bound in the cytoplasm due to the presence of the **G**TPase **A**ctivating **P**rotein (RanGAP). RanGTP binds to import complexes and dissociates them in the nucleus, terminating import. RanGTP also binds export factors, but in this case promotes their association to the cargo. Both these processes must occur only in the nucleus, and therefore eukaryotic cells position RanGAP in the cytoplasm to ensure that Ran hydrolyses its bound GTP as it enters the cytoplasm. In animals, RanGAP is attached to the cytoplasmic face of the NPC when modified by SUMO1 (**S**mall **U**biquitin-related **MO**difier 1) (Mahajan *et al.*, 1997). SUMO1 mediates the interaction of RanGAP with the cytoplasmic filament nucleoporin, Nup358. In plants there is no equivalent to Nup358, but RanGAP is nevertheless located to the outer nuclear membrane. In *Arabidopsis*, RanGAP contains a plant-specific WPP (Trp-Pro-Pro) domain, which interacts directly with the WIP (**W**PP-domain-**I**nteracting **P**rotein) and WIT (**W**PP-domain-**I**nteracting **T**ail-anchored protein) families of transmembrane proteins (Xu *et al.*, 2007; Zhao *et al.*, 2008). Therefore, WIPs and WITs are proteins that target RanGAP to the nuclear envelope. These proteins therefore appear to have a similar role to the Nup358/SUMO1 system, but no recognizable similarity.

4.5 Intermediate filaments first evolved in the nucleus

Interestingly, it is the slightly unusual intra-nuclear, type V intermediate filament protein family, the lamins, which appears to be the founder of this large protein family. With its extra 42 amino acids in the rod domain (Herrmann *et al.*, 2009), its complex post-translation lipid modifications that binds it to membrane, or which may be removed by cleaving off part of the protein (Young *et al.*, 2006), its unusual two-dimensional network organization on the surface of the inner nuclear membrane (Aebi *et al.*, 1986; Goldberg *et al.*, 2008; see also Figure 4.2), and its nuclear localization sequence, it does look a bit like the 'black sheep' that has been outcast from the cytoplasm to the confines of the nucleus. However, the nucleus appears to be where it all started. The simplest multicellular animals such as *Hydra attenuata* do not have recognizable intermediate filament proteins, other than a single lamin (Erber *et al.*, 1999). Insects such as *Drospohila melanogaster* (Melcer *et al.*, 2007) appear to have both A- and B-type lamins but no cytoplasmic intermediate filaments, although a new candidate was recently identified in *Isotomurus maculates* (Mencarelli *et al.*, 2011). The nematode worm, *Caenorhabditis elegans* has only a single lamin gene but does, however, have 11 genes coding for cytoplasmic intermediate filaments (Melcer *et al.*, 2007) and mammals have three lamin genes (LMNB1, LMNB2 and LMNA) (Burke and Stewart, 2006) and 70 non-nuclear intermediate filament protein genes (Herrmann *et al.*, 2009).

So, major components of the nucleoskeleton are the nuclear lamina at the nuclear periphery and the intra-nuclear lamin network, both of which have been shown to bind to certain DNA sequences known as Matrix (or Scaffold) Attachment Regions or MARs (SARs) (Ludérus *et al.*, 1994). We shall therefore start our exploration of the plant nucleoskeleton by considering whether plants contain a plant lamina, or 'plamina' as we like to call it (Fiserova *et al.*, 2009).

4.6 Plants require a rigid nuclear boundary

After initial failure to find a plant lamin equivalent, it was argued that plants may not require a nuclear lamina. At the time it was thought that the nuclear lamina was there to provide structural integrity to the nucleus and to protect it from physical stress. Such a view was reinforced when it was discovered that mutations in the lamin A/C gene in humans caused muscle degenerating diseases (Bonne *et al.*, 1999). Nuclei were shown to be misshapen and damaged (Sullivan *et al.*, 1999). It can be proposed that this is due to physical forces imposed by muscle contraction leading to cell death or dysfunction. Plant cells, on the other hand, are thought of as rather static with the contents protected by the cell wall. It is, of course, true that plant cells have a strong cell wall, which is made rigid by hydrostatic pressure. They are not, however, static. Plant nuclei move during growth and development (Hepler *et al.*, 2001). This means that nuclei can be subjected to considerable physical forces from which they must be protected. Moreover, the fact that they move at all, especially in a controlled manner such as during root hair tip growth (Ketelaar *et al.*, 2002) or repositioning during fungal infection (Gross *et al.*, 1993) means that there must be attachment points between the cytoskeleton and/or motor proteins and the outer nuclear membrane or possibly the nuclear pore complexes (Graumann *et al.*, 2010b). If these attachments were not anchored to a rigid underlying structure (such as the lamina, via the LINC complex, as in animals – Mellad *et al.*, 2011) movement would probably not be possible due to the fluidity of the membrane. Microtubule organizing sites containing γ-tubulin and associated proteins have been shown to exist at the nuclear envelope (Seltzer *et al.*, 2007), although actin is required for nuclear positioning during root hair tip growth (Ketelaar *et al.*, 2002). There is therefore clearly a requirement for a rigid structural network equivalent to the nuclear lamina.

4.7 Is there a trans-nuclear envelope complex in plants that links the nucleoskeleton to the cytoskeleton?

If there is a plamina, how is it linked to the actin/microtubule cytoskeleton? In animals the answer has been emerging in the last few years. Briefly, SUN-domain proteins are lamin-interacting (Haque *et al.*, 2006) integral membrane proteins of the inner nuclear membrane. The SUN domain is found in the nuclear envelope lumen where it interacts with the KASH (**K**larsicht, **A**NC-1, and **S**yne **H**omology) domain of nesprins (Haque *et al.*, 2006). Nesprins are integral membrane spectrin repeat containing proteins (mostly) of the outer nuclear membrane, which in turn interact with, and anchor, microtubules (via kinesins, dynein and dynactin) and directly to F-actin and intermediate filaments (Mellad *et al.*, 2011). Plants do contain SUN-domain proteins (Graumann *et al.*, 2010a; Murphy *et al.*, 2010). It was also recently shown that WIP proteins (mentioned above) possess KASH domain-like sequences and bind SUN proteins in *Arabidopsis* (Zhou *et al.*, 2012). So we know that there must be a link from the plamina to the inner nuclear membrane, to the outer nuclear membrane, to the cytoskeleton. However, currently we only know of one potential component of this linkage (inner to outer nuclear membrane).

4.8 Role of the nuclear lamina as part of the nucleoskeleton

The nuclear matrix is a stable residual nuclear structure remaining after the extraction of chromatin and other components by salt, nucleases and detergents (Berezney and Coffey, 1974; Penman, 1995). In animals, the lamina, and the nuclear lamins, are a part of this structure (Nakayasu and Berezney, 1991) and form a stable peripheral network (Hozak *et al.*, 1995). Contrary to the normal perception about nuclear lamins, there is good evidence for structures throughout the nuclear interior that also contain lamins (Hozak *et al.*, 1995; Muralikrishna *et al.*, 2001). Not only, therefore, do lamins constitute the peripheral skeleton but they also appear to be part of the internal skeleton. Currently we do not know what determines whether a lamin protein is incorporated into the periphery or the interior. B-type lamins can be found at internal structures in some cells (Foster *et al.*, 2007). This is surprising because B-type lamins, unlike A-type lamins, are permanently isoprenylated (Kitten and Nigg, 1991), which is a post-translational lipid modification that anchors the protein to membranes (Sorek *et al.*, 2009). In the case of lamins, which possess a NLS (Monteiro *et al.*, 1994), this would be the inner nuclear membrane, which of course is (usually) at the periphery of the nucleus. This membrane association is shown most strikingly during open mitosis of animal cells. When the nuclear membranes dissociate, B-type lamins depolymerize, but remain membrane associated (Gerace and Blobel, 1980). On the other hand, the oocyte specific lamin B3, which is also permanently isoprenylated, is detached from the membrane during meiotic metaphase (Firmbach-Kraft and Stick, 1993). Lamin B3 is also found in internal nucleoskeletal structures prepared from nuclei assembled in *Xenopus* egg extracts (Zhang *et al.*, 1996). Together these studies show that isoprenylated lamins can exist away from the membrane and suggest that membrane association may be controlled. The presence of A-type lamins in the nuclear interior is better studied and easier to understand intuitively. During the post-translational processing of lamin A, the isoprenylated C-terminal CAAX box is removed (Worman *et al.*, 2009), so that although lamin A is found at the nuclear periphery, and probably initially targeted to the inner nuclear membrane by the lipid modification, it does not appear to be directly associated with the membrane when fully processed. Indeed, when B-type lamins, or in fact any other lipidated NLS (**N**uclear **L**ocalization **S**equence) containing protein (Ralle *et al.*, 2004; Linde and Stick, 2010), are overexpressed in *Xenopus* oocytes there is a resultant proliferation of intranuclear membranes with a single lamin B layer assembled on it (Goldberg *et al.*, 2008). In contrast, lamin A overexpression has no effect on the membranes, but forms thick multiple layers on top of the endogenous lamina (Goldberg *et al.*, 2008). A-type lamins also clearly exist in distinct structures throughout the nucleus or in soluble pools as well as at the periphery (Markiewicz *et al.*, 2002). It appears likely that the peripheral location is stabilized by lamin-interacting proteins such as the integral membrane protein, emerin, because disruption of emerin results in internalization and destabilization of lamin A (Markiewicz *et al.*, 2002). Changes in lamin A/C mobility are also associated with the expression levels of other interacting proteins such as Lap2α, which alter during *in vitro* differentiation of muscle cells (Markiewicz *et al.*, 2005).

4.9 Structural evidence for the nucleoskeleton

The first evidence for a plant lamina and nucleoskeleton was from structural studies. The nucleoskeleton is not detectable by routine thin-section electron microscopy in any organism. However, when nuclei assembled in *Xenopus* egg extracts are critical point dried, fractured and viewed by field-emission scanning electron microscopy (Figure 4.3), smooth 10 nm filaments are observed amongst the chromatin. These are morphologically consistent with the filaments observed in nuclear matrix and nucleoskeleton preparations (Jackson and Cook, 1988) after extraction of cells with detergent and nucleases. Interestingly, when nuclei isolated from tobacco (*Nicotiana tabacum*) BY-2 culture cells are fractured and viewed in the scanning electron microscope, almost identical structures are observed (Figure 4.3 inset). Together these observations suggest that both animal and plant cells have a stable intra-nuclear filamentous network.

The nuclear lamina in some animal cells, on the other hand, is easily visible as a dark band between the inner nuclear membrane and the peripheral chromatin (Stick and Schwarz, 1983). In other cell types, however, it is not easily discernible. Using high-pressure freezing and low-temperature fixation and embedding to prepare tissue for thin-section electron microscopy can enhance nuclear lamina visibility (Senda *et al.*, 2005), but even then it is not easily detectable in all cells (Figure 4.4a). In ultra-thin sections of *Xenopus* oocyte nuclear envelopes, which possess a very simple single layer lamina that only contains B-type lamins (mainly lamin LIII – Stick and Hausen, 1985) and no A-type lamins, the lamina is not directly resolvable. However, the inner

Figure 4.3 Putative nucleoskeletal filaments seen in fractured nuclei assembled *in vitro* in *Xenopus* egg extracts (main image) and in fractured tobacco BY-2 cells (inset).

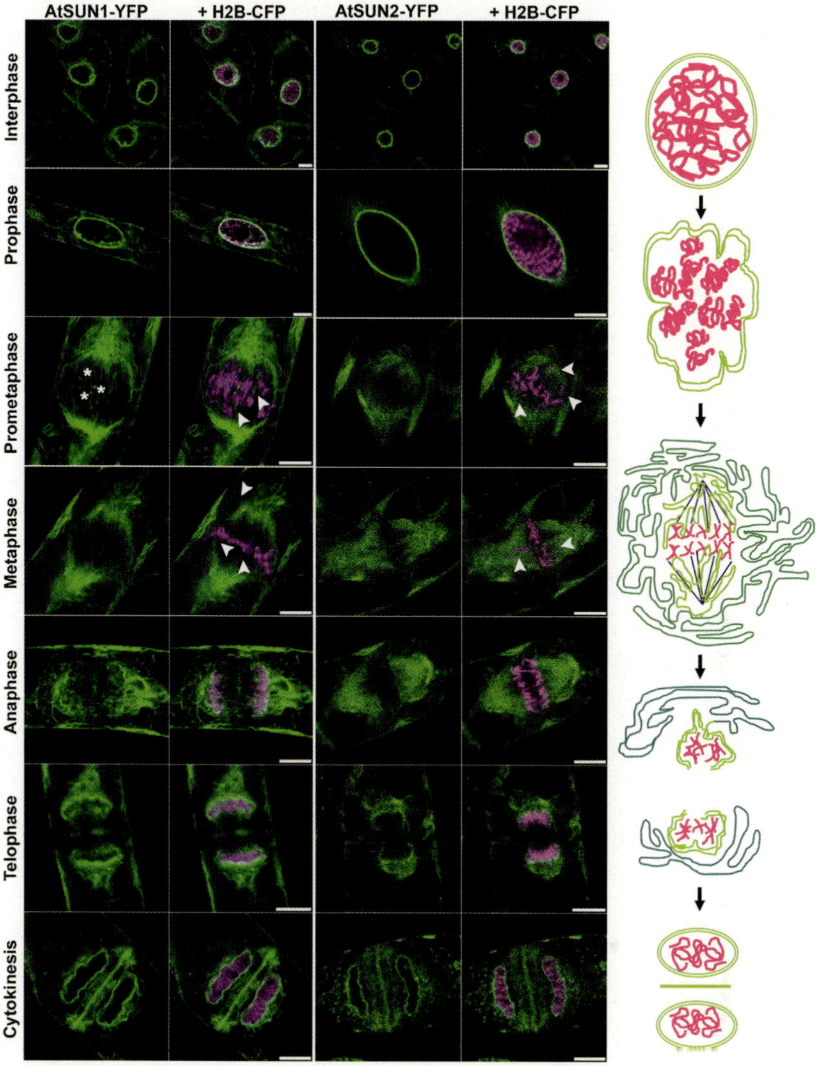

Plate 2.4 NE membrane dynamics during mitosis marked by SUN domain proteins. Stable transformed BY-2 cells co-expressing either AtSUN1-YFP (green in Plate 2.4) and chromatin marker histone H2B-CFP (magenta) or AtSUN2-YFP (green) and H2B-CFP were synchronized and living cells imaged by confocal microscopy. The two SUN proteins are present in the NE around chromatin in interphase and prophase. Upon NEBD, they distribute to mitotic ER and spindle membranes including tubules traversing the division zone. As the sister chromatids are separated, AtSUN1-YFP and AtSUN2-YFP accumulate in the reforming NE around chromatin first facing the spindle pole and finally proximal to the cell plate. In cytokinesis both fusion proteins are present in the expanding NE, phragmoplast and cell plate. AtSUN1-YFP is present in puncta in prometaphase (asterix) and both SUN-domain proteins are present in tubules in close proximity to chromatin (arrow heads). Scale bar = 10 μm (Graumann and Evans, 2011).

Annual Plant Reviews Volume 46: Plant Nuclear Structure, Genome Architecture and Gene Regulation, First Edition. Edited by David E. Evans, Katja Graumann and John A. Bryant.

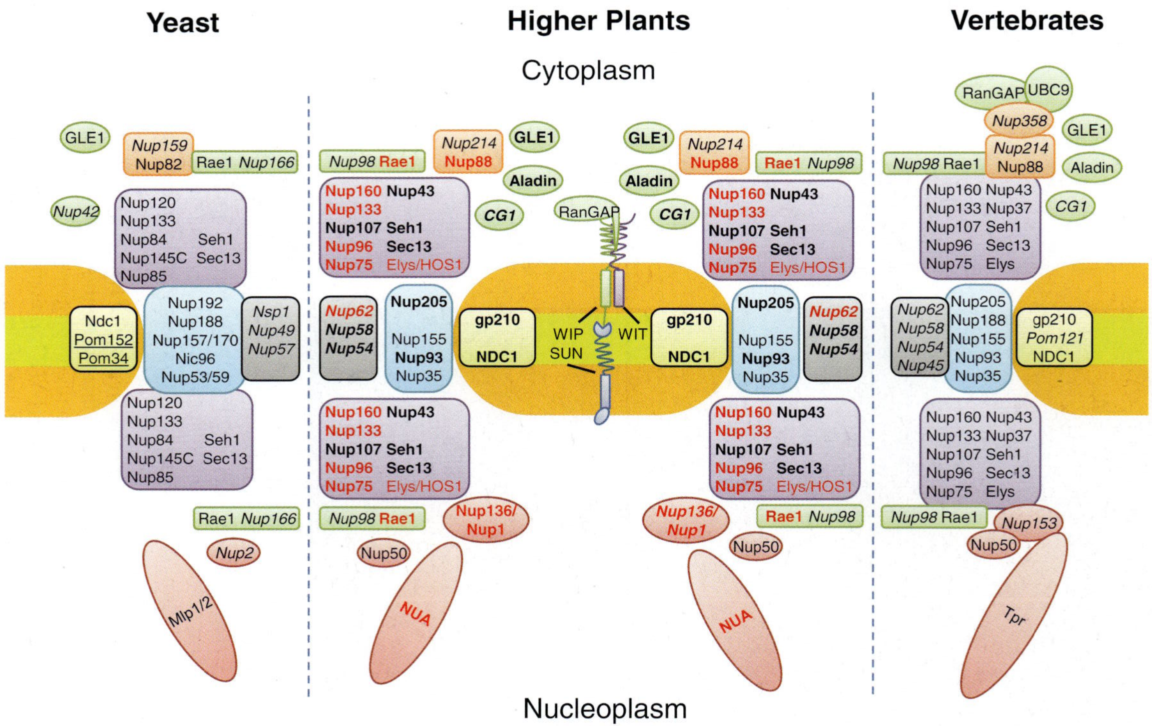

Plate 3.1 Comparison of the NPC components among higher plants, yeast and vertebrates. In Plate 3.1 protein complexes are grouped in single units, and the NPC subcomplexes are indicated in different colours – orange, cytoplasmic filaments; purple, Y-shape subcomplex; blue, Nup35-155/Nic96 subcomplex; grey, central FG subcomplex; and red, nuclear basket subcomplex. Other proteins or protein complexes are shown in green. FG Nups are indicated by italic font. Contacting units indicate confirmed interactions. In the higher plant NPC, bold protein names indicate confirmed NE localization, and mutant phenotypes have been reported for plant Nups indicated in red. In yeast, the underlined Pom152 and Pom34 have no homologues in vertebrates or higher plants. Positioning of plant Nups is based on their vertebrate counterparts.

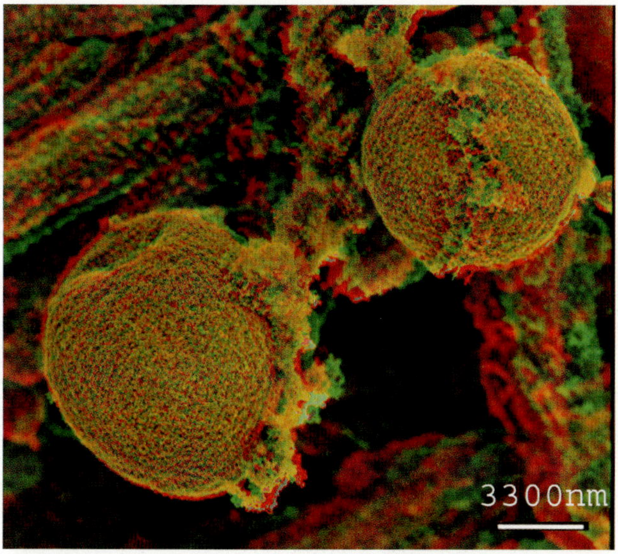

Plate 4.1 Stereo pair (use red/green glasses to view in Plate 4.1) of nucleoli from *Xenopus laevis* oocyte.

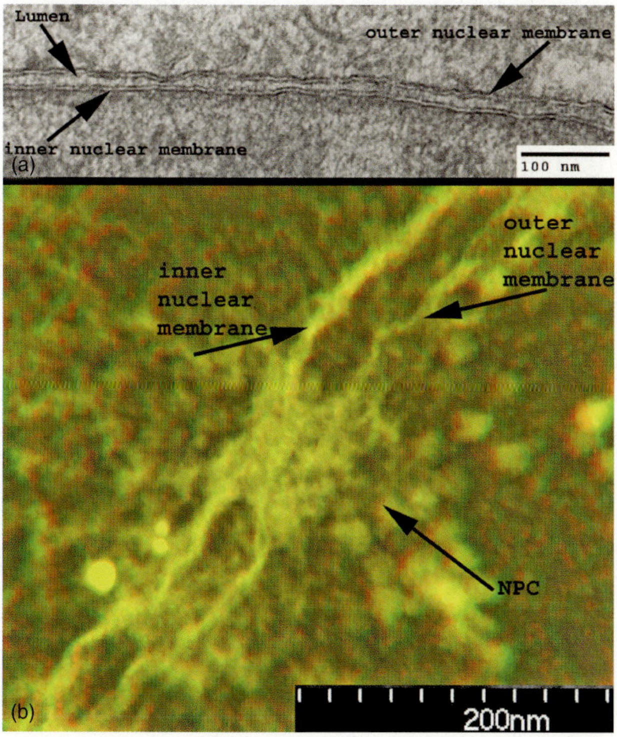

Plate 4.4 Thin sections through nuclear envelope of high pressure frozen/freeze substituted human dermal fibroblast cell (a), and *Xenopus* oocyte (b), imaged by transmission electron microscopy (a), and as a stereo pair (use red/green glasses to view) using field emission scanning electron microscopy (b).

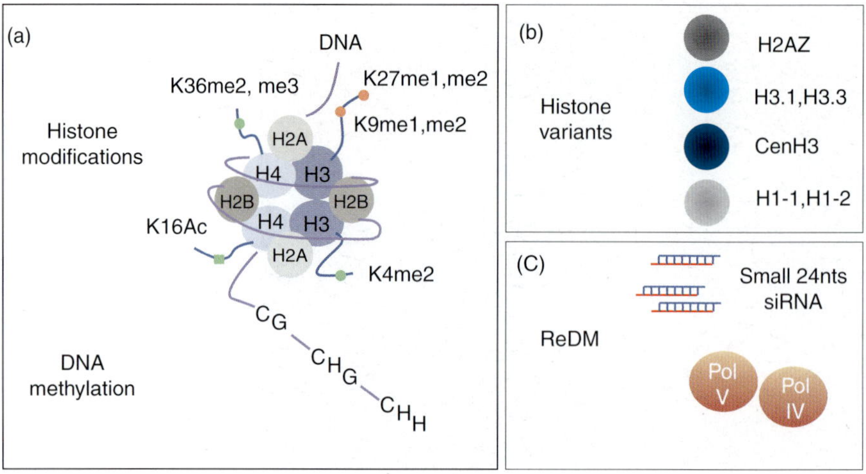

Plate 6.2 Heterochromatin features. a) The nucleosome is the basic structure of chromatin and includes histone octamer H2A, H2B, H3 and H4 as well as DNA. Lysine residues (K) lying within the histone tail at the N-terminal end of histone H3 and H4 can be modified either by methylation (me) or acetylation (Ac). Main repressive marks are indicated in red in Plate 6.2 (H3K9me1, me2 and H3K27me1, me2) whereas activating marks are indicated in green (H3K4me2, H4K36me2, me3 and H4K16Ac). DNA can also be subjected to DNA methylation at CG, CHG and CHH residues. Heterochromatin is enriched in methylated cytosine especially in CHH context, contains histone repressive marks and low levels of histone activating marks. b) Histone variants can substitute for canonical histones. H1-1, H1-2, H2AZ, H3.1, H3.3 and centromeric H3 (homologous to animal CenPA) are discussed in the frame of this review. H2AZ and H3.1 are enriched in actively transcribed regions while H3.3 and centromeric H3 are mostly found within heterochromatic regions. H1 variants are involved in chromosome segregation in plants. c) RNA Directed RNA Methylation (RdDM) occurs at heterochromatic loci and produces small non-coding RNA of 24nts called small interfering RNA (siRNA). RdDM involves plant specific polymerases called Polymerases IV and V (Pol IV and V) and is involved in the establishment of heterochromatin.

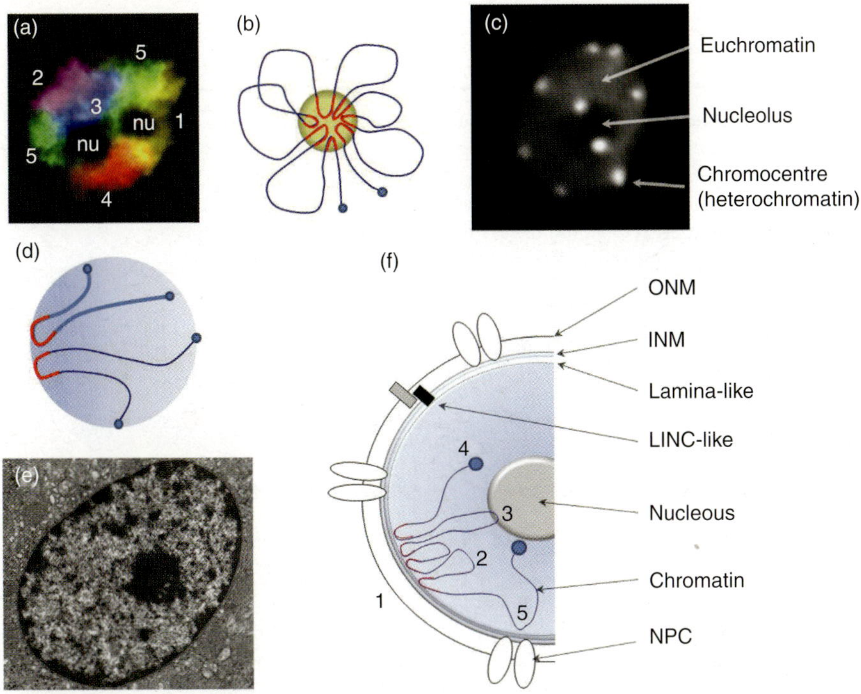

Plate 6.3 Chromatin organization in the nucleus. a) Chromosome Territories (CTs) were identified by FISH experiments using BAC probes against the five chromosomes from Arabidopsis with five differential labellings. Chromosome painting was then performed on a tetraploid nucleus including two nucleoli (nu) at interphase stage and revealed discrete CTs corresponding to the five chromosomes (1–5). With kind permission from Springer Science and Business media: Pecinka, A., Schubert, V., Meister, A., Kreth, A.G., Klatte, M., Lysak, M.A., Fuchs, J., Schubert, I. (2004) Chromosome territory arrangement and homologous pairing in nuclei of *Arabidopsis thaliana* are predominantly random except for NOR-bearing chromosomes. *Chromosoma* **113**, 258–269. b) The rosette-loop chromosome conformation as defined for Arabidopsis, from Fransz *et al.* (2002). Heterochromatic loci (in red in Plate 6.3) are grouped together at chromocentres and allow euchromatin to form loops. The chromocentre is indicated as a large yellow circle and corresponds to dense foci observed in 1C. Telomeres are indicated as blue circles. c) *Arabidopsis thaliana* nuclei at interphase from cotyledon stage stained with DAPI (kindly provided by Dr Sylvette Tourmente, http://www.gred-clermont.fr). Heterochromatin is more intensely stained due to its large amount of AT-rich sequences and appears as dense foci. Each spot represents a chromocentre although some chromocentres can merge and appear as a single spot. The nucleolus is seen as a non-stained region in the centre of the nucleus. Two chromocentres localize in the vicinity of the nucleolus whereas the six remaining chromocentres are located close to the nuclear periphery. d) The Rabl chromosome conformation is found in various genomes such as yeast (small genome) or wheat (large genome). Telomeres (blue circle at chromosome extremities in Plate 6.3) and centromeres (indicated in red) are located at opposite sites at the nuclear periphery. e) Electron microscopy image a of nucleus from a parietal stomach cell from *Rattus norvegicus*, showing the nuclear membrane, light euchromatin and dark heterochromatin, some of which is attached to the central nucleolus (Photo credit: image from Dr. Jastrow's electron microscopic atlas on the WWW http://www.drjastrow.de). f) Schematic of the chromatin organization in respect to

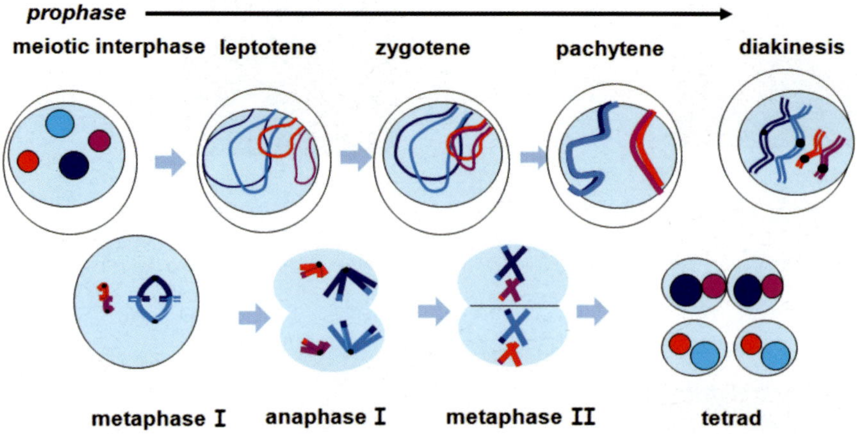

Plate 7.1 The behaviour of two pairs of chromosomes (red and blue in Plate 7.1) during meiosis.

Plate 6.3 (*Continued*) the Nuclear Envelope (NE). The NE is made of an inner nuclear membrane (INM), an outer nuclear membrane (ONM) and contains numerous NPC. A putative plant LINC-like complex is believed to connect the nucleoskeleton and the cytoskeleton. A lamina-like structure has been described in plants although its components remain to be described. Chromatin is represented as a blue line forming rosette-like structures. Heterochromatin (in red) is clustered at the NE (1) whereas euchromatic loops in blue (2) are at more internal locations within the nucleus. Various heterochromatic sequences, such as MARs and TFIIIC binding sites may participate in the recruitment of chromatin to the nuclear periphery. Loci encoding transfer RNA or ribosomal RNA (3) and telomeres (4) are brought into, or close to, the nucleolus (large grey circle). Data from other organisms suggest that other loci (5) may be recruited to the NPC where they may be actively transcribed. Panels D, E and F give insights about chromosome organization within the nucleus.

(a)

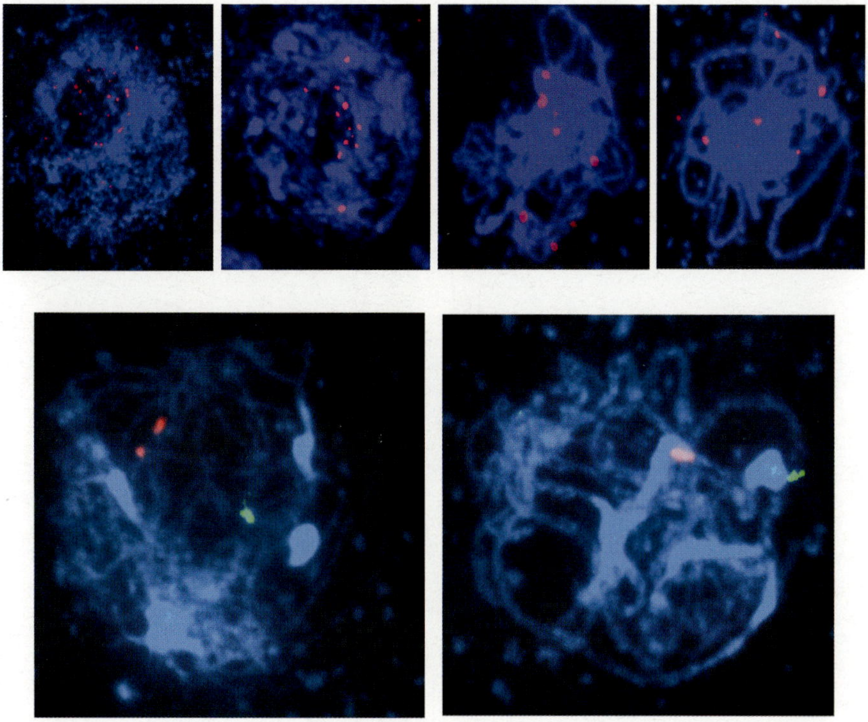

(b)

Plate 7.2 Panel a). FISH of WT Arabidopsis pollen mother cells probed with the telomere repeat probe (red), counterstained with DAPI, showing the dynamics and pairing of the telomeres. The first cell is in meiotic interphase, the telomeres are situated around the nucleolus and are unpaired (up to eighteen signals can be observed), in the second cell, at leptotene the telomeres start to pair in the nucleolar region (up to eleven signals observed). Following on from this paired telomeres (around eight signals) move to the nuclear periphery in the subsequent zyogotene and pachytene cells. Panel b). FISH with 2 BACs (subtelomeric probe detected with FITC, interstitial probe detected with Texas Red) of WT Arabidopsis pollen mother cells. In the first panel the cell is at leptotene the subterminal probe is paired (one signal) whereas the interstitial probe is unpaired at this stage, in the second cell at pachytene both probes are paired.

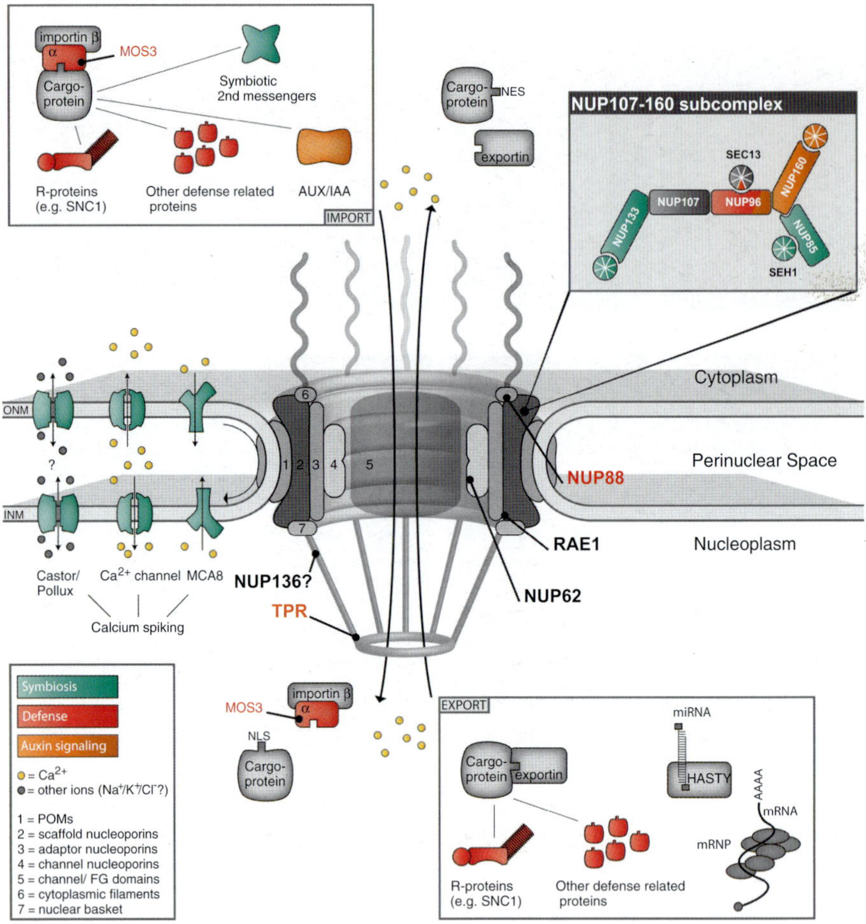

Plate 8.1 Model of the plant nuclear pore complex and its putative functions in plant signalling pathways. Plant nucleoporins are represented according to the position of the known homologues in the vertebrate NPC. The model of the NUP107-160 sub-complex in the top left panel is based on the structural characterization of the yeast NUP84 complex (Hoelz, Debler and Blobel, 2011). Mutations in plant nucleoporins affect various plant processes, including symbiotic signalling, plant defence and auxin signalling. The mutant phenotypes could be caused by deficient nucleo-cytoplasmic transport of various molecules. Potential targets in plant-defence and auxin signalling include R-proteins such as SNC1, other defense related proteins, AUX/IAA repressors, as well as certain mRNAs or miRNAs. In case of plant-microbe symbiosis, nuclear calcium spiking is abolished in the *ljnup85*, *ljnup133* and *nena/ljseh1* mutants. This might be due to structural changes in the NPC that either directly affect calcium homeostasis or indirectly impair proper transport of symbiotic components upstream of calcium spiking, such as calcium channels, the ion channels CASTOR and POLLUX, the calcium ATPase MAC8 or symbiotic second messengers. mRNP = messenger ribonucleoprotein; NLS = Nuclear localization signal; NES = Nuclear export signal; INM = Inner nuclear membrane; ONM = Outer nuclear membrane; POM = pore membrane protein.

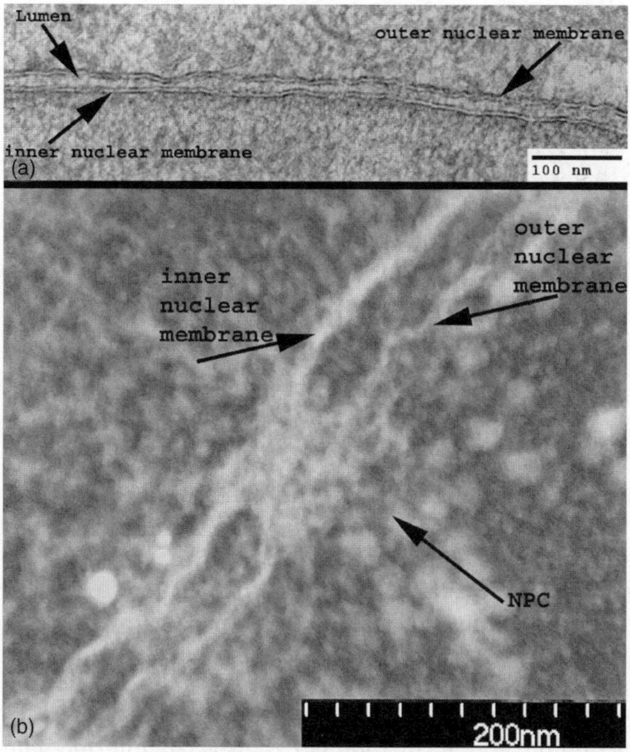

Figure 4.4 Thin sections through nuclear envelope of high pressure frozen/freeze substituted human dermal fibroblast cell (a), and *Xenopus* oocyte (b), imaged by transmission electron microscopy (a), and as a stereo pair (use red/green glasses to view) using field emission scanning electron microscopy (b). (*For colour details please see colour plate section.*)

membrane does appear thicker compared to the outer membrane (~10 nm versus ~6 nm, respectively – Figure 4.4b) It is likely that this extra thickness is due to the lamina. Indeed, if the same nuclear envelope is observed from the surface by feSEM (Goldberg *et al.*, 2008, Figure 4.2), then a regular array of parallel lamin filaments is clearly observed.

Examination of sections through plant nuclei does not usually reveal any electron-dense layer between the inner membrane and the chromatin. This does not mean that the lamina does not exist. As discussed above, it is not easily visible in many cell types that we know to have a nuclear lamina. Therefore the lamina could be very thin, and/or not stained sufficiently to be visualized. Moreover, there is probably insufficient data from the plant-thin section TEM literature to draw firm conclusions concerning the presence of a nuclear lamina, because high-resolution thin-section EM of plant cells presents many challenges due to the cell wall, vacuoles and air spaces, which can all adversely affect fixation methods. Careful examination of micrographs in the literature reveals some variation in material attached to the inner nuclear membrane in plants. In some micrographs a slight thickening of the inner membrane compared to the outer membrane is evident (De Zoeten and Gaard, 1969; Franke *et al.*, 1972). Although it is not striking, this thickening may be comparable to that seen in the single-layer lamina in *Xenopus* oocytes (Figure 4.4b). In tangential sections of the green alga *Acetabularia mediterranea* nuclear envelope, the nuclear pore complexes appear inter connected by very fine filaments (Franke *et al.*, 1975). In one study (Minguez *et al.*, 1994), thin sections of the marine alga, *Gymnodinium splendens*, do show a dark layer attached to the inner membrane (morphologically similar to the animal nuclear lamina) which is tantalizingly interrupted at the nuclear pores in exactly the same ways as in animal cells. More recently, it was shown that fracturing of tobacco BY-2 cells revealed the nuclear surface of the inner nuclear membrane (Fiserova *et al.*, 2010). Upon this surface there is clearly a filamentous structure that lies tightly on the membrane and inter-connects the NPCs. Strikingly, it was found to have an orthogonal-like organization that was almost identical to the B-type lamina found in *Xenopus* oocytes (Figure 4.2). From these studies it can be concluded that plants do contain a filamentous network, tightly apposed to the inner nuclear membrane and interconnecting the NPCs: a structure that is synonymous to the B-type lamina in animal cells.

Nucleoskeleton preparations of animal cells, followed by embedding in a removable resin, sectioning and resin removal reveals an intermediate filament-like network throughout the nucleus with a prominent peripheral structure of the lamina with embedded nuclear pore complexes (Jackson and Cook, 1988). The study of such residual structures in plants has not been extensive, but several studies have revealed similar structures. Nuclear matrix preparations of the dinoflagellate, *Amphidinium carterae*, showed a rather dense residual structure throughout the nucleus with a predominant peripheral 'lamina' (Minguez *et al.*, 1994). Immunofluorescence and immuno-blotting showed that these structures contained proteins that cross-reacted with certain antibodies that recognized (i) intermediate filaments in general, (ii) chicken and *Xenopus* lamins, and (iii) topoisomerase II, which is a possible component of the nucleoskeleton (Berrios *et al.*, 1985). Labelling was restricted to the nuclear periphery, or even to discrete structures in the nuclear interior, so it is uncertain if the 'lamina' contains proteins that are recognized by the anti-lamin antibodies. On the other hand, anti-lamin staining of isolated onion nuclei gave a peripheral staining, which extended to discrete sites in the nuclear interior when the chromatin was removed. Such staining could be due either to unmasking or reorganization of the epitopes after extraction. Isolation of nuclear matrix/lamina fractions from pea nuclei also lead to the enrichment of proteins that cross-reacted with other anti-lamin antibodies and had comparable molecular weights to lamins (60–70 kD) (McNulty and Saunders, 1992). Immuno-electron microscopy showed nuclear matrix labelling with the anti-lamin antibody. In another study, nuclear matrix structures, closely resembling animal nuclear matrices, were obtained from carrot (*Daucus carota* L.) cells (Wang *et al.*, 1996). In this case, anti-lamin labelling was more restricted to the periphery. Beven *et al.* (1991) raised monoclonal antibodies against nuclear matrix preparations of carrot cells and identified higher molecular weight proteins (~92 kD) that localized to the nuclear periphery. In contrast, the pan-intermediate filament 'IFA' antibody recognizes two lamin-sized proteins (~60 kD and ~65 kD) in carrot cells, and stains the nucleus, as well as the nuclear matrix (Frederick *et al.*, 1992). This suggests that carrot cells contain proteins that are immunologically related to intermediate filaments in the nucleus.

It is of potential interest that a protein recognized by an anti-lamin antibody is cleaved during induced apoptosis in tobacco protoplasts (Chen *et al.*, 2000). Cleavage of lamins during mammalian apoptosis is an important step, but, of course, many proteins are cleaved during apoptosis.

4.10 NuMA in plants

A consistent component of the vertebrate nuclear matrix is the large coiled-coil protein, NuMA (**Nu**clear **M**itotic **A**pparatus) (Radulescu and Cleveland, 2010). NuMA has defined roles in mitotic phosphorylation-dependent tethering of spindle microtubules to the spindle poles. It is also a component of the nuclear matrix (Kallajoki *et al.*, 1991). NuMA is cleaved during apoptosis (Gueth-Hallonet *et al.*, 1997) and mutation of the cleavage site results in resistance of the cell to apoptotic nuclear disintegration (Lin *et al.*, 2007). Conversely, RNAi silencing of NuMA accelerates caspase-induced apoptosis (Kivinen *et al.*, 2010). Together these results suggest that NuMA does have a role in maintaining the structural integrity of the nucleus and therefore is a structural component of the nucleoskeleton. NuMA can also self assemble into highly organized filament structures in the nucleus when overexpressed (Gueth-Hallonet *et al.*, 1998). It also appears to have a role in maintaining chromatin organization in differentiated mammary epithelial cells (Abad *et al.*, 2007) and interacts with transcription factors (Harborth *et al.*, 2000). Plants may contain proteins related to NuMA, but like many of the other nucleoskeleton-lamina fraction proteins, there is currently only immunological evidence. Yu and Moreno Diaz de la Espina (1999) showed that nuclear matrix preparations from onion cells contained large proteins of a similar size to human NuMA (~230kD) that were recognized by certain anti-NuMA antibodies. Immuno-EM showed that these were present in the residual filamentous nuclear matrix. Like human NuMA, the plant epitope appears to associate with the spindle, although the distribution is different. Therefore it seems that the proteins recognized by this antibody bare some similarities to NuMA, but to date no actual plant NuMA has been characterized and there are no obvious homologues in the Arabidopsis database.

4.11 Matrix attachment regions (MARs) and the role of the nucleoskeleton in chromatin organization

Although the nucleoskeleton may have important roles in maintaining the structural integrity of the nucleus, one of its better defined roles is in organizing chromatin. Preparation of nuclear matrices involves extensive endonuclease digestion of the DNA followed by elution of the resulting chromatin fragments. This leaves behind short DNA sequences that are attached to the nuclear matrix. These ~200bp DNA sequences are not particularly homologous but they are AT-rich and contain topoisomerase II consensus sequences. They bind to the nucleoskeleton, hence they are known as **M**atrix **A**ttachment **R**egions (MARs) or **S**caffold **A**ttachment **R**egions (SARs). When nuclei are incubated in low ionic strength buffers, the chromatin is dispersed out from the nucleus in loops (Figure 4.5). It is thought that the anchor for those loops is the attachment of MARs to the nuclear matrix (Smith *et al.*, 1984). Therefore one level of chromatin organization is into nucleoskeleton anchored loops, mediated by the MARs and nucleoskeletal proteins. Functionally related genes are sometimes organized into clusters (Loc and Strätling, 1988), which could facilitate co-ordinated chromatin decondensation and gene expression (Tikhonov *et al.*, 2000). MARs have been identified in plants and appear to

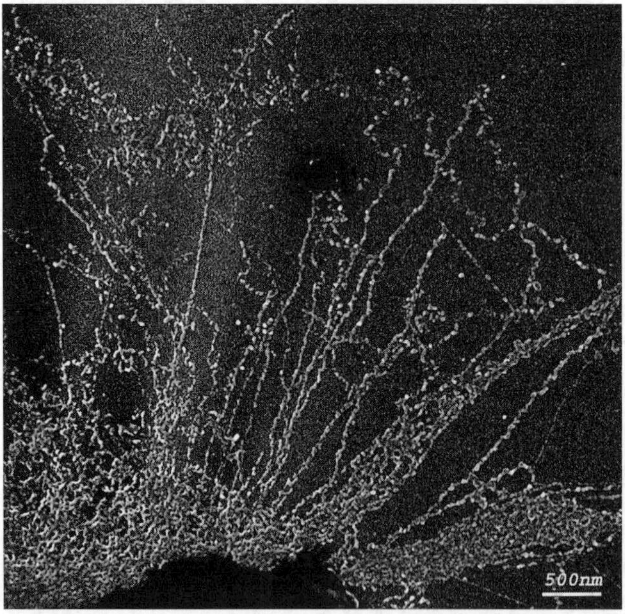

Figure 4.5 Chromatin 'loops' imaged by field emission scanning electron microscopy of isolated HeLa cell nucleus subjected to low ionic strength buffer and burst open by centrifugation.

have a role in defining the structural organization of chromatin (Hall *et al.*, 1991; Breyne *et al.*, 1992) and can be used to enhance or control the expression of transgenes (Van der Geest *et al.*, 2004; Wang *et al.*, 2007). Inclusion of tobacco MARs in Agrobacterium-mediated transformation of *Theobroma cacao* L. with green fluorescent protein (GFP) increased the level of GFP expression and also reduced the occurrence of gene silencing (Maximova *et al.*, 2003). Other studies (Li *et al.*, 2008), however, have found less dramatic or consistent effects of introducing MARs into transgenes.

Topoisomerase II is thought to be a component of the nucleoskeleton, located at positions where it interacts with DNA at the MARs (Razin *et al.*, 1991). MARs are cleaved in the presence of the topoisomerase II inhibitor VM-26 (Razin *et al.*, 1991). When Arabidopsis and maize nuclei are treated with VM-26, the genome is fragmented into specific sizes ranging from about 20 kbp to 100 kbp, suggesting that the plant genome is organized into MAR flanked loops of this size. However, Makarevitch and Somer (2006) found that topoisomerase II cleavage sites in Arabidopsis did not correlate with MARs.

4.12 Chromocentres and the plant nucleoskeleton

Further evidence for the organization of chromatin into loops in plants has come from studies into the spatial relationship between the condensed, methylated, inactive chromatin regions, called chromocentres, in comparison to active euchromatin on the same chromosome (Fransz *et al.*, 2002). Chromocentres occupy defined compact regions in the interphase nucleus and in Arabidopsis are located at the nuclear periphery or near the nucleoli. Using fluorescence *in situ* hybridization (FISH), it was shown that active chromosome regions loop away from the chromocentre. It is not known, however, if there is a relationship between the chromocentres and the nucleoskeleton.

4.13 Long coiled-coil proteins in plants and their role in nuclear organization: candidates for plamins and nucleoskeletal proteins?

Interestingly, there is a hint that there may be a chromocentre/nucleoskeleton relationship and this brings us back to possible candidates for proteins that may be components of the plamina. Masuda *et al.* (1997) used a monoclonal antibody raised against *Daucus carota L.* (carrot) nuclear matrix preparations to identify a protein called NMCP1 (nuclear matrix constituent protein 1). Immuno-fluorescence and immuno electron microscopy showed that the epitope was at the nuclear periphery. The antibody was used to screen a carrot expression library and the sequence of the resulting identified protein was deduced. From the sequence, NMCP1 appears to be a large coiled coil protein (with the diagnostic heptad repeats), it has nuclear localization signals and potential recognition sites for cdc2 kinase (lamins are phosphorylated by activated cdc2 kinase at the onset of mitosis and nuclear envelope breakdown – Peter *et al.*, 1990). It therefore has characteristics that are consistent with a putative plamina protein.

Recently, Dittmer *et al.* (2007) identified two NMCP1-related proteins in *Arabidopsis thaliana*, LITTLE NUCLEI1 (LINC1) and LINC2, which also had large coiled-coil domains with smaller N- and C-terminal regions. Each of these were fused to YFP (yellow fluorescent protein) and expressed in Arabidopsis plants. LINC1-YFP located to the nuclear periphery, whereas LINC2-YFP located to the nuclear interior, suggesting that LINC1 may be a plamina protein and LINC2 may be part of the nucleoskeleton. Single and double mutants were generated. Although single mutations did not have dramatic phenotypes, the linc1-1 linc2-1 double mutant produced small plants, with small cells and small nuclei. Interestingly the number of chromocentres (suggested to organize chromatin loops) is reduced significantly in the linc2-1 single mutant and to a much greater extent in the double mutant. Chromocentre number is related to the number of centromeres and hence to chromosome number. However the number can be reduced by aggregation of two or more chromocentres. Therefore disruption of the underlying structural framework of the nucleus, by mutation of one or more of its putative constituent proteins (LINC1/LINC2) could lead to chromocentre reduction.

The effects of LINC mutations on nuclear organization and plant development also offer some interesting parallels to rare human diseases where the nuclear lamins, or proteins that interact with them, are mutated. Such mutations lead to severe degenerative disorders and premature ageing (Worman, 2010), the so-called laminopathies or nuclear envelopathies. These diseases are not understood, despite several years of intense research and are rather diverse and complex. The phenotypes may be due to one or a combination of the following: nuclear fragility; changes to nuclear shape; changes to gene expression; and alterations of cytoskeletal and nucleoskeletal organization, which could have effects on signalling, cell polarity and migration. When expressed under the LINC1 native promoter, LINC1-GFP, like A-type lamins, is differentially expressed in different tissues (Dittmer and Richards, 2008). LINC1-GFP is found specifically in proliferating meristematic tissues such as the root tip, but not at all in differentiated root tissue. In contrast only B-type lamins are expressed in all cell types, whereas A-type lamins are not expressed in undifferentiated cells, such as germ cells (Lehner *et al.*, 1987). Interestingly, in human colonic crypts, proliferating cells do not express lamin A, but terminally differentiated cells as well as stem cells do express lamin A (Willis *et al.*, 2008). The expression status of LINC1 or LINC2 in stem cells has not been published but it does seem that whereas lamin A expression *in vivo* is associated with exit from the cell cycle; LINC1 is conversely required in dividing cells.

However, we do not fully understand the specific roles of lamin A in vertebrates, and we know almost nothing yet about the structural organization or interactions of LINC1 and LINC2. It could be for instance that LINC1 is not functionally equivalent to lamin A, but instead is more like lamin B. It is possible that the filaments that interconnect NPCs, contain LINC1, although this has not yet been tested. Interestingly, NPC-connecting filaments appear morphologically almost identical to the B-type lamina of Xenopus oocytes (Fiserova *et al.*, 2010 – compare with Goldberg *et al.*, 2008). It could therefore be speculated that proliferating cells require the nuclear organization achieved using either lamin B in vertebrates, or LINC1 in plants. This specific 'proliferation organization' is then altered to a differentiated state by expressing lamin A in vertebrates, or conversely in plants, by down-regulating LINC1 (and maybe expressing other proteins that are functionally equivalent to lamin A).

It is possibly relevant that B-type lamins and A-type lamins do not form heterodimers (Herrmann *et al.*, 2009) and therefore appear to assemble into separate structures. The B-type lamins are thought to be tightly apposed to the inner nuclear membrane (Goldberg *et al.*, 2008) due to the isoprenylation lipid modification at the C-terminus, which inserts into the lipid bilayer (Davies *et al.*, 2009). In fact overexpression of B-type lamins causes a proliferation of intra-nuclear membranes, suggesting that this modification must be embedded in a membrane (Linde and Stick, 2010). Overexpression of A-type lamins, which undergo a complex post-translational processing involving isoprenylation, followed by removal of the lipidated C-terminus (Davies *et al.*, 2009), on the other hand, does not affect the membranes but does result in a thickening of the lamina as layers of lamin filaments are overlaid onto the single B-type lamin layer (Goldberg *et al.*, 2008). Isoprenylation is a common and important protein modification in both plants and animals (Crowell and Huizinga, 2009), which requires the so-called CaaX box (C = Cysteine, a = aliphatic amino acid, X = any amino acid) at the C-terminus. None of the LINC proteins, or indeed any other nuclear envelope or nuclear matrix protein in plants, has a CaaX box and there is no other evidence for isoprenylation of any of these proteins. Yet, there does appear to be a protein filament system attached to the membrane. This then raises the question of how these proteins are attached to the membrane, or whether there are yet to be identified proteins that can be lipid modified.

The most obvious membrane association mechanism for a protein is to have a transmembrane domain. MAR binding filament like protein 1 (MFP1) was identified as a protein that was similar to large filamentous nuclear proteins in animals, and which could bind directly to DNA. MFP1 also has a predicted transmembrane domain, and therefore may be membrane associated (Meier *et al.*, 1996). Its DNA binding activity was also preferential for MARs. It was therefore looking a bit like a B-type lamin (a large coiled coil membrane associated protein that binds to MARs). Expression of MFP1-GFP constructs indicated that it was mainly located in the nuclear envelope. This was not an even distribution but rather in a speckled pattern (Gindullis and Meier, 1999). Although lamin localization is not speckled, it is discontinuous (Paddy *et al.*, 1990). MFP1 imunofluorescence of tobacco culture cells showed a similar nuclear peripheral speckled pattern. Nuclear matrix preparations, however, were stained on the interior as well (Gindullis and Meier, 1999). The presence of a transmembrane domain in MFP1 suggests that it may be redistributed from the nuclear envelope to the internal nuclear matrix during nuclear matrix preparation. However, in another study (Samaniego *et al.*, 2006), immunofluorescence of intact isolated nuclei using different antibodies in a variety of cells again showed an internal nuclear matrix pattern. Assuming that there are no internal membranes in the nucleus of these cells, and that the antibodies only recognize a single antigen within the MFP1 protein, of which there are no isoforms, then it can be suggested that MFP1 is either not membrane associated at all, or that the insertion of the hydrophobic N-terminal domain into the membrane is variable, and possibly could be controlled, not unlike the membrane association of lamins. It is uncertain from the sequence information how MFP1 is associated with the membrane (Gindullis and Meier, 1999). In plants, certain proteins can be post-translationally inserted into membranes (Schaad *et al.*, 1997), which would allow MFP1 to be either associated with the membrane or with the internal nuclear matrix. The function of MFP1 at the nuclear envelope is not known. MFP1 has also been shown to interact with MAF1 (MFP1 associated factor 1) (Gindullis *et al.*, 1999), which is a WPP domain containing protein. In plants, RanGAP, the GPTase activating protein of the small GTPase nuclear transport factor, Ran, also contains a WPP domain, which is responsible for targeting it to the nuclear envelope by binding to the transmembrane WIP and WIT proteins (Xu *et al.*, 2007; Zhao *et al.*, 2008). One function for MFP1 may therefore be to locate other proteins to specific locations on the nuclear envelope and nuclear matrix. Again there is a parallel here with lamins, which appear to have a role in locating certain transcription factors to specific locations (Markiewicz *et al.*, 2002; Andrés and González, 2009). It should be noted however that the location of MFP1 is controversial. In some studies MFP1 is not only located to the nuclear envelope and matrix, but there appear to be major components of this protein associated with chloroplast (Jeong *et al.*, 2003) and Golgi (Samaniego *et al.*, 2006), and there is evidence that the nuclear envelope location is involved in linking chloroplasts to the nucleus, rather than being involved in nuclear organization. What has not been shown unequivocally yet is whether MFP1 is located to the outer membrane, the inner membrane or both, which could be resolved by immuno-gold electron microscopy (Fiserova and Goldberg, 2010) or super-resolution light microscopy (Galbraith and Galbraith, 2011).

4.14 Actin and microtubules in the nucleus

One aspect of the nucleoskeleton that will not be covered in detail here is the role of the major cytoskeletal proteins tubulin and actin. To our knowledge, there is no evidence that tubulin has any role in the interphase nucleus, or at least not from the inside. In plants, however, the outer nuclear membrane appears to be the anchor point (or rather anchor surface) from which the microtubules are organized (Seltzer *et al.*, 2007). Therefore cytoplasmic microtubules may have important influences on nuclear architecture by their association with the nuclear surface.

The presence of actin in the nucleus, on the other hand, has been long debated but has recently become recognized as an important functional nuclear component. It has roles in gene expression and chromatin remodelling, and may even be important in large-scale chromatin organization (see Visa and Percipalle, 2010 for an extensive review). There are also a number of actin binding and regulating proteins in the nucleus, and therefore it seems likely that filamentous actin is an important component of the nucleoskeleton, although such filaments have not been visualized. The importance of nuclear actin in plants is still uncertain, but actin-related proteins appear to have roles in chromatin remodelling (Meagher *et al.*, 2007) and have, together with nuclear myosin I, been localized to transcription foci and nuclear matrix preparations (Cruz *et al.*, 2009).

4.15 Conclusions

So, do plants have a nucleoskeleton, and if so, what is it for? There is a 'smoking gun' and a few clues to both these questions respectively. However even in the animal field, where there is far more information, it is not universally agreed that the nucleoskeleton even exists, or if it does, what structural or functional form it could take. Electron microscopical evidence does point to the existence of a nucleoskeleton both in plants and animals. Antibodies against intermediate filaments, including lamins, and other nucleoskeletal proteins, stain plant nuclei and nuclear matrix preparations. Conversely there are no clearly identifiable equivalents for intermediate filament proteins, lamins or any other nuclear matrix proteins in the plant genome databases. However, there do exist proteins in plants that bear at least some of the features expected from these proteins, such as coiled-coil domains, large size, DNA binding properties and localization to the nuclear periphery and/or intra-nuclear networks. Currently, our understanding of the plant nucleoskeleton-lamin is really limited to a few candidate proteins of unknown function. However it is becoming increasingly clear that the structural organization of the nucleus is of fundamental importance to both gene expression and epigenetic control. Therefore the further characterization and identification of the proteins involved is key to understanding many fundamental and practical questions.

Acknowledgements

This work was supported by grants from the Biotechnology and Biological Sciences Research Council, UK, grant number BB/E015735/1 and BB/G011818/1. Thanks to Jindriska Fiserova for Figure 4.3.

References

Abad, P.C., Lewis, J., Mian, I.S. *et al.* (2007) NuMA influences higher order chromatin organization in human mammary epithelium. *Molecular Biology of the Cell* **18**, 348–361.

Aebi, U., Cohn, J., Buhle, L. and Gerace, L. (1986) The nuclear lamina is a meshwork of intermediate-type filaments. *Nature* **323**, 560–564.

Andrés, V. and González, J.M. (2009) Role of A-type lamins in signaling, transcription, and chromatin organization. *Journal of Cell Biology* **187**, 945–957.

Berezney, R. and Coffey, D.S. (1974) Identification of a nuclear protein matrix. *Biochemical and Biophysical Research Communications* **60**, 1410–1417.

Berrios, M., Osheroff, N., Fisher, P.A. (1985) In situ localization of DNA topoisomerase II, a major polypeptide component of the Drosophila nuclear matrix fraction. *Proceedings of the National Academy of Sciences of the USA* **82**, 4142–4146.

Beven, A., Guan, Y., Peart, J. *et al.* (1991) Monoclonal antibodies to plant nuclear matrix reveal intermediate filament related components within the nucleus. *Journal of Cell Science* **98**, 293–302

Blobel, G. (1985) Gene gating: a hypothesis. *Proceedings of the National Academy of Sciences of the USA* **82**, 8527–8529.

Blumenthal, S.S., Clark, G.B. and Roux, S.J. (2004) Biochemical and immunological characterization of pea nuclear intermediate filament proteins. *Planta* **218**, 965–975.

Bonne, G., Di Barletta, M.R., Varnous, S. *et al.* (1999) Mutations in the gene encoding lamin A/C cause autosomal dominant Emery-Dreifuss muscular dystrophy. *Nature Genetics* **21**, 285–288.

Braun, M. (2001) Association of spectrin-like proteins with the actin-organized aggregate of endoplasmic reticulum in the Spitzenkörper of gravitropically tip-growing plant cells. *Plant Physiology* **125**, 1611–1619.

Breyne, P., van Montagu, M., Depicker, N. and Gheysen, G. (1992) Characterization of a plant scaffold attachment region in a DNA fragment that normalizes transgene expression in tobacco. *Plant Cell* **4**, 463–471.

Brown, J.M., Green, J., das Neves, R.P. *et al.* (2008) Association between active genes occurs at nuclear speckles and is modulated by chromatin environment. *Journal of Cell Biology* **182**, 1083–1097.

Burke, B. and Stewart, C.L. (2006) The laminopathies: the functional architecture of the nucleus and its contribution to disease. *Annual Review of Genomics and Human Genetics* **7**, 369–405.

Cai, G. and Cresti, M. (2010) Microtubule motors and pollen tube growth – still an open question. *Protoplasma* **247**, 131–143.

Chamcheu, J.C., Siddiqui, I.A., Syed, D.N. *et al.* (2011) Keratin gene mutations in disorders of human skin and its appendages. *Archives of Biochemistry and Biophysics* **508**, 123–137.

Chang, L., Barlan, K., Chou, Y.H. *et al.* (2009) The dynamic properties of intermediate filaments during organelle transport. *Journal of Cell Science* **122**, 2914–2923.

Chen, H.M., Zhou, J. and Dai, Y.R. (2000) Cleavage of lamin-like proteins in in vivo and in vitro apoptosis of tobacco protoplasts induced by heat shock. *FEBS Letters* **480**, 165–168.

Cook, P.R. (2010) A model for all genomes: the role of transcription factories. *Journal of Molecular Biology* **395**, 1–10.

Cremer, T. and Cremer, M. (2010) Chromosome territories. *Cold Spring Harbor Perspectives in Biology* **2**, a003889.

Crick, F.H. (1952) Is alpha-keratin a coiled coil? *Nature* **170**, 882–883.

Crowell, D.N. and Huizinga, D.H. (2009) Protein isoprenylation: the fat of the matter. *Trends in Plant Science* **14**, 163–170.

Cruz, J.R. and Moreno Díaz de la Espina, S. (2009) Subnuclear compartmentalization and function of actin and nuclear myosin I in plants. *Chromosoma* **118**, 193–207.

Davies, B.S., Fong, L.G., Yang, S.H. *et al.* (2009) The posttranslational processing of prelamin A and disease. *Annual Review of Genomics and Human Genetics* **10**, 153–174.

Dechat, T., Pfleghaar, K., Sengupta, K. *et al.* (2008) Nuclear lamins: major factors in the structural organization and function of the nucleus and chromatin. *Genes and Development* **22**, 832–853.

De Zoeten, G.A. and Gaard, G. (1969) Possibilities for inter- and intracellular translocation of some icosahedral plant viruses. *Journal of Cell Biology* **40**, 814–823.

Dieppois, G. and Stutz, F. (2010) Connecting the transcription site to the nuclear pore: a multi-tether process that regulates gene expression. *Journal of Cell Science* **123**, 1989–1999.

Dittmer, T.A. and Richards, E.J. (2008) Role of LINC proteins in plant nuclear morphology. *Plant Signaling and Behavior* **3**, 485–487.

Dittmer, T.A., Stacey, N.J., Sugimoto-Shirasu, K. and Richards, E.J. (2007) LITTLE NUCLEI genes affecting nuclear morphology in Arabidopsis thaliana. *The Plant Cell* **19**, 2793–2803.

Emerson, L.J., Holt, M.R., Wheeler, M.A. *et al.* (2009) Defects in cell spreading and ERK1/2 activation in fibroblasts with lamin A/C mutations. *Biochimica et Biophysica Acta* **1792**, 810–821.

Erber, A., Riemer, D., Hofemeister, H. *et al.* (1999) Characterization of the Hydra lamin and its gene: A molecular phylogeny of metazoan lamins. *Journal of Molecular Evolution* **49**, 260–271.

Eriksson, J.E., Dechat, T., Grin, B. *et al.* (2009) Introducing intermediate filaments: from discovery to disease. *Journal of Clinical Investigation* **119**, 1763–1771.

Firmbach-Kraft, I. and Stick, R. (1993) The role of CaaX-dependent modifications in membrane association of Xenopus nuclear lamin B3 during meiosis and the fate of B3 in transfected mitotic cells. *Journal of Cell Biology* **123**, 1661–1670.

Fiserova, J., Kiseleva, E. and Goldberg, M.W. (2009) Nuclear envelope and nuclear pore complex structure and organization in tobacco BY-2 cells. *The Plant Journal* **59**, 243–255.

Fiserova, J., Richards,S.A., Wente, S.R. and Goldberg, M.W. (2010) Facilitated transport and diffusion take distinct spatial routes through the nuclear pore complex. *Journal of Cell Science* **123**, 2773–2780.

Fiserova, J., and Goldberg, M.W. (2010) Immunoelectron microscopy of cryofixed freeze-substituted Saccharomyces cerevisiae. *Methods in Molecular Biology* **657**, 191–204.

Foster, H.A., Stokes, P., Forsey, K. *et al.* (2007) Lamins A and C are present in the nuclei of early porcine embryos, with lamin A being distributed in large intranuclear foci. *Chromosome Research* **15**, 163–174

Franke, W.W., Scheer, U. and Fritsch, H. (1972) Intranuclear and cytoplasmic annulate lamellae in plant cells. *Journal of Cell Biology* **53**, 823–827.

Franke, W.W., Spring, H., Scheer, U. and Zerban, H. (1975) Growth of the nuclear envelope in the vegetative phase of the green alga Acetabularia. Evidence for assembly from membrane components synthesized in the cytoplasm. *Journal of Cell Biology* **66**, 681–689.

Fransz, P., De Jong, J.H., Lysak, M. *et al.* (2002) Interphase chromosomes in Arabidopsis are organized as well defined chromocenters from which euchromatin loops emanate. *Proceedings of the National Academy of Sciences of the USA* **99**, 14584–14589.

Frederick, S.E., Mangan, M.E., Carey, J.B. and Gruber, PJ. (1992) Intermediate filament antigens of 60 and 65 kDa in the nuclear matrix of plants, their detection and localization. *Experimental Cell Research* **199**, 213–222.

Galbraith, C.G. and Galbraith, J.A. (2011) Super-resolution microscopy at a glance. *Journal of Cell Science* **124**, 1607–1611.

Gerace, L. and Blobel, G. (1980) The nuclear envelope lamina is reversibly depolymerized during mitosis. *Cell* **19**, 277–287.

Gindullis, F. and Meier, I. (1999) Matrix attachment region binding protein MFP1 is localized in discrete domains at the nuclear envelope. *The Plant Cell* **11**, 1117–1128.

Gindullis, F., Peffer, N.J. and Meier, I. (1999) MAF1, a novel plant protein interacting with matrix attachment region binding protein MFP1, is located at the nuclear envelope. *The Plant Cell* **11**, 1755–1768.

Goldberg, M.W., Huttenlauch, I., Hutchison, C.J. and Stick, R. (2008) Filaments made from A- and B-type lamins differ in structure and organization *Journal of Cell Science* **121**, 215–225.

Goto, H., Tanabe, K., Manser, E. *et al.* (2002) Phosphorylation and reorganization of vimentin by p21-activated kinase (PAK). *Genes to Cells* **7**, 91–97.

Graumann, K., Runions, J. and Evans, D.E. (2010) Characterization of SUN-domain proteins at the higher plant nuclear envelope. *The Plant Journal* **61**, 134–144.

Graumann, K., Runions, J. and Evans, D.E. (2010) Nuclear envelope proteins and their role in nuclear positioning and replication. *Biochemical Society Transactions* **38**, 741–746.

Gross, P., Julius, C., Schmelzer, E. and Hahlbrock, K. (1993) Translocation of cytoplasm and nucleus to fungal penetration sites is associated with depolymerization of microtubules and defence gene activation in infected, cultured parsley cells. *The EMBO Journal* **12**, 1735–1744.

Gueth-Hallonet, C., Weber, K. and Osborn, M. (1997) Cleavage of the nuclear matrix protein NuMA during apoptosis. *Experimental Cell Research* **233**, 21–24.

Gueth-Hallonet, C., Wang, J., Harborth, J. *et al.* (1998) Induction of a regular nuclear lattice by overexpression of NuMA. *Experimental Cell Research* **243**, 434–452.

Hamada, T. (2007) Microtubule-associated proteins in higher plants. *Journal of Plant Research* **120**, 79–98.

Hall, G. Jr., Allen, G.C., Loer, D.S. *et al.* (1991) Nuclear scaffolds and scaffold-attachment regions in higher plants. *Proceedings of the National Academy of Sciences of the USA* **88**, 9320–9324.

Haque, F., Lloyd, D.J., Smallwood, D.T. *et al.* (2006) SUN1 interacts with nuclear lamin A and cytoplasmic nesprins to provide a physical connection between the nuclear lamina and the cytoskeleton. *Molecular and Cellular Biology* **26**, 3738–3751.

Harborth, J., Weber, K. and Osborn, M. (2000) GAS41, a highly conserved protein in eukaryotic nuclei, binds to NuMA. *Journal of Biological Chemistry* **275**, 31979–31985.

Hepler, P.K., Vidali, L. and Cheung, A.Y. (2001). Polarized cell growth in higher plants. *Annual Review of Cell and Developmental Biology* **17**, 159–187.

Herrmann, H. and Aebi, U. (2004) Intermediate filaments: molecular structure, assembly mechanism, and integration into functionally distinct intracellular Scaffolds. *Annual Review of Biochemistry* **73**, 749–789.

Herrmann, H., Strelkov, S.V., Burkhard, P. and Aebi, U. (2009) Intermediate filaments: primary determinants of cell architecture and plasticity.*Journal of Clinical Investigation* **119**, 1772–1783.

Hesse, M., Magin, T.M., Weber, K. (2001) Genes for intermediate filament proteins and the draft sequence of the human genome: novel keratin genes and a surprisingly high number of pseudogenes related to keratin genes 8 and 18. *Journal of Cell Science* **114**, 2569–2575.

Hozak, P., Sasseville, A.M., Raymond, Y. and Cook, P.R. (1995) Lamin proteins form an internal nucleoskeleton as well as a peripheral lamina in human cells. *Journal of Cell Science* **108**, 635–644.

Jackson, D.A. and Cook, P.R (1988) Visualization of a filamentous nucleoskeleton with a 23 nm axial repeat. *The EMBO Journal* **7**, 3667–3677.

Jackson, D.A. and Pombo, A. (1998) Replicon clusters are stable units of chromosome structure: evidence that nuclear organization contributes to the efficient activation and propagation of S phase in human cells. *Journal of Cell Biology* **140**, 1285–1295.

Jeong, S.Y., Rose, A. and Meier, I. (2003) MFP1 is a thylakoid-associated, nucleoid-binding protein with a coiled-coil structure. *Nucleic Acids Research* **31**, 5175–5185.

Joffe, B., Leonhardt, H. and Solovei, I. (2010) Differentiation and large scale spatial organization of the genome. *Current Opinion in Genetics and Development* **20**, 562–569.

Jost, W., Baur, A., Nick, P. *et al.* (2004) A large plant beta-tubulin family with minimal C-terminal variation but differences in expression. *Gene* **340**, 151–160.

Kallajoki, M., Weber, K., and Osborn, M. (1991). A 210 kDa nuclear matrix protein is a functional part of the mitotic spindle: A microinjection study using SPN monoclonal antibodies. *The EMBO Journal* **10**, 3351–3362.

Ketelaar, T., Faivre-Moskalenko, C., Esseling, J.J. *et al.* (2002) Positioning of nuclei in Arabidopsis root hairs: an actin-regulated process of tip growth. *The Plant Cell* **14**, 2941–2955.

Ketelaar, T., van der Honing, H.S. and Emons, A.M. (2010) Probing cytoplasmic organization and the actin cytoskeleton of plant cells with optical tweezers. *Biochemical Society Transaction* **38**, 823–828.

King, M.C., Lusk, C.P. and Blobel, G. (2006) Karyopherin-mediated import of integral inner nuclear membrane proteins. *Nature* **442**, 1003–1007.

Kitten, G.T., and Nigg, E.A. (1991) The CaaX motif is required for isoprenylation, carboxyl methylation, and nuclear membrane association of lamin B2. *Journal of Cell Biology* **113**, 13–23.

Kivinen, K., Taimen, P. and Kallajoki, M. (2010) Silencing of Nuclear Mitotic Apparatus protein (NuMA) accelerates the apoptotic disintegration of the nucleus. *Apoptosis* **15**, 936–945.

Lehner, C.F., Stick, R., Eppenberger, H.M. and Nigg, E.A. (1987) Differential expression of nuclear lamin proteins during chicken development. *Journal of Cell Biology* **105**, 577–587.

Li, J., Brunner, A.M., Meilan, R. and Strauss, S.H. (2008) Matrix attachment region elements have small and variable effects on transgene expression and stability in field-grown Populus. *Plant Biotechnology Journal* **6**, 887–896.

Lin, H.H., Hsu, H.L. and Yeh, N.H. (2007) Apoptotic cleavage of NuMA at the C-terminal end is related to nuclear disruption and death amplification. *Journal of Biomedical Science* **14**, 681–694.

Linde, N. and Stick, R. (2010) Intranuclear membranes induced by lipidated proteins are derived from the nuclear envelope. *Nucleus* **1**, 343–353.

Loc, P.V. and Strätling, W.H. (1988) The matrix attachment regions of the chicken lysozyme gene co-map with the boundaries of the chromatin domain. *The EMBO Journal* **7**, 655–664.

Ludérus, M.E., den Blaauwen, J.L., de Smit, O.J. *et al.* (1994) Binding of matrix attachment regions to lamin polymers involves single-stranded regions and the minor groove. *Molecular and Cellular Biology* **14**, 6297–6305.

Lusk, C.P., Blobel, G. and King, M.C. (2007) Highway to the inner nuclear membrane: rules for the road. *Nat Rev Molecular and Cellular Biology* **8**, 414–420.

Mahajan, R., Delphin, C., Guan, T. *et al.* (1997) A small ubiquitin-related polypeptide involved in targeting RanGAP1 to nuclear pore complex protein RanBP2. *Cell* **88**, 97–107.

Makarevitch, I. and Somers, D.A. (2006) Association of Arabidopsis topoisomerase IIA cleavage sites with functional genomic elements and T-DNA loci. *The Plant Journal* **48**, 697–709.

Maraldi, N.M., Capanni, C., Cenni, V. *et al.* (2011) Laminopathies and lamin-associated signaling pathways. *Journal of Cellular Biochemistry* **112**, 979–992.

Markiewicz, E., Ledran, M., Hutchison, C.J. (2005) Remodelling of the nuclear lamina and nucleoskeleton is required for skeletal muscle differentiation in vitro. *Journal of Cell Science* **118**, 409–420.

Markiewicz, E., Venables, R., Alvarez-Reyes, M. *et al.* (2002) Increased solubility of lamins and redistribution of lamin C in X-linked Emery-Dreifuss muscular dystrophy fibroblasts. *Journal of Structural Biology* **140**, 241–253.

Martelli, A.M., Falcieri, E., Zweyer, M. *et al.* (2002) The controversial nuclear matrix: a balanced point of view. *Histology and Histopathology* **17**, 1193–1205.

Masuda, K., Xu, Z.J., Takahashi, S. *et al.* (1997) Peripheral framework of carrot cell nucleus contains a novel protein predicted to exhibit a long alpha-helical domain. *Experimental Cell Research* **232**, 173–181.

Maximova, S., Miller, C., Antúnez de Mayolo, G. *et al.* (2003) Stable transformation of Theobroma cacao L. and influence of matrix attachment regions on GFP expression. *Plant Cell Reports* **21**, 872–883.

McNulty, A.K. and Saunders, M.J. (1992) Purification and immunological detection of pea nuclear intermediate filaments: evidence for plant nuclear lamins. *Journal of Cell Science* **103**, 407–414.

Meagher, R.B., Kandasamy, M.K., Deal, R.B. and McKinney, E.C. (2007) Actin-related proteins in chromatin-level control of the cell cycle and developmental transitions. *Trends in Cellular Biology* **17**, 325–332.

Melcer, S., Gruenbaum, Y. and Krohne, G. (2007) Invertebrate lamins. *Experimental Cell Research* **313**, 2157–2166.

Mellad, J.A., Warren, D.T. and Shanahan, C.M. (2011) Nesprins LINC the nucleus and cytoskeleton. *Current Opinion in Cellular Biology* **23**, 47–54.

Merkle, T. (2011) Nucleo-cytoplasmic transport of proteins and RNA in plants. *Plant Cell Reports* **30**, 153–176.

Mencarelli, C., Ciolfi, S., Caroti, D. *et al.* (2011) Isomin: a novel cytoplasmic intermediate filament protein from an arthropod species. *BMC Biology* **28**, 9–17.

Meier, I., Phelan, T., Gruissem, W. *et al.* (1996) MFP1, a novel plant filament-like protein with affinity for matrix attachment region DNA. *The Plant Cell* **8**, 2105–2115.

Mínguez, A., Franca, S. and Moreno Díaz de la Espina, S. (1994) Dinoflagellates have a eukaryotic nuclear matrix with lamin-like proteins and topoisomerase II. *Journal of Cell Science* **107**, 2861–2873.

Monteiro, M.J., Hicks, C., Gu, L. and Janicki, S. (1994) Determinants for intracellular sorting of cytoplasmic and nuclear intermediate filaments. *Journal of Cell Biology* **127**, 1327–1343.

Muralikrishna, B., Dhawan, J., Rangaraj, N. and Parnaik, V.K. (2001) Distinct changes in intranuclear lamin A/C organization during myoblast differentiation. *Journal of Cell Science* **114**, 4001–4011.

Murphy, S.P., Simmons, C.R. and Bass, H.W. (2010) Structure and expression of the maize (Zea mays L.) SUN-domain protein gene family: evidence for the existence of two divergent classes of SUN proteins in plants. *BMC Plant Biology* **10**, 269.

Nakamura, H., Morita, T. and Sato, C. (1986) Structural organizations of replicon domains during DNA synthetic phase in the mammalian nucleus. *Experimental Cell Research* **165**, 291–297.

Nakayasu, H., Berezney, R. (1991) Nuclear matrins: identification of the major nuclear matrix proteins. *Proceedings of the National Academy of Sciences of the USA* **88**, 10312–1036.

Neumann, U., Brandizzi, F. and Hawes, C. (2003) Protein transport in plant cells: in and out of the Golgi. *Annals of Botany* **92**, 167–180.

Nicolet, S., Herrmann, H., Aebi, U. and Strelkov, S.V. (2010) Atomic structure of vimentin coil 2. *Journal of Structural Biology* **170**, 369–376.

Paddy, M.R., Belmont, A.S., Saumweber, H. *et al.* (1990) Interphase nuclear envelope lamins form a discontinuous network that interacts with only a fraction of the chromatin in the nuclear periphery. *Cell* **62**, 89–106.

Pederson, T. (2011) The nucleolus. *Cold Spring Harbor Perspectives in Biology* **1**;3. pii: a000638. doi: 10.1101/cshperspect.a000638.

Pendle, A.F., Clark, G.P., Boon, R. *et al.* (2005) Proteomic analysis of the Arabidopsis nucleolus suggests novel nucleolar functions. *Molecular Biology of the Cell* **16**, 260–269.

Penman, S. (1995) Rethinking cell structure. *Proceedings of the National Academy of Sciences of the USA* **92**, 5251–5257.

Pérez-Munive, C. and Moreno Díaz de la Espina, S. (2011) Nuclear spectrin-like proteins are structural actin-binding proteins in plants. *Biology of the Cell* **103**, 145–157.

Perrin, B.J. and Ervasti, J.M. (2010) The actin gene family: function follows isoform. *Cytoskeleton (Hoboken)* **67**, 630–4.

Peter, M., Nakagawa, J., Dorée, M. *et al.* (1990. In vitro disassembly of the nuclear lamina and M phase-specific phosphorylation of lamins by cdc2 kinase. *Cell* **61**, 591–602.

Petrásek, J., Schwarzerová, K. (2009) Actin and microtubule cytoskeleton interactions. *Current Opinion in Plant Biology* **12**, 728–734.

Piette, B.M., Liu, J., Peeters, K. *et al.* (2009) A thermodynamic model of microtubule assembly and disassembly. *PLoS One* **4**:e6378.

Puckelwartz, M.J., Depreux, F.F. and McNally, E.M. (2011) Gene expression, chromosome position and lamin A/C mutations. *Nucleus* **2**, 162–167.

Radulescu, A.E. and Cleveland, D.W. (2010) NuMA after 30 years: the matrix revisited. *Trends in Cellular Biology* **20**, 214–222.

Ralle, T., Grund, C., Franke, W.W. and Stick, R. (2004) Intranuclear membrane structure formations by CaaX-containing nuclear proteins. *Journal of Cell Science* **117**, 6095–6104.

Razin, S.V., Vassetzky, Y.S. and Hancock, R. (1991) Nuclear matrix attachment regions and topoisomerase II binding and reaction sites in the vicinity of a chicken DNA replication origin. *Biochemical and Biophysical Research Communications* **177**, 265–270.

Reddy, A.S. (2001) Molecular motors and their functions in plants. *International Review of Cytology* **204**, 97–178.

Ridley, A.J. (2011) Life at the leading edge. *Cell* **145**, 1012–1022.

Rose, A., Manikantan, S., Schraegle, S.J. *et al.* (2004) Genome-wide identification of Arabidopsis coiled-coil proteins and establishment of the ARABI-COIL database. *Plant Physiology* **134**, 927–939.

Rose, A., Schraegle, S.J., Stahlberg, E.A. and Meier, I. (2005) Coiled-coil protein composition of 22 proteomes – differences and common themes in subcellular infrastructure and traffic control. *BMC Evolutionary Biology* **5**, 66.

Samaniego, R., Jeong, S.Y., Meier, I. and de la Espina, S.M. (2006) Dual location of MAR-binding, filament-like protein 1 in Arabidopsis, tobacco, and tomato. *Planta* **223**, 1201–1206.

Schaad, M.C., Jensen, P.E. and Carrington, J.C. (1997) Formation of plant RNA virus replication complexes on membranes: role of an endoplasmic reticulum-targeted viral protein. *The EMBO Journal* **16**, 4049–4059.

Schäpe, J., Prausse, S., Radmacher, M. and Stick, R. (2009) Influence of lamin A on the mechanical properties of amphibian oocyte nuclei measured by atomic force microscopy. *Biophysical Journal* **96**, 4319–4325.

Schneider, M., Lu, W., Neumann, S. *et al.* (2011) Molecular mechanisms of centrosome and cytoskeleton anchorage at the nuclear envelope. *Cellular and Molecular Life Sciences* **68**, 1593–1610.

Seltzer, V., Janski, N., Canaday, J. *et al.* (2007) Arabidopsis GCP2 and GCP3 are part of a soluble gamma-tubulin complex and have nuclear envelope targeting domains. *The Plant Journal* **52**, 322–331.

Senda, T., Iizuka-Kogo, A. and Shimomura, A. (2005) Visualization of the nuclear lamina in mouse anterior pituitary cells and immunocytochemical detection of lamin A/C by quick-freeze freeze-substitution electron microscopy. *Journal of Histochemistry and Cytochemistry* **53**, 497–507.

Shimi, T., Butin-Israeli, V., Adam, S.A. and Goldman, R.D. (2010) Nuclear lamins in cell regulation and disease. *Cold Spring Harbor Symposia on Quantitative Biology* **75**, 525–531.

Smith, H.C., Puvion, E., Buchholtz, L.A. and Berezney, R. (1984) Spatial distribution of DNA loop attachment and replicational sites in the nuclear matrix. *Journal of Cell Biology* **99**, 1794–1802.

Spector, D.L. and Lamond, A.I. (2011) Nuclear speckles. *Cold Spring Harbor Perspectives in Biology* **1**;3. pii: a000646. doi: 10.1101/cshperspect.a000646.

Sorek, N., Bloch, D. and Yalovsky, S. (2009) Protein lipid modifications in signaling and subcellular targeting. *Current Opinion in Plant Biology* **12**, 714–720.

Stick, R. and Hausen, P. (1985) Changes in the nuclear lamina composition during early development of Xenopus laevis. *Cell* **41**, 191–200.

Stick, R. and Schwarz, H. (1983) Disappearance and reformation of the nuclear lamina structure during specific stages of meiosis in oocytes. *Cell* **33**, 949–958.

Sullivan, T., Escalante-Alcalde, D., Bhatt, H. *et al.* (1999) Loss of A-type lamin expression compromises nuclear envelope integrity leading to muscular dystrophy. *Journal of Cell Biology* **147**, 913–920.

Tikhonov, A.P., Bennetzen, J.L. and Avramova, Z.V. (2000) Structural domains and matrix attachment regions along colinear chromosomal segments of maize and sorghum. *The Plant Cell* **12**, 249–264.

van der Honing, H.S., Emons, A.M. and Ketelaar, T. (2007) Actin based processes that could determine the cytoplasmic architecture of plant cells. *Biochimica et Biophysica Acta* **1773**, 604–614.

Van der Geest, A.H., Welter, M.E., Woosley, A.T. *et al.* (2004) A short synthetic MAR positively affects transgene expression in rice and Arabidopsis. *Plant Biotechnology Journal* **2**, 13–26.

Visa, N. and Percipalle, P. (2010) Nuclear functions of actin. *Cold Spring Harbor Perspectives in Biology* **2**:a000620.

Wang, J., Yang, C., Hans, R. and Zhai, Z. (1996) Nuclear lamina in plant cells. *Science in China Series C: Life Sciences* **39**, 449–457.

Wang, T., Xue, L., Hou, W. *et al.* (2007) Increased expression of transgene in stably transformed cells of Dunaliella salina by matrix attachment regions. *Applied Microbiology and Biotechnology* **76**, 651–657.

Wang, T.Y., Han, Z.M., Chai, Y.R. and Zhang, J.H. (2010) A mini review of MAR-binding proteins. *Molecular Biology Reports* **37**, 3553–3560.

Wente, S.R. and Rout, M.P. (2010) The nuclear pore complex and nuclear transport. *Cold Spring Harbor Perspectives in Biology* **2**:a000562.

Willis, N.D., Cox, T.R., Rahman-Casañs, S.F. *et al.* (2008) Lamin A/C is a risk biomarker in colorectal cancer. *PLoS One* **3**:e2988.

Worman, H.J., Fong, L.G., Muchir, A. and Young, S.G. (2009) Laminopathies and the long strange trip from basic cell biology to therapy. *Journal of Clinical Investigation* **119**, 1825–1836.

Worman, H.J. (2010) Prelamin A prenylation and the treatment of progeria. *Journal of Lipid Research* **51**, 223–225.

Xu, X.M., Meulia, T. and Meier, I. (2007) Anchorage of plant RanGAP to the nuclear envelope involves novel nuclear-pore-associated proteins. *Current Biology* **17**, 1157 1163.

Yoon, K.H., Yoon, M., Moir, R.D. *et al.* (2001) Insights into the dynamic properties of keratin intermediate filaments in living epithelial cells. *Journal of Cell Biology* **153**, 503–516.

Young, S.G., Meta, M., Yang, S.H. and Fong, L.G. (2006) Prelamin A farnesylation and progeroid syndromes. *Journal of Biological Chemistry* **281**, 39741–39745.

Yu, W. and Moreno Díaz de la Espina, S. (1999) The plant nucleoskeleton: ultrastructural organization and identification of NuMA homologues in the nuclear matrix and mitotic spindle of plant cells. *Experimental Cell Research* **246**, 516–526.

Zhang, C., Jenkins, H., Goldberg, M.W. *et al.* (1996) Nuclear lamina and nuclear matrix organization in sperm pronuclei assembled in Xenopus egg extract. *Journal of Cell Science* **109**, 2275–2286.

Zhang, H. and Dawe, R.K. (2011) Mechanisms of plant spindle formation. *Chromosome Research* **19**, 335–344.

Zhao, Q., Brkljacic, J., Meier, I. (2008) Two distinct interacting classes of nuclear envelope-associated coiled-coil proteins are required for the tissue-specific nuclear envelope targeting of Arabidopsis RanGAP. *The Plant Cell* **20**, 1639–1651.

Zhou, X., Graumann, K., Evans, D.E. and Meier, I. (2012) Novel plant SUN-KASH bridges are involved in RanGAP anchoring and nuclear shape determination. *Journal of Cell Biology* **196**, 203–211.

Zuleger, N., Kelly, D.A., Richardson, A.C. *et al.* (2011) System analysis shows distinct mechanisms and common principles of nuclear envelope protein dynamics. *Journal of Cell Biology* **193**, 109–123.

Annual Plant Reviews (2013) **46**, 123–156
doi: 10.1002/9781118472507.ch5
http://onlinelibrary.wiley.com

Chapter 5

GENOMICS AND CHROMATIN PACKAGING

Eugenio Sanchez-Moran
School of Biosciences, University of Birmingham, Birmingham B15 2TT, UK

Abstract: The role of DNA is to store an individual's genetic information so that it can be used during development and can be copied accurately during the divisions of the cell. DNA has to be packed within a small nucleus in an organized manner to be accessible for a variety of vital cell processes (transcription, DNA replication, repair, mitosis and meiosis). This is achieved by DNA associating with different proteins to form chromatin.

Different levels of compaction are involved, from the nucleosome fibre to higher order chromatin structures. All the chromatin components involved at different levels of DNA compaction are highly dynamic, allowing the required accessibility.

The understanding of chromatin components, their dynamics and interactions in plants would allow us to provide plant breeders with new tools to help fulfil the increased demand for food that will be required in future decades of increased global population and climate change.

Keywords: chromatin; chromosomes; histones; scaffold; axis; DNA; genomics

5.1 Chromatin components and structure in higher eukaryotes

The key role of DNA is to store the genetic information of an individual so that it can be used during normal growth and development and so that it can be accurately copied during cell division. The genomic DNA of higher eukaryotes is divided into individual molecules constituting chromosomes. If we could line up the genomic DNA of all the chromosomes present in a diploid human cell (ca six Gb), it would be about 2 m long. This 2 m of DNA is packed within the cell nucleus, which is only 0.01 mm (10 μm) in

Annual Plant Reviews Volume 46: Plant Nuclear Structure, Genome Architecture and Gene Regulation, First Edition. Edited by David E. Evans, Katja Graumann and John A. Bryant.

diameter. If the cell nucleus were as big as a football (22.5 cm in diameter) then the extended DNA would cover more than 450 times the length of a football pitch (about 100 m long). Although the DNA is highly packed in the nucleus, this packaging must be organized in such a way that it is readily accessible for a variety of crucial processes. The information it contains must be easily read (transcription) so that the cell can rapidly produce proteins and enzymes. It must be readily duplicated (DNA replication) and accurately separated during cell division (mitosis) and gamete production (meiosis). It is also essential that any break, knot or tangle that might occur can be repaired (DNA repair). The way that higher eukaryotes manage the packaging and accessibility of the DNA in the nucleus is by association with different proteins forming a dynamic nucleo-protein complex called chromatin. This enables the compaction necessary to fit the naked DNA inside the cell nucleus whilst maintaining access to the genetic information to carry out the vital biological functions of the DNA (Manuelidis, 1990; Haaf and Schmid, 1991; Wolffe, 1998; Belmont *et al.*, 1999).

There are different levels of compaction involved in packaging DNA into chromosomes (Figure 5.1). These levels of compaction build the hierarchical structure of the chromatin in higher eukaryotes. The first level of compaction is the nucleosomal fibre (diameter ca. 10 nm with fivefold compaction). The basic structure of chromatin is the nucleosome, formed by wrapping naked DNA around a core of proteins known as histones. The nucleosomes are arranged along the DNA forming the nucleosomal fibre, also known as 'beads on a string' because of their appearance in an electron microscope. Despite the old impression that nucleosomes were static structures, nowadays a nucleosome is considered to be a highly dynamic assemblage. Changes to this structure are facilitated through histone modifications, histone variants, modelling factors and exchange of histone proteins. The second level of compaction is the higher order chromatin fibre (diameter ~ 30 nm/ ~ 40-fold compaction). The nucleosomal fibre is further compacted by winding itself into a higher order chromatin fibre whose formation and structure *in vivo* remains controversial. The third level of compaction is the arrangement of

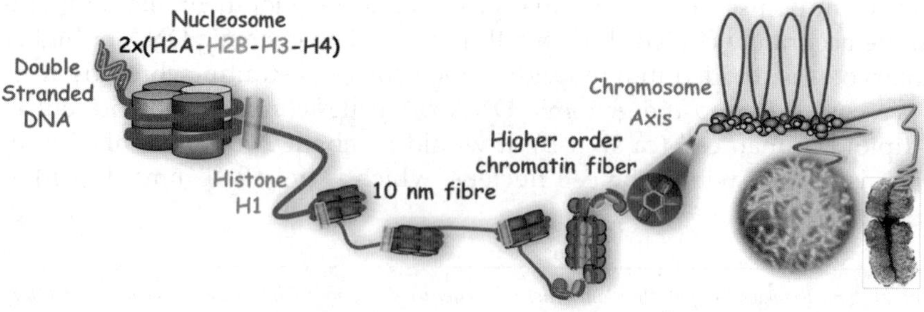

Figure 5.1 Classical model of chromatin structure.

the higher order chromatin fibre into loops attached to a chromosome scaffold (diameter ~300 nm/ ~200-fold compaction). The higher order chromatin fibre is additionally arranged into loops (30–100 kb/loop) that are attached to a multi-protein axis called the chromosome scaffold or axis. This seems to be the basic organization of the chromatin in an interphase nucleus. Furthermore, during the process of cell division when the individual chromosomes have been duplicated, chromosome condensation is necessary to ensure their accurate distribution. After replication the chromatin is organized in a manner identical to interphase nuclei but as it goes through prophase it is then helically folded into radial loops to form the final metaphase chromosome (diameter ~ 700 nm/chromatid/ 10 000 to 20 000-fold compaction) (Figure 5.1). Misregulation of chromosome condensation can lead to cell death, cancer and improper chromosome segregation during cell cycle or during the production of gametes.

Although the biochemistry of histones and other chromosome-associated proteins has been studied intensively, their interactions to achieve the chromatin structure and chromosome condensation are still poorly understood. Andrew Belmont (2006) said that the mitotic chromosome structure is like 'a riddle, wrapped in a mystery, inside an enigma'. This statement was used to emphasize the fact that mutations in different key chromatin proteins cause only slight perturbations in the global chromosome condensation. Thus, the classical model of chromatin organization and structure is undergoing continuous re-evaluation.

The key components involved in chromatin structure and chromosome condensation are conserved throughout eukaryotic evolution indicating that their roles are so fundamental that they are species independent. Study of plant systems make more detailed analysis of the chromatin structure in higher eukaryotes possible. Several plant genomes have been, or are being, sequenced (Table 5.1), allowing us to compare the level of compaction needed

Table 5.1 Haploid genome size and estimated length for the genomic DNA of a diploid cell in various species (NCBI genome data: www.ncbi.nlm.nih.gov/genome)

Species	Haploid Genome Size	Genomic DNA Length in a Diploid Cell
Saccharomyces cerevisiae	~12 Mb	~7.7 mm
Caenorhabditis elegans	~100.2 Mb	~6.4 cm
Drosophila melanogaster	~139.4 Mb	~8.9 cm
Mus musculus	~2.7 Gb	~1.7 cm
Homo sapiens	~3.1 Gb	~2 m
Arabidopsis thaliana	~120 Mb	~7.7 cm
Brassica oleracea[a]	~650Mb	~41.9 cm
Oryza sativa	~389 Mb	~25 cm
Triticum aestivum (6X)[a]	~16 Gb	~10.3 m

[a]Not yet sequenced

for the DNA into the nucleus in each species. *Arabidopsis thaliana* haploid genome contains about 120 Mb. If we could line up the genomic DNA of all the chromosomes present in a diploid Arabidopsis cell (ca. 240 Mb) it would be about 7.7 cm long. Although much shorter than the human DNA, this 7.7 cm of DNA also has to be closely packed within the cell nucleus, which is only 0.01 mm (10 μm) in diameter. Obviously, the Arabidopsis genome is one of the smallest in the plant kingdom. In other species, for example in bread wheat (*Triticum aestivum*), the genome size is even bigger than in humans. If we could extend the total genomic DNA from a diploid cell of bread wheat (ca. 32 Gb), its length would be about 10 m. Using the football comparison again, if the plant cell nucleus were as big as a football then the extended DNA could cover more than 2250 times the length of a football pitch, a staggering 225 km or the distance from London to Sheffield as the crow flies. The compaction of the genomic DNA into chromatin is very important to allow it to be accommodated in the nucleus but it is also crucial for the biological processes of the DNA. Not only does the 10 m of genomic DNA in bread wheat have to be packed into a 10 μm diameter nucleus in the cell but it has to be done in such a way that any specific region can be found and readily accessed at any time. This is a colossal task.

5.2 Histones and nucleosome fibre

The German biochemist Ludwig Karl Martin Leonhard Albrecht Kossel (1853–1927) investigated a substance isolated from pus cells called 'nuclein', isolated by Freidrich Meischer, which was highly acidic and rich in phosphorus (Jones, 1953). Kossel showed that the substance consisted of proteins and a non-protein component that he later further isolated. He characterized its five basic compounds, named nucleobases: adenine, cytosine, guanine, thymine and uracil (Jones, 1953). 'Nuclein' has now become known as nucleic acids, and this discovery won Kossel the Nobel Prize in 1910. Furthermore, he was very interested in the chemical composition of proteins, and his research led to the discovery of the polypeptide nature of the proteins. Interestingly, during the last years of his career, Kossel discovered a very abundant class of proteins in the nucleus: the histones (Jones, 1953). Therefore, Kossel not only explored the nucleic acids (DNA) providing the stepping stones that led to the double-helix model of the DNA (Watson and Crick, 1953), but also discovered the basic elements of the chromatin.

It was not until 1973 that chromatin was shown to be a repeating subunit structure (Hewish and Burgoyne, 1973). The authors observed that digestion of chromatin by a nuclease resulted in a DNA ladder of discrete series of DNA fragments of sizes which are multiples of 180-200 bp (Williamson, 1970; Hewish and Burgoyne, 1973). Nowadays, we know that these multiples of 180-200 bp represent the DNA associated with the nucleosome. Soon afterwards, the DNA fragments were isolated as a complex with protein (Sahasrabuddhe and van Holde, 1974). The analysis of these DNA-protein complexes by electron microscopy showed a structure like 'beads on a string', delivering further confirmation for a repetitive subunit structure (Woodcock, 1973; Olins and Olins, 1974). Kornberg and Thomas (1974) characterized the different oligomers of histones. All these investigations led Kornberg (1974) to propose the nucleosome model in which a nucleosome consists of DNA wrapped around an octamer of histones. Each octamer is built by a tetramer of histones ($(H3/H4)^2$) and two dimers of histones (H2A/H2B). Different crystal structures of the histone core and its attachment to the DNA substantially clarified the nucleosome structure (Richmond *et al.*, 1984; Arents *et al.*, 1991; Arents and Moudrianakis, 1993; Luger *et al.*, 1997). Thus, the tetramer $(H3/H4)^2$ seems to organize the nucleosome, whereas the dimers H2A/H2B interact with the tetramer and the ends of the DNA. The nucleosome is the basic repeating structural and functional unit of the chromatin and is composed of two pairs of positively charged histone proteins (H2A, H2B, H3 and H4) with about 146 bp of DNA wrapped around them 1.67 times in a left-handed super-helical turn (Luger et al., 1997).

Histone proteins have a relatively small molecular weight of approximately 11–16 kDa, and contain large amounts of basic amino acids (more than 20% of lysine and arginine residues) (van Holde, 1988). Histone protein structure contains an extended histone-fold domain at the carboxyl terminal end (Ct) and a charged tail at the amino terminal end (Nt) (Arents *et al.*, 1991). The Ct histone folding domain consists of three α-helices (α1-3) joined by two linkers (L1-2). This Ct is thought to play an important role in the interaction between the histones themselves and with the DNA (Alva, Ammelburg and Lupas, 2007). Histones dimerize along their α2 helices. The H2A/H2B dimer binds onto the H3/H4 tetramer by H4 and H2B hydrophobic interactions (Luger *et al.*, 1997). Histone octamer stability depends on the presence of DNA and the basic charge of the four core histones. The histones appear to organize the DNA around the nucleosome through electrostatic interactions between arginine residues and the phosphodiester backbone as shown by histone removal from DNA by high salt concentrations (Wolffe, 1998). Furthermore, the charged Nt histone tails protrude from the histone octamer and are the target sites for post-translation modifications (Strahl and Allis, 2000; Jenuwein and Allis, 2001) and interaction with other structural components of chromatin. The structure of the Nt tails has not been characterized by crystallography due to their high flexibility (Zheng and Hayes, 2003). Histone amino acid sequences are evolutionarily conserved, with H3 and H4 being the most highly conserved. Thus, bovine histone H4 is only different in two residues of the 102 amino acids from histone H4 in pea (De Lange et al., 1969a, b). Similarly, histone H4 in humans differs only in two amino acids from that in Arabidopsis (Figure 5.2). Even then, the changes involve amino acids with similar characteristics: a valine (V) for an isoleucine (I), both nonpolar and

(a)

ALIGN (BLASTp): Method: Compositional matrix adjust.
Score = 200 bits (508), **Expect** = 6e-57,
Identities = 101/103 (98%), **Positives** = 103/103 (100%), **Gaps** = 0/103 (0%)

```
H4Hs  MSGRGKGGKGLGKGGAKRHRKVLRDNIQGITKPAIRRLARRGGVKRISGLIYEETRGVLKVFLENVIRDAVTYTEHAKRKTVTAMDVVYALKRQGRTLYGFGG
      MSGRGKGGKGLGKGGAKRHRKVLRDNIQGITKPAIRRLARRGGVKRISGLIYEETRGVLK+FLENVIRDAVTYTEHA+RKTVTAMDVVYALKRQGRTLYGFGG
H4At  MSGRGKGGKGLGKGGAKRHRKVLRDNIQGITKPAIRRLARRGGVKRISGLIYEETRGVLKIFLENVIRDAVTYTEHARRKTVTAMDVVYALKRQGRTLYGFGG
```

(b)

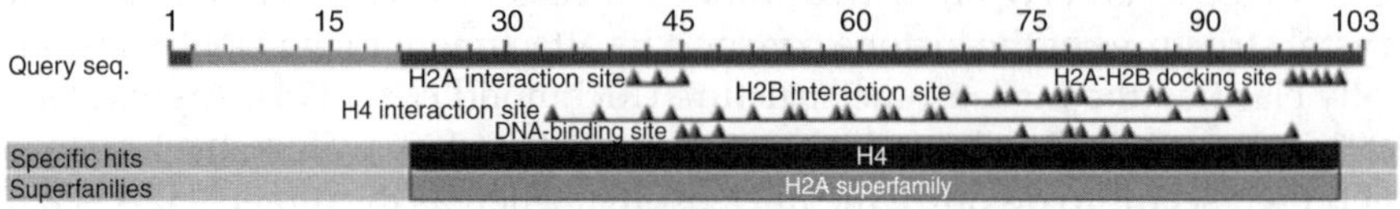

Figure 5.2 Histone core H4 sequence has remained nearly the same during evolution. (a) BLAST alignment of H4 sequences in human ($H4^{Hs}$) and Arabidopsis ($H4^{At}$). (b) Conserved Domain for H4 proteins (CD00076). Marchler-Bauer *et al.* (2011).

hydrophobic amino acids, and, a lysine (K) for an arginine (R), both basic amino acids.

The binding of nucleosomes to DNA along the genome is non-sequence specific, although some DNA sequences are more prolific in their attachment to nucleosomes than others (Segal *et al.*, 2006). There are over 120 histone-DNA interactions in a nucleosome together with several hundred water-mediated ones (Davey *et al.*, 2002). There are two differentiated sites in the octamer to bind to the DNA: the α1α1 site, formed by the two α1-helices from two different histones; and the L1L2 site, formed by the loops L1 and L2. Water-mediated interactions include salt linkages and hydrogen bonding through basic and hydroxyl groups, and amide groups with phosphate groups (Davey *et al.*, 2002). Furthermore, other types of interactions of histones with DNA exist, for example, non-polar interactions between histones and the deoxyribose sugars on DNA or insertions into two minor grooves of the DNA by histones H3 and H2B Nt tails (Davey *et al.*, 2002). The arrangement of nucleosomes along the genomic DNA sequence, nucleosome positioning, could be a factor in regulation of gene expression, DNA repair or DNA recombination (see review by Radman-Livaja and Rando, 2010). The development of high-throughput genomic scale and microarray techniques has permitted the mapping of nucleosomes across genomes from different organisms (Albert *et al.*, 2007; Ozsolak *et al.*, 2007; Valouev *et al.*, 2008; Mavrich *et al.*, 2008a,b). These genome-wide studies have shown that promoter regions are relatively free of nucleosomes compared to transcribed regions. The 'nucleosome-free regions' (NFRs) are found just upstream of the transcription start sites (TSS). The localization of the first nucleosome after the TSS (+1 nucleosome) tends to be located at a specific position but this position varies among organisms, reflecting perhaps differences in the machinery involved in the regulation of transcription amongst species. This canonical chromatin structure of promoters is generally found in housekeeping genes.

Nucleosome positioning has been determined in two species of Nicotiana at the 5S ribosomal DNA (5S rDNA) repeat units (Fulnecek *et al.*, 2006). The data suggested that 5S rDNA does not have fixed position sites for the nucleosomes and that the units could be wrapped in different alternative frames. A genome-wide analysis of nucleosome positioning in the Arabidopsis genome has been carried out in combination with the DNA methylation pattern at single-base resolution (Chodavarapu *et al.*, 2010). The results suggested that nucleosome positioning influences the genomic DNA methylation profile. Furthermore, the authors showed that DNA methyltransferases preferentially target DNA bound to the nucleosome as also seen in humans (Chodavarapu *et al.*, 2010). Thus, it seems that the relationships between DNA methyltransferases and nucleosomes are conserved.

Histone families include multiple histone isoforms, with several copies for each histone. The genome of *Saccharomyces cerevisiae* contains two identical genes encoding each of the core histones: H2A (HTA1 & HTA2), H2B (HTB1 & HTB2), H3 (HHT1 & HHT2) and H4 (HHF1 & HHF2). The result of this gene redundancy is that mutation of one of the copies is not lethal (Dollard *et al.*, 1994). In multicellular eukaryotes the number of copies for each individual canonical histone is increased. The Arabidopsis genome encodes many different histone isoforms: 13 isoforms for H2A (HTA1-13), 11 isoforms for H2B (HTB1-11), 15 isoforms for H3 (HHT1-15) and eight identical isoforms for H4 (The Arabidopsis Information Resource (TAIR): http://www.arabidopsis.org/). In humans, there are 22 H2A isoforms, 20 H2B isoforms, 12 H3 isoforms and 13 H4 isoforms. The four core histones are synthesized and mostly deposited during DNA replication (Bryant and Dunham, 1988; Jin *et al.*, 2005). In addition, there are some histone isoforms that are synthesized and assembled independently of DNA replication; these are referred to as histone variants (Malik and Henikoff, 2003). So histone-core isoforms can be defined as histones deposited during DNA replication which have identical or very similar sequences. On the other hand, histone variants can be defined as histones that are deposited throughout the cell cycle, with similar sequences but with specific differences that give them a specific role and/or location on the chromatin and chromosome structure. Histone H4 isoforms in Arabidopsis are identical with no change in their amino acid sequence. This is a clear case of gene redundancy in chromatin components and denotes the importance of histone duplications in evolution. However, H2A, H2B and H3 isoforms have some differences in their amino acid sequence. Arabidopsis possesses 11 isoforms for histone H2B, which can be classified in different groups according to their sequence similarity (Table 5.2). Similarly, H2A and H3 histone isoforms in Arabidopsis can be classified into different groups. Different histone variants have been described for histones H2A and H3 in different organisms and some of them are present in plants. Histone sequence heterogeneity is exploited to regulate a wide range

Table 5.2 Histone core isoforms in *Arabidopsis thaliana*. Alignments obtained using WU-Blast (TAIR)

Histone		Locus		Alignments
H2A	HTA12	AT5G02560	***SPKK*** DNA bind	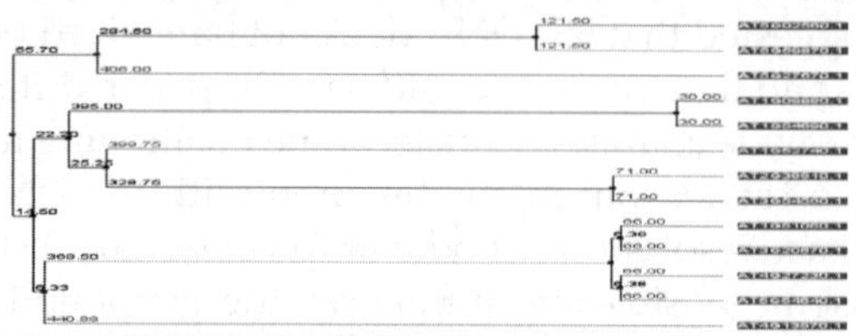
	HTA6	AT5G59870	***SPKK*** DNA bind	
	HTA7	AT5G27670	***SPKK*** DNA bind	
	HTA5	AT1G08880	H2AX-2	
	HTA3	AT1G54690	H2AX-1	
	HTA9	AT1G52740	H2AZ	
	HTA8	AT2G38810	H2AZ	
	HTA11	AT3G54560	H2AZ	
	HTA10	AT1G51060		
	HTA13	AT3G20670		
	HTA2	AT4G27230		
	HTA1	AT5G54640	RAT5	
	HTA4	AT4G13570		
H2B	HTB8	AT1G08170		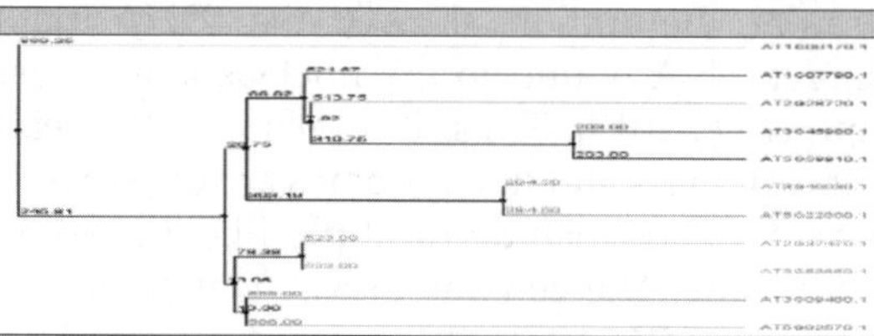
	HTB1	AT1G07790		
	HTB3	AT2G28720		
	HTB9	AT3G45980		
	HTB4	AT5G59910		
	HTB11	AT3G46030		
	HTB2	AT5G22880		
	HTB5	AT2G37470		
	HTB6	AT3G53650		
	HTB7	AT3G09480		
	HTB10	AT5G02570		
H3	HTR2	AT1G09200	H3.1	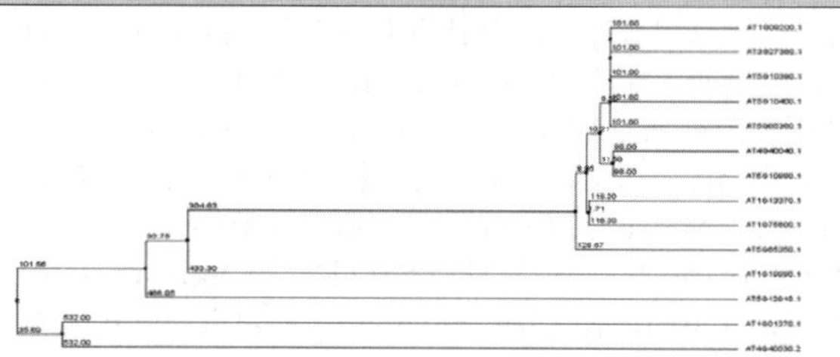
	HTR3	AT3G27360	H3.1	
	HTR13	AT5G10390	H3.1	
	HTR9	AT5G10400	H3.1	
	HTR1	AT5G65360	H3.1	
	HTR5	AT4G40040	H3.3	
	HTR8	AT5G10980	H3.3	
	HTR14	AT1G75600	H3.3 LIKE	
	HTR6	AT1G13370	H3.3 LIKE	
	HTR11	AT5G65350	H3.1 LIKE	
	HTR15	AT5G12910	H3.3 LIKE	
	HTR10	AT1G19890	H3.3 like AtMGH3	
	HTR4	AT4G40030	H3.3	
	HTR12	AT1G01370	CENH3	
	HTR7	AT1G75610	Pseudogene	
H4		AT1G07660		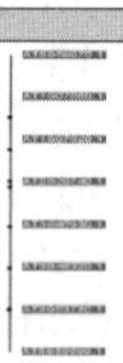
		AT1G07820		
		AT2G28740		
		AT3G45930		
		AT3G53730		
		AT3G46320		
		AT5G59690		
		AT5G59970		

of nuclear functions, and evidence is accumulating that histone variants do indeed have distinct functions.

Interestingly, Arabidopsis histone H2A1 (HTA1) has been found to be involved in the efficient transformation of roots by *Agrobacterium tumefaciens* (Mysore *et al.*, 2000). Further, overexpression of other H2A histones could compensate for the loss of HTA1 in Arabidopsis (Yi *et al.*, 2006). This shows a functionally redundancy of histones H2A for *Agrobacterium*-mediated transformation. Nevertheless, the fact that the lack of HTA1 produced a mutant phenotype shows that the levels and/or patterns of expression of the other histones H2A cannot compensate for HTA1 lack of function. Furthermore, overexpression of histones H4 (HFO) and H3-11(HTR11) led to increased transformation susceptibility suggesting that some other histones can increase the efficiency of *Agrobacterium*-mediated transformation (Tenea *et al.*, 2009). T-DNA targeting into the genomic plant chromatin is a consequence of the interaction between VIP1 (VirE2 interacting protein 1) and histones (Li *et al.*, 2005c; Loyter *et al.*, 2005). Thus it has been proposed that the disruption of *HTA1* gene could lead to changes in the chromatin structure and chromatin remodelling (Tenea *et al.*, 2009). Similar results have been obtained in tobacco (*Nicotiana benthamiana*) where H2A and H3 depletion produces a decrease in *Agrobaterium*-mediated transformation (Anand *et al.*, 2007). In rice, the overexpression of Arabidopsis histone H2A1 increases transformation efficiencies (Zheng *et al.*, 2009).

5.2.1 Histone variants

H2A and H3 histone variants have shown diversification of histone function. These histone variants have different roles and are usually located at specific positions in chromatin, related to their detailed function.

Nucleosomal histone core H2A can be replaced by histone variant H2AZ, which can modulate gene transcription. Nucleosomes have a functionally active role in controlling gene expression by modulating access of transcription factors and associated 'machinery' to DNA (Segal and Widom, 2009). The modulation occurs by the wrapping and unwrapping of the DNA from the nucleosomes. Supporting this, it has been shown that RNA Pol II does not remove nucleosomes by itself but waits for the local unwrapping of the DNA from nucleosomes before transcription can start (Hodges *et al.*, 2009). Nucleosomes containing histone variant H2A.Z display a much tighter wrapping of their DNA (Thambirajah *et al.*, 2006) and using reconstituted nucleosomes Fan *et al.* (2002) observed that, in nucleosomes containing H2A.Z, the intranucleosomal interactions are stronger.

H2AZ plays an essential role in higher eukaryotes development as its absence in early mouse embryos leads to lethality (Faast *et al.*, 2001) and genome instability in cell culture (Rangasamy *et al.*, 2004). H2A.Z-depleted cells have shown different nuclear and chromosomal abnormalities, including the existence of multiple nuclei, micronuclei, lagging chromosomes and anaphase bridges in mouse and *Xenopus* (Rangasamy *et al.*, 2004). H2A.Z controls the proper localization of HP1α (essential for constitutive heterochromatin) and its chromatin interactions at the pericentromeric heterochromatic regions of the chromosomes. This in turn leads to an impairment of chromosome segregation in H2A.Z-depleted cells (Rangasamy *et al.*, 2004). In budding yeast H2A.Z seems to have a role in the regulation of gene expression (Santisteban *et al.*, 2000). In addition, H2A.Z in *Drosophila melanogaster* also carries out the role of histone variant H2AX (Madigan *et al.*, 2002).

In Arabidopsis, histone variant H2AZ has been associated with plant thermosensory perception (Kumar and Wigge, 2010). The Arabidopsis genome contains two H2A.Z histone variants (HTA8 and HTA11). The double mutant *hta9 hta11* shows acceleration of the transition to reproductive development (early flowering) and architectural responses, mainly that the seedlings showed elongation of the hypocotyls and petioles when growing at 17 °C (Kumar and Wigge, 2010). Wild-type plants at an increased ambient temperature can have identical dramatic effects in development, with also an increase in the elongation of plant axes and early flowering. Furthermore, the double *hta9 hta11* mutant displays an increased expression of heat-shock protein HSP70, which is strongly up-regulated at high temperatures. High temperature exposure can activate the synthesis of heat-shock proteins (HSPs) whose role is to defend proteins against denaturation and maintain the cell's metabolism. Kumar and Wigge (2010) used the thermal response of *HSP70* gene expression and forward genetic screen analysis to find mutants displaying thermosensitivity at cool temperatures. The authors identified and isolated different multiple mutant alleles for the gene *ARP6*. The *arp6* mutants phenocopy the double *h2a.z* mutant, showing the same phenotype as wild-type plants growing at higher temperatures. The *ARP6* gene encodes for a subunit of the SWR1 chromatin remodelling complex that is necessary for the deposition of H2A.Z in the nucleosomes (Li *et al.*, 2005b). Nucleosomes containing H2A.Z occupying promoter regions repress transcription of those genes in Arabidopsis and yeast. Higher temperatures facilitate the eviction of H2A.Z from chromosomes and therefore, gene transcription (Kumar and Wigge, 2010). These findings are extremely exciting when climate change is a certainty and we are unable to predict how plants are going to respond to further increases in temperature. Further knowledge in understanding the basic mechanisms of temperature perception and plant adaptability will allow us to breed crops able to endure climate change.

H2AX is a highly conserved H2A variant, which constitutes 2–25% of the whole H2A histone pool present on the chromatin (Rogakou *et al.*, 1998; Redon *et al.*, 2002). Histone variant H2AX possesses an unique Ct tail with a serine residue at position 139, which is very rapidly phosphorylated around a DNA double-strand break (DSB) (Redon *et al.*, 2002). Double-strand breaks can be generated by environmental sources such as ionizing radiation (IR) or genotoxic chemicals or by endogenous factors such as stalled replication forks and metabolism intermediates. Furthermore, DSBs are naturally programmed during meiosis and immunoglobulin V(D)J gene rearrangement. Incorrect repair of DSBs creates genome instability and thus can lead to apoptosis, senescence and even to cancer. It has been suggested that the phosphorylation of H2AX (to γH2AX) could produce chromosome structural changes that facilitate repair of DSBs in DNA (Fernandez-Capetillo *et al.*, 2004). Nevertheless, its direct role is still unknown. There are different pathways to repair DSBs: error free (Homologous Recombination –HR-) and some error prone (Non-Homologous End Joining -NHEJ- and Single-Strand Annealing –SSA-) (Bonner *et al.*, 2008; Shrivastav *et al.*, 2008). Homologous Recombination is the only error-free DNA repair pathway and always requires a homologous copy of the affected DNA sequence. In somatic cells, HR uses the sister chromatid and in meiocytes, the homologous chromosome, allowing the formation of crossovers (visualized as chiasmata at metaphase I). Surprisingly, the different DNA repair pathways seem to compete for DSBs and their balance differs widely among species, between different cell types of a single species and during different cell cycle phases of a single cell type. Remarkably, NHEJ is predominantly used to repair DSBs in mammals and plants (Shrivastav *et al.*, 2008). In budding yeast HR is the main DSB repair pathway indicating that the structural integrity of genomic DNA is far more important than the preservation of the local DNA sequence in multicellular organisms. The histone variant H2AX is phosphorylated very rapidly, in minutes, around the DSBs and seems to facilitate DNA repair by the retention of repair complexes and signalling factors at the DSBs (Bonner *et al.*, 2008). The phosphorylation of H2AX to produce γH2AX seems to be dependent on PI3K-like protein serine/threonine kinases ATM (ataxia telangiectasia mutated), ATR (ataxia Rad3-related) and DNAPKcs (DNA-dependent protein kinase catalytic subunit) (Durocher and Jackson, 2001; Falck *et al.*, 2005). There is a controversy about which PI3K-related kinase phosphorylates H2AX in mammalian cells. Similar results have been obtained in Arabidopsis with the difference that DNAPKcs have not been found yet in plants and that ATM and ATR-dependent H2AX phosphorylation is less than one-third of the rate of γH2AX foci in mammalian cells. This suggests a lower rate of DSB formation in plants and, perhaps, a higher resistance to radiation in plants compare to yeast and mammals (Friesner *et al.*, 2005). Interestingly, the Arabidopsis genome possesses two copies for H2AX that are 98% identical at the amino acid level with the evolutionarily conserved SQ∗F motif at the Ct (139-142 residues). This duplication of histone variant H2AX is conserved among other plant species including rice (*Oryza sativa*), grapevine (*Vitis vinifera*), *Brachypodium distachyon* and black cottonwood poplar (*Populus trichocarpa*) suggesting that H2AX duplication might have an evolutionary significance (Figure 5.3). Plants as sedentary organisms are constantly exposed to environmental radiation (IR) and chemicals that induce DSBs in their genome so a robust DNA repair system must be necessary at different levels.

The histone H3 centromeric variant known as CENH3, originally called CENP-A in humans (Earnshaw and Rothfield, 1985) and HTR12 in Arabidopsis (Talbert *et al.*, 2002) replaces the canonical histone H3 in the nucleosome core at the centromere regions of chromosomes at the inner plate of the kinetochore. The kinetochore constitutes the anchoring point for the spindle microtubules allowing proper segregation of the chromosomes during mitosis and meiosis. All eukaryotic centromeres contain CENH3 even though centromeric DNA sequences are exceptionally diverse (Black and Bassett, 2008). CENH3 is required for the recruitment and assembly of the kinetochore protein complex, mitotic progression and chromosome segregation (Howman *et al.*, 2000; Oegema *et al.*, 2001; Blower *et al.*, 2006). CENH3 serves as an epigenetic mark that propagates centromere identity (Allshire and Karpen, 2008). Remarkably, CENH3 is not well conserved amongst species in contrast

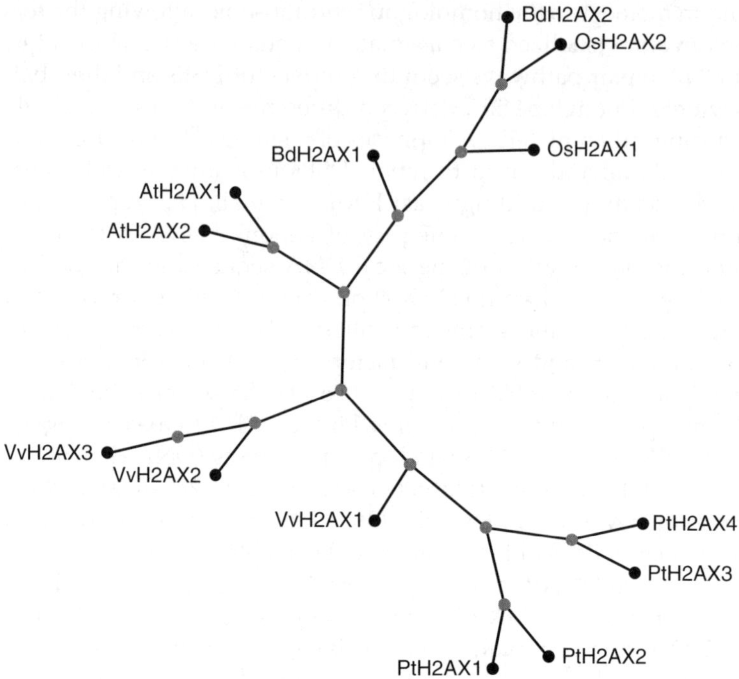

Figure 5.3 H2AX is duplicated in plants. At: *Arabidopsis thaliana*, Os: *Oryza sativa*, Vv: *Vitis vinifera*, Pt: *Populus trichocarpa* Bd: *Brachypodium distachyon*.

to 'conventional' histones (Malik and Henikoff, 2003). CENH3 is evolutionarily divergent and complementation by the CENH3 protein from different species is not possible (Vermaak *et al.*, 2002; Wieland *et al.*, 2004; Baker and Rogers, 2006; Ravi *et al.*, 2010). The CENH3 Nt domain is more variable than the Ct that contains the histone-fold domain. The Nt domain shares almost no similarity amongst different species, even between plant species, for example comparing *Arabidopsis thaliana* with *Zea mays* (Malik and Henikoff, 2003; Ravi *et al.*, 2010). This Nt tail is not needed for the kinetochore localization of CENH3 but is necessary for plant fertility (Ravi and Chan, 2010). These authors replaced the Nt tail of CENH3 with the tail of a conventional H3 and tagged it with a GFP protein in Arabidopsis. With this mutated CENH3 (GFP-tailswap, Copenhaver and Preuss, 2010) plants presented apparently normal viable phenotypes but were sterile suggesting that the Nt tail might have a specific function in gamete fertility (for example in meiosis). Indeed, the authors observed that these GFP-tailswap plants were sterile due to chromosome mis-segregation at metaphase I. This mis-segregation was caused by the lack of CENH3 variant at the meiotic kinetochores and the compromised recruitment of other essential kinetocore proteins such as Mis12. Thus, centromeres must have a meiosis-specific structure that is dependent on the Nt

CENH3 tail domain. This Nt tail is the target for multiple post-translational modifications and might be involved in the regulation of centromere structure. The CENH3 GFP-tailswap plants retain normal somatic function for proper mitotic segregation (viable plants) but are unable to form the proper meiotic centromere structure. It is worth mentioning that the result of the chromosome mis-segregation in this mutation produces uniparental chromosome elimination in the zygote. Uniparental chromosome elimination has been found in different interspecific hybrids in plants (Devaux and Pickering, 2005). Recently, it has been demonstrated that, in barley hybrid (*Hordeum vulgare x Hordeum bulbosum*) embryos, the elimination of *H. bulbosum* chromosomes is a result of the disappearance of the CENH3 protein from the centromeres of these chromosomes (Sanei *et al.*, 2011). In these hybrids not all CENH3 variants present are incorporated in all the parental species triggering the mitosis-dependent process of chromosome elimination of *H. bulbosum* (Sanei *et al.*, 2011). Ravi and Chan (2010) showed that modifying CENH3 in one parent could induce targeted elimination of the chromosomes of that parent. The distinction between maternal or paternal CENH3 is lost after the first few divisions and the remaining divisions are able to divide normally throughout the plant development, resulting in a haploid plant (Ravi and Chan, 2010). These haploids could be used to obtain doubled-haploid progeny through somatic chromosome doubling or non-reductional meiosis (Copenhaver and Preuss, 2010). Incorporating doubled haploids into plant breeding would accelerate the development of new inbred lines by saving time in achieving homozygosity (Copenhaver and Preuss, 2010; Dunwell, 2010).

Histone variant 3.3 differs in only four to five amino residues from the canonical histone H3, and both are expressed in multicellular eukaryotes including plants (Ahmad and Henikoff, 2002). H3.3 localization is enriched in chromatin regions that are transcriptionally active (Ahmad and Henikoff, 2002; Schwartz and Ahmad, 2005). The deposition of histone H3.3 is mainly by replacing the canonical histone H3 from the nucleosomes (Tagami *et al.*, 2004) This replacement of histone H3 by H3.3 is carried out through different chaperones including histone regulator A (HIRA; Phelps-Durr *et al.*, 2005), ATRX (Wong *et al.*, 2010), death-associated protein DAXX (Drané *et al.*, 2010) and DEK (Sawatsubashi *et al.*, 2010). H3.3 plays an important role in chromosome condensation in spermatocytes (Sakai *et al.*, 2009), in zygotic epigenetic reprogramming in plants (Ingouff *et al.*, 2010) and epigenetic memory (Ng and Gurdon, 2008). Remarkably, the H3.3 mutation can be rescued by over-expression of canonical histone H3 in *Drosophila melanogaster* (Sakai *et al.*, 2009). The Arabidopsis genome encodes six histones H3.3 or H3.3-like (HTR4, HTR5, HTR6, HTR8, HTR14 and the male gamete-specific HTR10/AtMGH3) (Okada *et al.*, 2005). Histone H3.3 (HTR4) differs from the canonical histone H3 (HTR1) in four amino acids in Arabidopsis (residues 31, 41, 87 and 90). Recently, Shi *et al.* (2011) observed that residue 87 and to some extent 90 guide H3.3 nucleosome assembly, whereas residues 31 and 41 are critical for H3.3 nucleosome disassembly in nucleolar rDNA regions of Arabidopsis nuclei. The male gamete-specific histone H3 is expressed in pollen in *Lilium longiflorum* (Okada *et al.*, 2005), and more specifically in the generative and sperm cells in Arabidopsis (Okada *et al.*, 2005; Brownfield *et al.*, 2009). An AtMGH3 T-DNA insertion mutant was analysed and showed that other H3 histones could compensate for the loss of AtMGH3 as the insertion line had normal development and fertility (Okada *et al.*, 2005).

5.2.2 Histone modifications

Histone post-translational modifications were discovered in the 1960s and even then they were predicted to have an effect on transcriptional regulation (Allfrey *et al.*, 1964). In the early stages, it was thought that modifications such as acetylation or phosphorylation could loosen the association of the DNA-histone by lowering the charge of the histone (Fenley *et al.*, 2001). Nevertheless, different findings led the authors to propose the histone code hypothesis, which postulates that post-translational histone modifications allow the specific recognition of other proteins that could modify chromatin structure and promote or inhibit transcription (Turner *et al.*, 1992; Strahl and Allis, 2000). Over a decade of experimental observations have proven this suggestion right by identifying specific proteins that recognize histone modifications leading to functional outcomes for the cell (Margueron *et al.*, 2005; Nightingale *et al.*, 2006).

More than 60 different amino acid residues of the four core histones have already been shown to be modified. Most of these residues are found at the Nt tails of histones H3 and H4. The Nt tails are exposed, protruding through the DNA wrapped around the nucleosome, so they are highly accessible. The modifications include acetylation of lysines, methylation of lysines and arginines, phosphorylation of serines and threonines, and ubiquitinylation and sumoylation of lysines (Kouzarides, 2007). The different histone modifications are functionally associated with a variety of processes and the histone code is highly complex. The nature of the modification is crucial for the response but also for its position in the protein. For example, histone H3 lysine 9 methylation is linked to silenced promoters whereas H3 lysine 4 methylation can be found in active genes (Boggs *et al.*, 2002). Histone modifications can also have longer term effects on chromatin and nuclear processes as with constitutive heterochromatin (Mateescu *et al.*, 2004; Fischle *et al.*, 2005). However, in some instances the same modification can correlate with different chromatin states; for example H3 serine 10 phosphorylation can be found in active chromatin and in condensed chromosomes (Cheung *et al.*, 2000). Another important point is that the order of histone modification also contributes to the response. Histone H4 arginine 3 methylation precedes and promotes H4 lysine 8 acetylation to stimulate a transcriptionally active state (Wang *et al.*, 2001). Thus, a single histone modification should not be considered in isolation but within a wider picture. The increasing variety of histone modifications, their functionality and interactive properties could be a chapter in itself, so I will not go into more detail here. So far, in plants, the genomic distribution patterns of histone H3 methylation have been determined by the combination of microarray analysis and chromatin immunoprecipitation (ChIP-chip) (Zhang, 2008). Integrative epigenomic mapping for 12 different histone modifications has defined four main chromatin states (CS) in the *Arabidopsis thaliana* genome (Roudier *et al.*, 2011). CS1 (54% of genes) corresponds to transcriptionally active genes (enriched in H3K4 and H3K36 trimethylated forms), CS2 (23% of genes) is mainly associated with genes under PRC2-mediated repression (enriched in trimethylated H3K27), CS3 (83% of transposable elements – TEs) corresponds to classical heterochromatin and is mostly located over silent TEs (enriched in H3K9, dimethylated and H4K20 methylated forms), CS4 (10% of genes) does not contain any prevalent histone modification mark and is associated with weakly expressed genes and intergenic regions (Roudier *et al.*, 2011).

Histone modifications can be considered as a signalling pathway that allows organisms to respond to environmental stimuli (Schreiber and Bernstein, 2002). There is an increase in histone acetylation and methylation at transcriptionally active promoters and this can change rapidly in response to external stimuli (Turner *et al.*, 1992; Hazzalin and Mahadevan, 2005; Metivier *et al.*, 2003). We have already seen a similar effect in the thermosensory response of histone variant H2AZ in Arabidopsis (Kumar and Wigge, 2010) and how the phosphorylation of H2AX variant signals DSBs on chromosomes to attract DNA repair complexes (Fernandez-Capetillo *et al.*, 2004). Modifications seem to turn over quickly to allow such rapid responses (Hazzalin and Mahadevan, 2005). The information accumulated by histone modifications is considered epigenetic, because it is not encoded in the DNA but is inherited by the daughter cells (Turner, 2007). An epigenetic code, then, has been based on the different roles of histone modifications and how the environment can regulate them through specific enzyme-based modifications (Turner, 2007, 2009). Thus, chromatin components (histone modifications) can be sensitive to environmental and metabolic agents that alter gene expression, which can be passed on to subsequent cells in an individual. These gene expression changes occurring in germ cells may be passed on to the offspring and, thus, constitute a potential route through which the environment might directly influence evolution (Turner, 2009).

5.2.3 Nucleosome dynamics

The first impression is that the nucleosome is a static structure but although the histone-DNA interaction is very stable, the nucleosome is highly dynamic. Interestingly, Moehs *et al.* (1992) observed that the histone octamer of bread wheat (*T. aestivum*) was much more stable than in vertebrate animals. This enhanced stability of the octamer seems to be achieved through histones H2A and H2B (Moehs *et al.* 1992). The DNA is in equilibrium between wrapping and unwrapping the histone octamer (Li *et al.*, 2005c). Studies using fluorescence resonance energy transfer (FRET) revealed that the nucleosomal DNA remains wrapped for about 250 ms and then is unwrapped for 10–50 ms and again rewrapped. The nucleosomal DNA is exposed by gradually unwrapping from the histone core.

As discussed previously, different post-translational modifications, together with the incorporation of histone variants, make the nucleosome a very dynamic structure with the potential to alter chromatin structurally and functionally (Malik and Henikoff, 2003; Kouzarides, 2007). Furthermore, nucleosomes can be mobilized by ATP-dependent chromatin remodelling enzymes. These remodelling factors possess a variety of distinct reactions dependent on ATP and chromatin that can slide nucleosomes, disrupt histone-DNA interactions, generate negative DNA torsion, introduce histone variants and affect chromatin structure in general (Whitehouse *et al.*, 1999; Havas *et al.*, 2000; Bruno *et al.*, 2003; Kassabov *et al.*, 2003; Mizuguchi *et al.*, 2004). Eukaryotes have developed such a great variety of resources to modulate chromatin structure and function, which allow DNA accessibility to be altered locally. This chromatin remodelling allows the high level of control required to organize the different biological processes of the nucleus such as DNA transcription and DNA repair in higher eukaryotes. Recently, in plants, it has been shown that the progression of pluripotent stem cells to differentiated cell lineages requires chromatin remodelling enzymes to shift the cell differentiation programmes as well as to maintain the undifferentiated nature of the stem cells (for review see Kornet and Scheres, 2008). It is interesting to note that chromatin remodelling factor mutations are normally lethal in metazoans but are viable in plants. Plant chromatin remodelling factor mutants have revealed their control in a wide variety of developmental processes (Wagner, 2003).

5.3 Linker histone and the higher order chromatin-order fibre

Higher eukaryotes possess a fifth histone called the linker histone. The most common histone linker is histone H1, which is a highly basic protein, very rich in lysine and with a molecular weight just a little higher than those of the core histones. H1 Histone H1 is loosely bound to the linker DNA. Linker histones can be dissociated by moderate-strength ionic solutions (slighter higher than 0.35 M NaCl). The structure of linker histones contains highly charged Nt and Ct tails, and a central domain which crystallographic analysis revealed as a winged helix motif (Ramakrishnan *et al.*, 1993). The Nt domain is about 45 amino acids long and is rich in basic amino acids; the central globular domain (~75 amino acids) is highly conserved among H1 histones and the Ct domain (~100 amino acids) is highly enriched in lysine, serine and proline. The linker histones stabilize the histone-DNA interactions within the nucleosome and the linker histone tails interact with the linker DNA between the nucleosomes. Linker DNA length varies from 20bp up to 90bp (Wong *et al.*, 2007), usually in steps of 10bp that correspond to the number of base pairs involved in one rotational turn of the DNA double helix (Widom, 1992). The average length of the DNA linker shows species specificity: 20 bp in yeast (Lee *et al.*, 2007), 30bp in Drosophila (Mavrich *et al.*, 2008b), 40 bp in mammals (Schones *et al.*, 2008) and 30bp in the model plant Arabidopsis (Chodavarapu *et al.*, 2010). The linker histone binds around 20 bp of the linker DNA between the nucleosomes. Originally, the nucleosome associated with histone linker H1 was termed a chromatosome (Simpson, 1978). Thus, a nucleosome could be defined as a nucleosomal core plus about 20 bp of DNA and one H1 histone and denotes that the histone H1 must be associated with the DNA at the entry and exit site of the DNA associated with the histone octamer of the nucleosome. *In vitro* analysis, crystallographic and computational docking data support the idea that the globular domain of linker histones is in contact with the nucleosomal DNA (Clark *et al.*, 1993; Zhou *et al.*, 1998; Fan and Roberts, 2006). Brown and collaborators (2006) analysed the interaction sites between the globular domain of linker histones and the DNA by combining Fluorescence Recovery After Photobleaching (FRAP) assays *in vivo* and molecular modelling. The authors proposed a model in which H1 binds near the nucleosomal dyad by the winged helix motif of the globular domain, and has a second DNA interacting site on the linker DNA about 15 bp away from the histone octamer (Brown *et al.*, 2006).

Histone H1 interaction with chromatin limits the mobility of nucleosomes and DNA accessibility. Nevertheless, histone H1 is not a static structural component acting as a repressor but a highly dynamic component. H1 linker histone interaction with a nucleosome is transient with a residence time of three to four minutes (Misteli *et al.*, 2000; Phair *et al.*, 2004). Furthermore, the kinetics of histone linker binding is influenced by the competition with other chromatin binding proteins such as the high mobility group (HMG) (Catez *et al.*, 2004). Histone linker H1 can be targeted by several post-translational modifications that include phosphorylation, acetylation, methylation and ubiquitination. H1 phosphorylation has been implicated in chromatin decondensation during S phase, and condensation during M phase of mitosis (Bradbury *et al.*, 1974; Baatout and Derradji, 2006).

The linker histone H1 family includes a considerable diversity of isoforms, which can be differentially expressed in different tissues and developmental stages within the same species. Furthermore, it seems that histone H1 sequences and even their functions are not evolutionarily conserved amongst eukaryotes (Steinbach *et al.*, 1997; Khochbin, 2001; De *et al.*, 2002; Godde and Ura, 2008; Izzo *et al.*, 2008; Happel and Doenecke, 2009). The linker histone H1 family in humans consists of eleven different H1 proteins: seven somatic isoforms (H1.0, H1.1-5, and H1x) (Albig *et al.*, 1997; Parseghian and Hamkalo, 2001; Ausio and Abbott, 2004; Happel *et al.*, 2005) and four germ cell isoforms (H1t, H1T2, HILS1 and $H1^{oo}$) (Seyedin and Kistler, 1980; Iguchi *et al.*, 2003;

Yan *et al.*, 2003; Martianov *et al.*, 2005). The analysis of H1 functionality in mammals and plants is difficult because of the redundancy of the different isoforms. Deletion of H1hop in yeast inhibits DNA repair by homologous recombination (Downs *et al.*, 2003). In *Ascoblus immersus* (a fungus), the deletions of H1 reduced its life span and global DNA methylation (Barra *et al.*, 2000). Histone *H1.0* gene disruption in mouse resulted in a non-apparent phenotype (Sirotkin *et al.*, 1995). Interestingly, the stoichiometry of histone H1 to nucleosome did not change, suggesting compensation by other H1 histones. The same was observed for disruption of *H1t* (Drabent *et al.*, 2000; Lin *et al.*, 2000) and *H1.1* (Rabini *et al.*, 2000). The sequential mutations of *H1.2, H1.3* and *H1.4* genes in mouse resulted in embryo lethality. The chromatin in these mutated embryos had around half the amount of H1 histone (to nucleosome ratio) proving that a correct stoichiometry of linker histone is essential (Fan *et al.*, 2003). Furthermore, it has been demonstrated that H1 depletion leads to chromatin reorganization such as decondensation, suggesting a role of H1 histones in stabilizing and even organizing a higher level of chromatin compaction (Fan *et al.*, 2005). The Arabidopsis genome contains ten different histone *H1* and *H1-like* genes. In Arabidopsis, more than a 90% reduction in *H1* expression by RNAi constructs resulted in different developmental phenotypes (Wierzbicki and Jerzmanoski, 2005). Mutant phenotypes included different shape and size of leaves and flowers, and different apical dominance and flowering time. Some of these phenotypes were similar to that observed in DNA hypomethylation mutants, but this down-regulation of H1 genes did not change the general methylation of genomic DNA (Wierzbicki and Jerzmanoski, 2005). Nevertheless, significant minor changes in the methylation patterns of repetitive and single-copy sequences were found in these mutants revealing a histone H1 role in DNA methylation (Wierzbicki and Jerzmanoski, 2005). Furthermore, different novel roles have been associated with different histone H1 plant isoforms. H1 histone from wheat enhances DNA hydrolysis of WEN1 and WEN2 endonucleases from the coleoptiles (Fedoreyeva *et al.*, 2008). In tobacco, H1 histone has been involved in the organization of radial microtubules (Nakayama *et al.*, 2008). Plant cells lack the centrosome and the surface of the nucleus acts as an organizing centre for microtubules. A histone H1-related protein abundant on nuclear surfaces was isolated and *in vitro* analysis showed that, if incubated together with tubulin, it could form ring-shaped complexes that could nucleate and elongate the radial microtubules (Hotta *et al.*, 2007; Nakayama *et al.*, 2008).

5.3.1 The elusive higher order chromatin fibre

A higher order chromatin fibre (ca. 30 nm) was obtained when histone H1 was added *in vitro* to a chain of nucleosomes (10 nm fibre) (Thoma *et al.*, 1979). The authors described this as a solenoidal arrangement of the nucleosomal fibre or solenoid fibre. Nevertheless, a similar structure was obtained in absence of histone H1 at increased ionic strength, although not as regular as the previous one. Much has been discussed about the structure of the 30 nm fibre and its structure but it is still controversial. Different models have been proposed with two major types of organization that have been described according to the orientation of the nucleosomes within the fibre. The one-start helix or the original solenoid model is where the sequential nucleosomes follow each other (Thoma *et al.*, 1979; Robinson *et al.*, 2006), and the two-start helix or 'zigzag' model is where H1 connects two rows of nucleosomes back and forth (Bednar *et al.*, 1998; Schalch *et al.*, 2005). In fact, it could be considered that different levels of chromatin compaction exist and therefore different higher order chromatin structures could co-exist *in vivo* in higher eukaryotes (Robinson and Rhodes, 2006). The crystal structure of a tetranucleosome array allowed the positions of the linker DNA and nucleosomes to be defined showing that the chromatin acquires a zigzag two-start helix model of the 30 nm fibre (Schalch *et al.*, 2005). It also revealed different features such as the interaction of acidic amino acid residues from histone H2A present in a nucleosome with the Nt region of histone H4 of a different nucleosome (Schalch *et al.*, 2005). Interactions like this one could be functionally relevant for the formation of the high-order chromatin, independently of histone linker H1. Thus, the precise role of linker histones in the formation of higher order chromatin fibre remains to be determined. Recently, data suggested that different higher order chromatin models could be present simultaneously in eukaryotes forming a heteromorphic chromatin fibre with a uniform 30 nm diameter. The linker histones, nucleosome interactions and divalent cations like Mg^{2+} could be involved in the formation of these higher order structures along the genomic DNA of higher eukaryotes (Grigoryeva *et al.*, 2009). Nevertheless, *in vivo* analysis of *Xenopus* cells with either histone linker H1 depletion or overexpression has shown unpredicted effects on chromosome condensation. Histone H1 depletion caused elongation of the mitotic chromosome (about twofold) (Maresca *et al.*, 2005), whereas overexpression of histone linker H1 resulted in a higher compaction of interphase chromatin (Freedman and Heald, 2010).

Furthermore, whether the higher order chromatin fibre structure even exists *in vivo* is itself uncertain. The reasons underlying this uncertainty include technical limitations to detect these fibre structures in nuclei (Tremethick, 2007).

5.4 Chromatin loops and chromosome axis

The model for the structure of chromatin loops arranged radially along the chromosome axis came from different observations including examination of the morphology of lampbrush chromosomes and the structure of nuclei and chromosomes treated with high salt solutions to extract the histones. Lampbrush chromosomes are found in oocytes at the meiotic stage of diplotene at prophase I. They have been described mostly in amphibians, birds and insects but they cannot be found in mammals (Callan, 1986; Morgan, 2002; Gaginskaya *et al.*, 2009). The lampbrush chromosomes contain a series of loops covered with RNA polymerases emerging from a single continuous chromosome axis, indicating active transcription (Callan, 1986). Each individual loop always contains the same DNA sequence (Morgan, 2002; Gaginskaya *et al.*, 2009). The axes of lampbrush chromosomes contain compacted regions of inactive chromatin, which correspond to the majority of the DNA, at the base of the loops on the axis. Each chromatin loop is about 100 kb in size. Interphase chromosomes are expected to be similarly arranged in loops. Although chromatin loops have not been directly observed in interphase nuclei, fluorescence *in situ* hybridization (FISH) has shown that the loci are paired with each other along the interphase chromosome giving evidence of a similar structure. DNase I digestion of intact chromosomes from *Drosophila* resulted in chromosomal fragments of different sizes suggesting a chromatin organization in domains of around 100 kb of DNA (Benyajati and Worcel, 1976). Observation of chromosomes and nuclei from which the histones were extracted by high salt solutions made it possible to observe loop sizes of 40–90 kb of DNA attached to a residual chromosome scaffold (Cook and Brazell, 1975; Paulson and Laemmli, 1977; Jackson *et al.*, 1990). A second model of chromosome organization has been proposed in which the higher order chromatin fibre folds without specific attachments to an undisrupted chromosome axis (Belmont *et al.*, 1987, 1989). Observations on mitotic chromosomes suggest a hierarchy of higher order chromatin folding patterns where the loops are not orientated consistently in radial loops about any given axis (Belmont and Bruce, 1994). Analysis *in vivo* of repeats of the *lac* operator introduced into chromosomes and GFP-tagged *lac* repressor protein has allowed the observation of a large-scale chromatin folding above the 30 nm fibre (about up to 100 nm diameter fibre) (Robinett *et al.*, 1996). All these observations are consistent with the anchoring of loops into a chromosome axis but without a strictly radial symmetry. Although the radial arrangement of 30 nm chromatin loops is not consistent with microscopy observations, in reality the chromosome could be formed by irregular chromatin loops toward the chromosome centre formed by different networks of chromosome axis proteins (Maeshima and Eltsov, 2008). Current thinking is that the chromatin loops are constantly changing and reorganizing their structure according to the different DNA processes that are required. Chromatin loops have been associated with functional 'transcription factories' through the spatial association of transcriptional protein complexes (polymerase plus transcription factors) and the different chromatin loops (Cook, 2010; Dean, 2011). The control of gene transcription involves different regulatory elements that can interact over large genomic distances; the chromatin loop structure can provide a physical organization of these elements. Recently, it has been observed that the immunoglobulin heavy-chain (*IgH*) gene locus contains two levels of chromosomal compaction that is required for its repositioning within the nucleus and locus contraction in preparation for recombination

(Guo *et al.*, 2011). Two forms of loops generate the chromatin conformation in the *IgH* locus. Furthermore, there is evidence for an epigenetic memory at higher order folding of the chromatin fibre to maintain active/repressed chromatin activity (Deng and Blobel, 2010). Chromosomal rearrangements or illegitimate recombination are very important for genomic evolution and have a tendency to occur at specific genomic sites that are normally separated by 50–100 kb regions that are very seldom involved in rearrangements. There is an obvious correlation between these distances and the chromatin loop organization and spacing. Furthermore, it has been shown that DNA sequences attached to the chromosome axis contain DNA recombination hot spots and this could be one plausible source for species evolution (see review by Kantidze and Razin, 2009).

Higher order chromatin degradation (HOCD) mediated by DNA digestion at the anchored site of chromatin loops to the chromosome axis is a hallmark of programmed cell death. The result is an excision of the chromatin loops from chromosomes leading to somatic mutations and cell death. Hydrogen peroxide (H_2O_2) rapidly induces HCOD and can produce neurodegenaration (Konat, 2002).

The DNA sequences attached to the chromosome axis have been named as scaffold-associated regions (SARs) or matrix-attachment regions (MARs) and they form the bases of chromatin loops in eukaryotic chromatin (Laemmli *et al.*, 1992). Scaffold-associated regions seem to play an important role in facilitating chromatin dynamics for DNA processes such as gene transcription and replication (Hart and Laemmli, 1998). Scaffold-associated regions are highly AT-rich DNA regions and a continuous rearrangement of these SARs can generate AT-queue chromosome regions. Chromosomal banding patterns arise from the irregular folding of the AT-queue chromosome regions. Similarly, SARs have been shown to have an important structural and functional role in organizing plant chromatin (see review by Breyne *et al.*, 1994). Scaffold-associated regions have also been shown to possess a domain-defining and regulatory role in maize and Sorghum chromosomes (Tikhonov *et al.*, 2000). Petersen *et al.* (2002) suggest that SARs also increase transformation frequencies (up to twofold) and reduce transgene expression in barley. Intragenic SARs have been suggested to be important for spatiotemporal regulation of gene expression in Arabidopsis (Tetko *et al.*, 2006).

Originally, two main components of the chromosome scaffold Sc1 (Topoisomerase II) and Sc2 (condensin complex: SMC2-SMC4) were described (Lewis and Laemmli, 1982). Topoisomerase II is able to pass a double DNA strand through another by generating a double-strand break and rejoining it (see review by Nitiss, 2009a and b). Topoisomerase II is a mitotic chromosome axis component (Earnshaw and Heck, 1985; Earnshaw *et al.*, 1985). Mutations in topoisomerase II are lethal for the cell as the protein is necessary for proper chromosome segregation (DiNardo *et al.*, 1984). Topoisomerase II expression has been correlated with cell proliferation in Arabidopsis (Xie and Lam, 1994). The stability and maintenance of chromosomes (SMC) family of proteins are highly conserved in structure amongst eukaryotes with two coiled-coil protein-protein interaction domains. The proteins are essential for the viability of the cell with the chromosomes not segregating properly when these proteins are absent (Samejima *et al.*, 1993; Strunnikov *et al.*, 1995), similar to topoisomerase II mutants. In plants, SMC protein complexes are essential for sister chromatid cohesion, chromosome condensation, DNA repair and recombination (see review by Schubert, 2009).

Other scaffold components have been identified, for example the high mobility group (HMG) proteins. HMGs are the second most abundant chromosomal protein after histones and have an architectural role in chromatin (Catez and Hock, 2010). High mobility groups are estimated to bind to up to 10% of the nucleosomes (Johns, 1982). HMG proteins were named according to their high mobility when run in a polyacrylamide gel (less than 30 kDa) (Goodwin *et al.*, 1973). High mobility group proteins can be divided into three distinct families according to their DNA binding motifs: HMGA proteins (formerly HMGI/Y), which contain at least one AT-hook DNA binding motif, HMGB proteins (formerly HMG1/2), which contain HMG-box domains, and HMGN proteins (formerly HMG14/17), which contain a nucleosome binding domain (Bustin, 2001). High mobility groups participate in a variety of DNA-dependent process such as chromatin remodelling, regulation of gene expression, DNA repair and recombination and apoptosis; they even have a role in epigenetic changes (Catez and Hock, 2010). HMGA proteins have a variable number of AT-hook motifs, the motifs attached to AT-rich regions of the DNA like those found in SARs. These AT-hook motifs might play a crucial role in modelling chromatin structure (Sgarra *et al.*, 2004). Although one AT-hook motif is able to provide a binding site to DNA, more AT-hook motifs in the protein would significantly increase the affinity of the protein for the DNA (Catez and Hock, 2010). Each AT-hook domain consists of a highly conserved amino acid sequence based on positively charged amino acids: proline, arginine, glycine, arginine and proline (PRPGRP) (Sgarra *et al.*, 2004). Different HMG proteins belonging to the HMGA and HMGB families have been characterized in plants (Spiker, 1988; Grasser, 1995) whereas HMGNs are not found in plants. Arabidopsis HMGB14, which has the closest sequence homology to HMGN does not interact with nucleosomes (Grasser *et al.*, 2006). Plant HMGA proteins have an Nt domain with a similar sequence to the globular domain of histones H1 and can contribute to the regulation of target genes (Grasser, 2003; Klosterman and Hadwiger, 2002).

Space does not permit the enumeration of the different chromosome axis-associated proteins that have been identified to date in different organisms, including plants. However, I do need to mention one example of these components that has a very specific role during gametogenesis. ASY1, is a meiosis-specific Arabidopsis protein with some sequence homology to yeast protein HOP1 which acts as a meiosis-specific barrier to sister-chromatid repair. ASY1 plays a crucial role in co-ordinating the activity of a key member of the homologous recombination machinery, AtDMC1 in Arabidopsis (see review by Sanchez-Moran *et al.*, 2008). This example shows that there is a huge variety of chromosome axis-associated proteins with different special nuclear roles like the case of ASY1.

5.5 Conclusions and future prospects

Chromatin structure and chromatin components are evolutionarily conserved in higher eukaryotes including plants. The different levels of DNA compaction and the structural components have a biological significance relating to chromosome dynamics, chromosome condensation and segregation and to different DNA processes such as replication, transcription, repair, recombination. There is a wealth of information available on the different chromatin components but we are still trying to understand their dynamics and how they all interact to be able to carry out all these important biological processes.

It is predicted that as a result of population increase, industrialization and climate change the global demand for food will double by 2050 (Royal Society, 2009). New crop varieties that are improved in different ways will be required to meet this challenge. Classical plant breeding methods will continue to play a major part in meeting these aims but more new technologies will be required. The knowledge of chromatin structure components and their role in DNA repair and meiotic recombination will produced a template for these new technologies.

References

Ahmad, K., and Henikoff, S. (2002) The histone variant H3.3 marks active chromatin by replication-independent nucleosome assembly. *Molecular Cell* **9**, 1191–1200.

Albert, I., Mavrich, T.N., Tomsho, L.P. *et al.* (2007) Translational and rotational settings of H2A.Z nucleosomes across the Saccharomyces cerevisiae genome. *Nature* **446**, 572–576.

Albig, W., Meergans, T. and Doenecke, D. (1997) Characterisation of the H1.5 gene completes the set of H1 subtype genes. *Gene* **184**, 141–148.

Allfrey, V.G., Faulkner, R. and Mirsky, A.E. (1964) Acetylation and methylation of histones and their possible role in the regulation of RNA synthesis. *Proceedings of the National Academy of Sciences of the USA* **51**, 786–794.

Allshire, R.C. and Karpen, G.H. (2008) Epigenetic regulation of centromeric chromatin: old dogs, new tricks? *National Review of Genetics* **9**, 923–937.

Alva, V., Ammelburg, M. and Lupas, A.N. (2007) On the origin of the histone fold. *BMC Structural Biology* **7**, 17.

Anand, A., Zarir, V., Ryu, C.M. *et al.* (2007) Identification of plant genes involved in *Agrobacterium*-mediated transformation by using virus induced gene silencing as a functional genomics tool. *Molecular Plant-Microbe Interactions* **20**, 41–52.

Arents, G., Burlingame, R.W., Wang, B.W. *et al.* (1991) The nucleosomal core histone octamer at 3.1 Å resolution: a tripartite protein assembly and a left-handed superhelix. *Proceedings of the National Academy of Sciences of the USA* **88**, 10148–10152.

Arents, G. and Moudrianakis, E.N. (1993) Topography of the histone octamer surface: repeating structural motifs utilized in the docking of nucleosomal DNA. *Proceedings of the National Academy of Sciences of the USA* **90**, 10489–10493.

Ausio, J., and Abbott, D.W. (2004) The role of histone variability in chromatin stability and folding. *New Comprehensive Biochemistry* **39**, 241–290.

Baatout, S. and Derradji, H. (2006) About histone H1 phosphorylation during mitosis. *Cell Biochemistry and Function* **24**, 93–94.

Baker, R.E., and Rogers, K. (2006) Phylogenetic analysis of fungal centromere H3 proteins. *Genetics* **174** (3), 1481–1492.

Barra, J.L., Rhounim, L., Rossignol, J.L. and Faugeron, G. (2000) Histone H1 is dispensable for methylation-associated gene silencing in *Ascobolus immersus* and essential for long life span. *Molecular and Cellular Biology* **20** (1), 61–69.

Bednar, J., Horowitz, R.A., Grigoryev, S.A. *et al.* (1998) Nucleosomes, linker DNA, and linker histone form a unique structural motif that directs the higher-order folding and compaction of chromatin. *Proceedings of the National Academy of Sciences of the USA* **95**, 14173–14178.

Belmont, A.S. (2006) Mitotic chromosome structure and condensation. *Current Opinion in Cell Biology* **18** (6), 632–638.

Belmont, A.S., Braunfeld, M.B., Sedat, J.W. and Agard, D.A. (1989) Large scale chromatin structural domains within mitotic and interphase chromosomes in vivo and in vitro. *Chromosoma* **98**, 129–143.

Belmont, A.S. and Bruce, K. (1994) Visualization of G1 chromosomes: a folded, twisted, supercoiled chromonema model of interphase chromatid structure. *Journal of Cellular Biology* **127**, 287–302.

Belmont, A.S., Dietzel, S., Nye, A.C. *et al.* (1999) Large-scale chromatin structure and function. *Current Opinion in Cell Biology* **11**, 307–311.

Belmont, A.S., Sedat, J.W. and Agard, D.A. (1987) A three dimensional approach to mitotic chromosome structure: evidence for a complex hierarchical organization. *Journal of Cellular Biology* **105**, 77–92.

Benyajati, C. and Worcel, A. (1976) Isolation, characterization and structure of the folded interphase genome of *Drosophila melanogaster*. *Cell* **9**, 393–407.

Black, B.E., and Bassett, E.A. (2008) The histone variant CENP-A and centromere specification. *Current Opinion in Cell Biology* **20**, 91–100.

Blower, M.D., Daigle, T., Kaufman, T. and Karpen, G.H. (2006) Drosophila CENP-A mutations cause a BubR1-dependent early mitotic delay without normal localization of kinetochore components. *PLOS Genetics* 2, e110.

Boggs, B.A., Cheung, P., Heard, E. *et al.* (2002) Differentially methylated forms of histone H3 show unique association patterns with inactive human X chromosomes. *Nature Genetics* **30**, 73–76.

Bonner, W.M., Redon, C.E., Dickey, J.S. *et al.* (2008) γH2AX and cancer. *Nature Reviews* **8**, 957–967.

Bradbury, E.M., Inglis, R.J. and Matthews, H.R. (1974) Control of cell division by Very Lysine Rich Histone (F1) phosphorylation. *Nature* **247**, 257–261.

Breyne, P., Van Montagu, M., and Gheysen, G. (1994) The role of scaffold attachment regions in the structural and functional organization of plant chromatin. *Transgenic Research* **3** (3), 195–202.

Brown, D.T., Izard, T. and Mistelli, T. (2006) Mapping the interaction surface of linker histone H1(0) with the nucleosome of native chromatin in vivo. *Nature Structural and Molecular Biology* **13**, 250–255.

Brownfield, L., Hafidh, S., Borg, M. *et al.* (2009) A plant germline-specific integrator of sperm specification and cell cycle progression. *PLoS Genetics* **5** (3), e1000430.

Bruno, M., Flaus, A., Stockdale, C. *et al.* (2003) Histone H2A/H2B dimer exchange by ATP-dependent chromatin remodeling activities. *Molecular Cell* **12** (6), 1599–1606.

Bryant, J.A. and Dunham, V.L. (1988) The biochemistry of DNA replication. In J.A. Bryant and V.L. Dunham (eds), *DNA Replication in Plants*, CRC Press, Boca Raton, FL, pp. 17–54.

Bustin, M. (2001) Revised nomenclature for high mobility group (HMG) chromosomal proteins. *Trends in Biochemical Sciences* **26**, 152–153.

Callan, H.G. (1986) *Lampbrush Chromosomes*, Springer-Verlag, Berlin.

Catez, F., and Hock, R. (2010) Binding and interplay of HMG proteins on chromatin: Lessons from live cell imaging. *Biochimica et Biophysica Acta – Gene Regulatory Mechanisms* **1799** (1–2), 15–27.

Catez, F., Yang, H., Tracey, K.J. *et al.* (2004) Network of dynamic interactions between histone H1 and high-mobility-group proteins in chromatin. *Molecular and Cellular Biology* **24**, 4321–4328.

Cheung, P., Allis, C.D. and Sassone-Corsi, P. (2000) Signaling to chromatin through histone modifications. *Cell* **103**, 263–271.

Chodavarapu, R.K., Feng, S, Bernatavichute, Y.V. *et al.* (2010) Relationship between nucleosome positioning and DNA methylation. *Nature* **466**, 388–392.

Clark, K.L., Halay, E.D., Lai, E. *et al.* (1993) Co-crystal structure of the HNF-3/fork head DNA recognition motif resembles histone H5. *Nature* **264**, 412–420.

Copenhaver, G.P. and Preuss, D. (2010) Haploidy with histones. *Nature Biotechnology* **28** (5), 423–424.

Cook, P.R. (2010) A model for all genomes: the role of transcription factories. *Journal of Molecular Biology* **395** (1), 1–10.

Cook, P.R. and Brazell, I.A. (1975) Supercoils in human DNA. *Journal of Cellular Science* **19**, 261–279.

Davey, C.A., Sargent, D.F., Luger, K. *et al.* (2002) Solvent mediated interactions in the structure of the nucleosome core particle at 1.9 a resolution. *Journal of Molecular Biology* **319** (5), 1097–1113.

De, S., Brown, D.T., Lu, Z.H. *et al.* (2002) Histone H1 variants differentially inhibit DNA replication through an affinity for chromatin mediated by their carboxyl-terminal domains. *Gene* **292** (1–2), 173–181.

Dean, A. (2011) In the loop: long range chromatin interactions and gene regulation. *Briefings in Functional Genomics* **10** (1), 3–10.

De Lange, R.J., Farnbrough, D.M., Smith, E.L. and Bonner, J. (1969a) Calf and pea histone IV: the complete amino acid sequence of calf thymus histone IV; presence of N-acetyllysine. *Journal of Biological Chemistry* **244**, 319–334.

De Lange, R.J., Farnbrough, D.M., Smith, E.L. and Bonner, J. (1969b) Calf and pea histone W: complete amino acid sequence of pea seedling histone IV; comparison with the homologous calf thymus histone. *Journal of Biological Chemistry* **244**, 5669–5679.

Deng, W. and Blobel, G.A. (2010) Do chromatin loops provide epigenetic gene expression states? *Current Opinion in Genetic Development* **20** (5), 548–554.

Devaux, P and Pickering, R. (2005) Haploids in the improvement of Poaceae. In: C.E. Palmer, W.A. Keller and K.J. Kasha (eds) *Haploids in Crop Improvement II*, Springer-Verlag, Berlin, pp. 215–242.

DiNardo, S., Voelkel, K. and Sternglanz, R. (1984) DNA topoisomerase II is required for segregation of daughter molecules at the termination of DNA replication. *Proceedings of the National Academy of Sciences of the USA* **81**, 2616–2620.

Dollard, C., Ricupero-Hovasse, S.L., Natsoulis, G. *et al.* (1994) SPT10 and SPT21 are required for transcription of particular histone genes in Saccharomyces cerevisiae. *Molecular and Cellular Biology* **14** (8), 5223–5228.

Downs, J.A., Kosmidou, E., Morgan, A. and Jackson, S.P. (2003) Suppression of homologous recombination by the Saccharomyces cerevisiae linker histone. *Molecular Cell* **11**, 1685–1692.

Drabent, B., Saftig, P., Bode, C. *et al.* (2000) Spermatogenesis proceeds normally in mice without linker histone H1t. *Histochemistry and Cell Biology* **113**, 433–442.

Drané, P., Ouararhni, K., Depaux, A. *et al.* (2010) The death-associated protein DAXX is a novel histone chaperone involved in the replication-independent deposition of H3.3. *Genes and Development* **24**, 1253–1265.

Dunwell, J.M. (2010) Haploids in flowering plants: origins and exploitation. *Plant Biotechnology Journal* **8**, 377–424.

Durocher, D. and Jackson, S.P. (2001) DNA-PK, ATM and ATR as sensors of DNA damage: variations on a theme? *Current Opinion in Cell Biology* **13** (2), 225–231.

Eamshaw, W.C. and Heck, M.M.S. (1985) Localization of topoisomerase II in mitotic chromosomes. *Journal of Cell Biology* **100**, 1716–1725.

Earnshaw, W.C. and Rothfield, N. (1985) Identification of a family of human centromere proteins using autoimmune sera from patients with *Scleroderma*. *Chromosoma* **91**: 313–321.

Earnshaw, W.C., Halligan, B., Cooke, C.A. *et al.* (1985) Topoisomerase II is a structural component of mitotic chromosome scaffolds. *Journal of Cell Biology* **100**, 1706–1715.

Faast, R., Thonglairoam, V., Schulz, T.C. *et al.* (2001) Histone variant H2A.Z is required for early mammalian development. *Current Biology* **11**, 1183–1187.

Falck, J., Coates, J. and Jackson, S.P. (2005) Conserved modes of recruitment of ATM, ATR and DNA-PKcs to sites of DNA damage. *Nature*, **434** (7033), 605–611.

Fan, J.Y., Gordon, F., Luger, K. *et al.* (2002) The essential histone variant H2A.Z regulates the equilibrium between different chromatin conformational states. *Nature Structural Billogy* **9**, 172–176.

Fan, L., and Roberts, V.A. (2006) Complex of linker histone H5 with the nucleosome and its implications for chromatin packing. *Proceedings of the National Academy of Sciences of the USA* **103**, 8384–8389.

Fan, Y., Nikitina, T., Morin-Kensicki, E.M. *et al.* (2003) H1 linker histones are essential for mouse development and affect nucleosome spacing in vivo. *Molecular and Cellular Biology* **21**, 7933–7943.

Fan, Y., Nikitina, T., Zhao, J. *et al.* (2005) Histone H1 depletion in mammals alters global chromatin structure but causes specific changes in gene regulation. *Cell* **123**, 1199–1212.

Fedoreyeva, L.I., Smirnova, T.A., Kolomijtseva, G.Y. *et al.* (2008) H1 histone modulates DNA hydrolysis with WEN1 and WEN2 endonucleases from wheat coleoptiles. *Biochemistry* **74** (2), 145–152.

Fenley, A.T., Adams, D.A. and Onufriev, A.V. (2001) Charge state of the globular histone core controls stability of the nucleosome. *Biophysical Journal* **99** (5), 1577–1585.

Fernandez-Capetillo, O., Lee, A., Nussenzweig, M. and Nussenzweig, A. (2004) H2AX: the histone guardian of the genome. *DNA Repair (Amsterdam)* **3** (8–9), 959–967.

Fischle, W., Tseng, B.S., Dormann, H.L. *et al.* (2005) Regulation of HP1-chromatin binding by histone H3 methylation and phosphorylation. *Nature* **438**, 1116–1122.

Freedman, B.S., and Heald, R. (2010) Functional comparison of H1 histones in *Xenopus* reveals isoform-specific regulation by Cdk1 and RanGTP. *Current Biology* **20** (11), 1048–1052.

Friesner, J.D., Liu, B., Culligan, K. and Britt, A.B. (2005) Ionizing radiation–dependent γ-H2AX focus formation requires Ataxia telangiectasia mutated and Ataxia telangiectasia mutated and Rad3-related. *Molecular Biology of the Cell* **16**, 2566–2576.

Fulnecek, J., Matyasek, R and Kovarik, A. (2006) Plant 5S rDNA has multiple alternative nucleosome positions. *Genome* **49**, 840–850.

Gaginskaya, E., Kulikova, T., and Krasikova, A. (2009) Avian Lampbrush Chromosomes: a Powerful Tool for Exploration of Genome Expression. *Cytogenetic and Genome Research* **124**, 251–267.

Godde, J.S., and Ura, K. (2008) Cracking the enigmatic linker histone code. *Journal of Biochemistry* **143** (3), 287–293.

Goodwin, G.H., Sanders, C. and Johns, E.W. (1973) A new group of chromatin-associated proteins with a high content of acidic and basic amino acids. *European Journal of Biochemistry* **38**, 14–19.

Grasser, K.D. (1995) Plant chromosomal high mobility group (HMG) proteins. *Plant Journal* **7**, 185–192.

Grasser, K.D. (2003) Chromatin-associated HMGA and HMGB proteins; versatile co-regulators of DNA-dependent processes. *Plant Molecular Biology* **53**, 281–295.

Grasser, K.D., Lentz, A., Lichota, J. *et al.* (2006) The Arabidopsis genome encodes structurally and functionally diverse HMGB-type proteins. *Journal of Molecular Biology* **358**, 654–664.

Grigoryeva, S.A., Aryab, G., Correlia, S. *et al.* (2009) Evidence for heteromorphic chromatin fibers from analysis of nucleosome interactions. *Proceedings of the National Academy of Sciences of the USA* **106**, 13317–13322.

Guo, C., Gerasimova, T., Hao, H. *et al.* (2011) Two forms of loops generate the chromatin conformation of the immunoglobulin heavy-chain gene locus. *Cell* **147** (2), 332–343.

Haaf, T. and Schmid, T. (1991) Chromosome topology in mammalian interphase nuclei. *Experimental Cell Research* **192**, 325–332.

Happel, N., and Doenecke, D. (2009) Histone H1 and its isoforms: contribution to chromatin structure and function. *Gene* **431** (1–2), 1–12.

Happel, N., Schulze, E. and Doenecke, D. (2005) Characterisation of human histone H1x. *Journal of Biological Chemistry* **386**, 541–551.

Hart, C.M., and Laemmli, U.K. (1998) Facilitation of chromatin dynamics by SARs. *Current Opinion in Genetics and Development* **8** (5), 519–525.

Havas, K., Flaus, A., Phelan, M. *et al.* (2000) Generation of superhelical torsion by ATP-dependent chromatin remodeling activities. *Cell* **103** (7), 1133–1142.

Hazzalin, C.A., and Mahadevan, L.C. (2005) Dynamic acetylation of all lysine 4-methylated histone H3 in the mouse nucleus: analysis at c-fos and c-jun. *PLOS Biology* **3**, e393.

Hewish, D.R. and Burgoyne, U.A. (1973) Chromatin sub-structure: the digestion of chromatin DNA at regularly spaced sites by a nuclear deoxyribonuclease. *Biochemical and Biophysical Research Communication* **52**, 504–510.

Hodges, C., Bintu, L., Lubkowska, L. *et al.* (2009) Nucleosomal fluctuations govern the transcription dynamics of RNA polymerase II. *Science* **325**, 626–628.

Hotta, T., Haraguchi, T. and Mizuno, K. (2007) A novel function of plant histone H1: microtubule nucleation and continuous plus end association. *Cell Structure and Function* **32** (2), 79–87.

Howman, E.V., Fowler, K.J., Newson, A.J. *et al.* (2000) Early disruption of centromeric chromatin organization in centromere protein A (Cenpa) null mice. *Proceedings of the National Academy of Sciences of the USA* **97**, 1148–1153.

Iguchi, N., Tnaka, H., Yomogida, K. *et al.* (2003) Isolation and characterization of a novel cDNA-binding protein (HILS1) specifically expressed in testicular haploid germ cells. *International Journal of Andrology* **26**, 354–365.

Ingouff, M., Rademacher, S., Holec, S. *et al.* (2010) Zygotic resetting of the HISTONE 3 variant repertoire participates in epigenetic reprogramming in Arabidopsis. *Current Biology* **20**, 2137–2143.

Izzo, A., Kamieniarz, K. and Schneider, R. (2008) The histone H1 family: specific members, specific functions? *Journal of Biological Chemistry* **389** (4), 333–343.

Jackson, D.A., Dickinson, P. and Cook, P.R. (1990) The size of chromatin loops in HeLa cells. *The EMBO Journal* **9**, 567–571.

Jenuwein, T. and Allis, C.D. (2001) Translating the histone code. *Science* **293**, 1074–1080.

Jin, J., Cai, Y., Li, B. *et al.* (2005) In and out: histone variant exchange in chromatin. *Trends in Biochemical Sciences* **30**, 680–687.

Johns, E.W (ed.) (1982) *The HMG Chromosomal Proteins*, Academic Press, London.

Jones, M.E. (1953) Albrecht Kossel, A Biographical Sketch. *Yale Journal of Biology and Medicine (National Center for Biotechnology Information)* **26** (1), 80–97.

Kantidze, O.L., and Razin, S.V. (2009) Chromatin loops, illegitimate recombination, and genome evolution. *Bioessays* **31** (3), 278–286.

Kassabov, S.R., Zhang, B., Persinger, J. and Bartholomew, B. (2003) SWI/SNF unwraps, slides, and rewraps the nucleosome. *Molecular Cell* **11** (2): 391–403.

Khochbin, S. (2001) Histone H1 diversity: bridging regulatory signals to linker histone function. *Gene* **271** (1), 1–12.

Klosterman, S.J., and Hadviger, L.A. (2002) Plant HMG proteins bearing the AT-hook motif. *Plant Science* **162**, 855–866.

Konat, G.W. (2002) Higher order chromatin degradation: implications for neurodegeneration. *Neurochemical Research* **27** (11), 1447–1451.

Kornberg, R. (1974) Chromatin structure: a repeating unit of histones and DNA. *Science* **184**, 868–871.

Kornberg, R. and Thomas, J.O. (1974) Chromatin structure: oligomers of histones. *Science* **184**, 865–868.

Kornet, N. and Scheres, B. (2008) Stem cell factors in plants: chromatin connections. *Cold Spring Harbor Symposia on Quantitative Biology* **73**, 235–242.

Kouzarides, T. (2007) Chromatin modifications and their function. *Cell* **128**, 693–705.

Kumar, S.V. and Wigge, P.A. (2010) H2A.Z-containing nucleosomes mediate the thermosensory response in Arabidopsis. *Cell* **140** (1), 136–147.

Laemmli, U.K., Käs, E., Poljak, L. and Adachi, Y. (1992) Scaffold-associated regions: cis-acting determinants of chromatin structural loops and functional domains. *Current Opinion in Genetics and Development* **2** (2), 275–285.

Lee, W., Tillo, D., Bray, N. *et al.* (2007) A high-resolution atlas of nucleosome occupancy in yeast. *Nature Genetics* **39**, 1235–1244.

Lewis, C.D. and Laemmli, U.K. (1982) Higher order metaphase chromosome structure: evidence for metalloprotein interactions. *Cell* **29**, 171–181.

Li, B., Pattenden, S.G., Lee, D. *et al.* (2005b) Preferential occupancy of histone variant H2AZ at inactive promoters influences local histone modifications and chromatin remodeling. *Proceedings of the National Academy of Sciences of the USA* **102**, 18385–18390.

Li, G., Levitus, M., Bustamante, C. and Widom, J. (2005c) Rapid spontaneous accessibility of nucleosomal DNA. *Nature Structural and Molecular Biology* **12** (1), 46–53.

Li, J., Krychevsky, A., Vaidya, M. *et al.* (2005a) Uncoupling of the function of the Arabidopsis VIP1 protein in transient and stable plant genetic transformation by *Agrobacterium*. *Proceedings of the National Academy of Sciences of the USA* **102**, 5733–5738.

Lin, Q., Sirotkin, A. and Skoultchi, A.L. (2000) Normal spermatogenesis in mice lacking the testis-specific linker histone H1t. *Molecular and Cellular Biology* **20**, 2122–2128.

Loyter, A., Rosenbluh, J., Zakai, N. *et al.* (2005) The plant VirE2 interacting protein 1. A molecular link between the *Agrobacterium* T-complex and the host cell chromatin? *Plant Physiology* **138**, 1318–1321.

Luger, K., Mader, A.W., Richmond, R.K. *et al.* (1997) Crystal structure of the nucleosome core particle at 2.8 Å resolution. *Nature* **389**, 251–260.

Madigan, J.P., Chotkowski, H.L. and Glaser, R.L. (2002) DNA double-strand break-induced phosphorylation of *Drosophila* histone variant H2Av helps prevent radiation-induced apoptosis. *Nucleic Acids Research* **30**, 3698–3705.

Maeshima, K., and Eltsov, M. (2008) Packaging the genome: the structure of mitotic chromosomes. *Journal of Biochemistry* **143** (2), 145–153.

Malik, H.S. and Henikoff, S. (2003) Phylogenomics of the nucleosome. *Nature Structural and Molecular Biology* **10**, 882–891.

Manuelidis, L. (1990) A view of interphase chromosomes. *Science* **250**, 1533–1540.

Marchler-Bauer, A., Lu, S., Anderson, J.B. *et al.* (2011) CDD: a conserved domain database for the functional annotation of proteins. *Nucleic Acids Research* **39**: D225–9.

Maresca, T.J., Freedman, B.S. and Heald, R. (2005) Histone H1 is essential for mitotic chromosome architecture and segregation in *Xenopus laevis* egg extracts. *Journal of Cell Biology* **169** (6): 859–869.

Margueron, R., Trojer, P. and Reinberg, D. (2005) The key to development: interpreting the histone code? *Current Opinion in Genetic Development* **15**, 163–176.

Martianov, I., Brancorsini, S., Catena, R. *et al.* (2005) Polar nuclear localization of H1T2, a histone H1 variant, required for spermatid elongation and DNA condensation during spermiogenesis. *Proceedings of the National Academy of Sciences of the USA* **102**, 2808–2813.

Mateescu, B., England, P., Halgand, F. *et al.* (2004) Tethering of HP1 proteins to chromatin is relieved by phosphoacetylation of histone H3. *EMBO Reports* **5**, 490–496.

Mavrich, T.N., Ioshikhes, I.P., Venters, B.J. *et al.* (2008a) A barrier nucleosome model for statistical positioning of nucleosomes throughout the yeast genome. *Genome Research* **18**, 1073–1083.

Mavrich, T.N., Jiang, C., Ioshikhes, I.P. *et al.* (2008b) Nucleosome organization in the *Drosophila* genome. *Nature* **453**, 358–362.

Metivier, R., Penot, G., Hübner, M.R. *et al.* (2003) Estrogen receptor-α directs ordered, cyclical, and combinatorial recruitment of cofactors on a natural target promoter. *Cell* **115**, 751–763.

Misteli, T., Gunjan, A., Hock, R. *et al.* (2000) Dynamic binding of histone H1 to chromatin in living cells. *Nature* **408**, 877–881.

Mizuguchi, G., Shen, X., Landry, J. *et al.* (2004) ATP-driven exchange of histone H2AZ variant catalyzed by SWR1 chromatin remodeling complex. *Science* **303** (5656): 343–348.

Moehs, C.P., Baxevanis, A.D., Moudrianakis, E.N. and Spiker, S. (1992) Enhanced stability of histone octamers from plant nucleosomes: Role of H2A and H2B histones. *Biochemistry* **31**, 10844–10851.

Morgan, G.T. (2002) Lampbrush chromosomes and associated bodies: new insights into principles of nuclear structure and function. *Chromosome Research* **10**, 177–200.

Mysore, K.S., Nam, J. and Gelvin, S.B. (2000) An *Arabidopsis* histone H2A mutant is deficient in *Agrobacterium* T-DNA integration. *Proceedings of the National Academy of Sciences of the USA* **97** (2), 948–953.

Nakayama, T., Ishii, T., Hotta, T. *et al.* (2008) Radial microtubule organization by histone H1 on nuclei of cultured tobacco BY-2 cells. *Journal of Biological Chemistry* **283** (24), 16632–16640.

Ng, R.K., and Gurdon, J.B. (2008) Epigenetic memory of an active gene state depends on histone H3.3 incorporation into chromatin in the absence of transcription. *Nature Cell Biology* **10**, 102–109.

Nightingale, K.P., O'Neill, L.P. and Turner, B.M. (2006) Histone modifications: signalling receptors and potential elements of a heritable epigenetic code. *Current Opinion in Genetics and Development* **16**, 125–136.

Nitiss, J.L. (2009a) DNA Topoisomerase II and its growing repertoire of biological functions. *Nature Reviews Cancer* **9**, 327–337.

Nitiss, J.L. (2009b) Targeting DNA topoisomerase II in cancer chemotherapy. *Nature Reviews Cancer* **9**, 338–350.

Oegema, K., Desai, A., Rybina, S. *et al.* (2001) Functional analysis of kinetochore assembly in *Caenorhabditis elegans*. *Journal of Cellular Biology* **153**, 1209–1226.

Okada, T., Bhalla, P.L. and Singh, M.B. (2005) Transcriptional activity of male gamete-specific histone gcH3 promoter in sperm cells of *Lilium longiflorum*. *Plant and Cell Physiology* **46** (5), 797–802.

Okada, T., Endo, M., Singh, M.B. and Bhalla, P.L. (2005) Analysis of the histone H3 gene family in Arabidopsis and identification of the male-gamete-specific variant AtMGH3. *Plant Journal* **44**, 557–568.

Olins, A.L. and Olins, D.E. (1974) Spheroid chromatin units (V bodies). *Science* **183**, 330–332.

Ozsolak, F., Song, J.S., Liu, X.S. and Fisher, D.E. (2007) High-throughput mapping of the chromatin structure of human promoters. *Nature Biotechnology* **25**, 244–248.

Parseghian, M.H. and Hamkalo, B.A. (2001) A compendium of the histone family of somatic subtypes: an elusive cast of characters and their characteristics. *Biochemistry and Cell Biology* **79**, 289–304.

Paulson, J.R. and Laemmli, U.K. (1977) The structure of histone-depleted metaphase chromosomes. *Cell* **12**, 817–828.

Petersen, K., Leah, R., Knudsen, S. and Cameron-Mills, V. (2002) Matrix attachment regions (MARs) enhance transformation frequencies and reduce variance of transgene expression in barley. *Plant Molecular Biology* **49** (1), 45–58.

Phair, R.D., Scaffidi, P., Elbi, C. *et al.* (2004) Global nature of dynamic protein-chromatin interactions in vivo: three-dimensional genome scanning and dynamic interaction networks of chromatin proteins. *Molecular and Cellular Biology* **24**, 6393–6402.

Phelps-Durr, T.L., Thomas, J., Vahab, P. and Timmermans, M.C. (2005) Maize rough sheath2 and its Arabidopsis orthologue ASYMMETRIC LEAVES1 interact with HIRA, a predicted histone chaperone, to maintain knox gene silencing and determinacy during organogenesis. *Plant Cell* **17**, 2886–2898.

Rabini, S., Franke, L., Saftig, P. *et al.* (2000) Spermatogenesis in mice is not affected by histone H1.1 deficiency. *Experimental Cell Research* **255**, 114–124.

Radman-Livaja, M., and Rando, O.J. (2010) Nucleosome positioning: how is it established, and why does it matter? *Developmental Biology* **339**, 258–266.

Ramakrishnan, V., Finch, J.T., Graziano, V. *et al.* (1993) Crystal structure of globular domain of histone H5 and its implications for nucleosome binding. *Nature* **362** (6417), 219–223.

Rangasamy, D., Greaves, I and Tremethick, D.J. (2004) RNA interference demonstrates a novel role for H2A.Z in chromosome segregation. *Nature Structural and Molecular Biology* **11**, 650–655.

Ravi, M., and Chan, S.W. (2010) Haploid plants produced by centromere-mediated genome elimination. *Nature* **464** (7288), 615–618.

Ravi, M., Kwong, P.N., Menorca, R.M. *et al.* (2010) The Rapidly Evolving Centromere-specific Histone Has Stringent Functional Requirements in *Arabidopsis thaliana*. *Genetics* **186**, 461–471.

Redon, C., Pilch, D., Rogakou, E. *et al.* (2002) Histone H2A variants H2AX and H2AZ. *Current Opinion in Genetics and Development* **12**, 162–169.

Richmond, T.J., Finch, J.T., Rushton, B. *et al.* (1984) Structure of the nucleosome core particle at 7 Å resolution. *Nature* **311**, 532–537.

Robinett, C.C., Straight, A.A., Li, G. *et al.* (1996) In vivo localization of DNA sequences and visualization of large-scale chromatin organization using lac operator repressor recognition. *The Journal of Cell Biology* **135**, 1685–1700.

Robinson, P.J., Fairall, L., Huynh, V.A. *et al.* (2006) EM measurements define the dimensions of the '30-nm' chromatin fiber: evidence for a compact, interdigitated structure. *Proceedings of the National Academy of Sciences of the USA* **103**, 6506–6511.

Robinson, P.J., and Rhodes, D. (2006) Structure of the '30-nm' fibre: a key role for the linker histone. *Current Opinion in Structural Biology* **16**, 336–343.

Rogakou, E.P., Pilch, D.R., Orr, A.H. *et al.* (1998) DNA double-stranded breaks induce histone H2AX phosphorylation on serine 139. *Journal of Biological Chemistry* **273**, 5858–5868.

Roudier, F., Ahmed, I., Berard, C. *et al.* (2011) Integrative epigenomic mapping defines four main chromatin states in Arabidopsis. *The EMBO Journal* **30** (10), 1928–1938.

Royal Society (2009) *Reaping the Benefits: Science and the Sustainable Intensification of Global Agriculture*. The Royal Society Policy Reports, Statements and Responses, http://royalsociety.org/policy/publications/2009/reaping-benefits/ (accessed 21 November 2012).

Sahasrabuddhe, C.G. and van Holde, K.E. (1974) The effect of trypsin on nuclease-resistant chromatin fragments. *Journal of Biological Chemistry* **249**, 152–156.

Sakai, A., Schwartz, B.E., Goldstein, S. and Ahmad, K. (2009) Transcriptional and developmental functions of the H3.3 histone variant in *Drosophila*. *Current Biology* **19**, 1816–1820.

Samejima, I., Matsumoto, T., Nakaseko, Y. *et al.* (1993) Identification of seven new cut genes involved in *Schizosaccharomyces pombe* mitosis. *Journal of Cell Science* **105**, 135–143.

Sanchez-Moran, E., Osman, K., Higgins, J.D. *et al.* (2008) ASY1 coordinates early events in the plant meiotic recombination pathway. *Cytogenet. Genome Research* **120** (3–4):302–12.

Sanei, M., Pickering, R., Kumkea, K. *et al.* (2011) Loss of centromeric histone H3 (CENH3) from centromeres precedes uniparental chromosome elimination in interspecific barley hybrids. *Proceedings of the National Academy of Sciences of the USA* **108**: E498–505.

Santisteban, M.S., Kalashnikova, T. and Smith, M.M. (2000) Histone H2A.Z regulates transcription and is partially redundant with nucleosome remodelling complexes. *Cell* **103**, 411–422.

Sawatsubashi, S., Murata, T., Lim, J. *et al.* (2010) A histone chaperone, DEK, transcriptionally coactivates a nuclear receptor. *Genes and Development* **24**, 159–170.

Schalch, T., Duda, S., Sargent, D.F. *et al.* (2005) X-ray structure of a tetranucleosome and its implications for the chromatin fibre. *Nature* **436**, 138–141.

Schones, D., Cui, K., Cuddapah, S. *et al.* (2008) Dynamic regulation of nucleosome positioning in the human genome. *Cell* **132**, 887–898.

Schreiber, S.L., and Bernstein, B.E. (2002) Signaling network model of chromatin. *Cell* **111**, 771–778.

Schubert, V. (2009) SMC proteins and their multiple functions in higher plants. *Cytogenetic and Genome Research* **124** (3–4), 202–214.

Schwartz, B.E., and Ahmad, K. (2005) Transcription triggers deposition and removal of the histone variant H3.3. *Genes and Development* **19**, 804–814.

Segal, E., and Widom, J. (2009) From DNA sequence to transcriptional behaviour: a quantitative approach. *Nature Reviews Genetics* **10**, 443–456.

Seyedin, S.M., and Kistler, W.S. (1980) Isolation and characterization of rat testis H1t. An H1 histone variant associated with spermatogenesis. *Journal of Biological Chemistry* **255**, 5949–5954.

Sgarra, R., Rustighi, A., Tessari, M.A. *et al.* (2004) Nuclear phosphoproteins HMGA and their relationship with chromatin structure and cancer. *FEBS Letters* **574** (1–3), 1–8.

Shi, L., Wang, J., Hong, F. *et al.* (2011) Four amino acids guide the assembly or disassembly of Arabidopsis histone H3.3-containing nucleosomes. *Proceedings of the National Academy of Sciences of the USA* **108** (26), 10574–10578.

Shrivastav, M., De Haro, L.P. and Nickoloff, J.A. (2008) Regulation of DNA double-strand break repair pathway choice. *Cell Research* **18**, 134–147.

Simpson, R.T. (1978) Structure of the chromatosome, a chromatin particle containing 160 base pairs of DNA and all the histones. *Biochemistry* **17**, 5524–5531.

Sirotkin, A.M., Edelmann, W., Chen, G. *et al.* (1995) Mice develop normally without the H1° linker histone. *Proceedings of the National Academy of Sciences of the USA* **92**, 6434–6438.

Spiker, S. (1988) Histone variants and high mobility group non-histones chromosomal proteins of higher plants; their potential for forming a chromatin structure that is either poised for transcription or transcriptionally inert. *Physiolia Plantarum* **74**, 200–213.

Steinbach, O.C., Wolffe, A.P. and Rupp, R.A. (1997) Somatic linker histones cause loss of mesodermal competence in *Xenopus*. *Nature* **389** (6649), 395–399.

Strahl, B.D. and Allis, C.D. (2000) The language of covalent histone modifications. *Nature* **403** (6765): 41–45.

Strunnikov, A.V., Hogan, E. and Koshland, D. (1995) SMC2, a *Saccharomyces cerevisiae* gene essential for chromosome segregation and condensation defines a subgroup within the SMC-family. *Genes and Development* **9**, 587–599.

Tagami, H., Ray-Gallet, D., Almouzni, G. and Nakatani, Y. (2004) Histone H3.1 and H3.3 complexes mediate nucleosome assembly pathways dependent or independent of DNA synthesis. *Cell* **116**, 51–61.

Talbert, P.B., Masuelli, R., Tyagi, A.P. *et al.* (2002) Centromeric localization and adaptive evolution of an Arabidopsis histone H3 variant. *Plant Cell* **14**, 1053–1066.

Tenea, G.N., Spantzel, J., Lee, L.-Y. *et al.* (2009) Overexpression of several Arabidopsis histone genes increases agrobacterium-mediated transformation and transgene expression in plants. *Plant Cell* **21** (10), 3350–3367.

Tetko, I.V., Haberer, G., Rudd, S. *et al.* (2006) Spatiotemporal expression control correlates with intragenic scaffold matrix attachment regions (S/MARs) in *Arabidopsis thaliana. PLoS Computational Biology* **2** (3), e21.

Thambirajah, A.A., Dryhurst, D., Ishibashi, T. *et al.* (2006) H2A.Z stabilizes chromatin in a way that is dependent on core histone acetylation. *Journal of Biological Chemistry* **281**, 20036–20044.

Thoma, F., Koller, T. and Klug, A. (1979) Involvement of histone H1 in the organization of the nucleosome and of the salt-dependent superstructures of chromatin. *Journal of Cell Biology* **83**, 403–427.

Tikhonov, A.P., Bennetzen, J.L. and Avramova, Z.V. (2000) Structural domains and matrix attachment regions along colinear chromosomal segments of maize and sorghum. *Plant Cell* **12** (2), 249–264.

Tremethick, D.J. (2007) Higher-order structures of chromatin: the elusive 30 nm fiber. *Cell* **128**, 651–654.

Turner, B.M. (2007) Defining an epigenetic code. *Nature Cell Biology* **9** (1), 2–6.

Turner, B.M. (2009) Epigenetic responses to environmental change and their evolutionary implications. *Philosophical Transactions of the Royal Society B* **364**, 3403–3418.

Turner, B.M., Birley, A.J. and Lavender, J. (1992) Histone H4 isoforms acetylated at specific lysine residues define individual chromosomes and chromatin domains in *Drosophila* polytene nuclei. *Cell* **69**, 375–384.

Valouev, A., Ichikawa, J., Tonthat, T. *et al.* (2008) A high-resolution, nucleosome position map of *C. elegans* reveals a lack of universal sequence-dictated positioning. *Genome Research* **18**, 1051–1063.

Vermaak, D., Hayden, H.S. and Henikoff, S. (2002) Centromere targeting element within the histone fold domain of Cid. *Molecular Cell Biology* **22** (21), 7553–7561.

Van Holde, K.E. (1988) *Chromatin: Springer Series in Molecular Biology*, New York, Springer-Verlag.

Wagner, D. (2003) Chromatin regulation of plant development. *Current Opinion in Plant Biology* **6** (1), 20–28.

Wang, H., Huang, Z.Q., Xia, L. *et al.* (2001) Methylation of histone H4 at arginine 3 facilitating transcriptional activation by nuclear hormone receptor. *Science* **293**, 853–857.

Watson, J.D. and Crick, F. (1953) A structure for deoxyribose nucleic acid. *Nature* **171** (4356), 737–738.

Whitehouse, I., Flaus, A., Cairns, B.R. *et al.* (1999) Nucleosome mobilization catalysed by the yeast SWI/SNF complex. *Nature* **400** (6746): 784–787.

Widom, J. (1992) A relationship between the helical twist of DNA and the ordered positioning of nucleosomes in all eukaryotic cells. *Proceedings of the National Academy of Sciences of the USA* **89**, 1095–1099.
Wieland, G., Orthaus, S., Ohndorf, S. *et al.* (2004) Functional complementation of human centromere protein A (CENP-A) by Cse4p from *Saccharomyces cerevisiae*. *Molecular Cell Biology* **24** (15), 6620–6630.
Wierzbicki, A.T. and Jerzmanowski, A. (2005) Suppression of histone H1 genes in Arabidopsis results in heritable developmental defects and stochastic changes in DNA methylation. *Genetics* **169** (2), 997–1008.
Williamson, R. (1970) Properties of rapidly labelled deoxyribonucleic acid fragments isolated from the cytoplasm of primary cultures of embryonic mouse liver cells. *Journal of Molecular Biology* **51**, 157–168.
Wolffe, A. (1998) *Chromatin Structure and Function*, 3rd edn, Academic Press, London.
Wong, H., Victor, J.M. and Mozziconacci, J. (2007) An all-atom model of the chromatin fiber containing linker histones reveals a versatile structure tuned by the nucleosomal repeat length. *PLOS One* **2** (9), e877.
Wong, L.H., McGhie, J.D., Sim, M. *et al.* (2010) ATRX interacts with H3.3 in maintaining telomere structural integrity in pluripotent embryonic stem cells. *Genome Research* **20**, 351–360.
Woodcock, C.L.F. (1973) Ultrastructure of inactive chromatin. *Journal of Cell Biology* **59**, 368a.
Xie, S., and Lam, E. (1994) Abundance of nuclear DNA topoisomerase II is correlated with proliferation in *Arabidopsis thaliana*. *Nucleic Acids Research* **22** (25), 5729–5736.
Yan, W., Ma, L., Burns, K.H. and Matzuk, M.M. (2003) HILS1 is a spermatid-specific linker histone H1-like protein implicated in chromatin remodelling during mammalian spermatogenesis. *Proceedings of the National Academy of Sciences of the USA* **100**, 10546–10551.
Yi, H., Sardesai, N., Fujinuma, T., Chan, C.-W. *et al.* (2006) Constitutive expression exposes functional redundancy between the Arabidopsis histone H2A gene HTA1 and other H2A gene family members. *Plant Cell* **18** (7), 1575–1589.
Zhang, X. (2008) The epigenetic landscape of Plants. *Science* **320**, 489–492.
Zheng, C. and Hayes, J.J. (2003) Structures and interactions of the core histone tail domains. *Biopolymers* **68** (4), 539–546.
Zheng, Y., He, X.-W., Ying, Y.-H. *et al.* (2009) Expression of the Arabidopsis thaliana histone gene AtHTA1 enhances rice transformation efficiency. *Molecular Plant* **2** (4), 832–837.
Zhou, Y.-B., Gerchman, S.E., Ramakrishnan, V. and Muyldermans, S.V. (1998) Position and orientation of the globular domain of linker histone H5 on the nucleosome. *Nature* **395**, 402–405.

Annual Plant Reviews (2013) **46**, 157–190 http://onlinelibrary.wiley.com
doi: 10.1002/9781118472507.ch6

Chapter 6

HETEROCHROMATIN POSITIONING AND NUCLEAR ARCHITECTURE

Emmanuel Vanrobays, Mélanie Thomas and Christophe Tatout

GReD Laboratory, UMR CNRS 6293, INSERM U1103, University Blaise Pascal, Aubière, France

Abstract: Heitz (1928) first described the two states of chromatin known as *euchromatin* and *heterochromatin*. Heterochromatin is often described as the gene-poor part of the genome associated with a silent and condensed state of chromatin inaccessible to transcription factors. However, this simple view has been challenged many times as heterochromatin is indeed transcribed and differences between the two states of chromatin depend on many other criteria. Data collected from the model species *Arabidopsis thaliana* indicate that heterochromatin relies on the repetitiveness of specific DNA sequences such as satellites, transposable elements and ribosomal DNA (rDNA) but also on epigenetic marks specifically associated with these repeated arrays. Recent studies on small RNA pathways have highlighted the central role of the RNA-directed DNA methylation pathway in heterochromatin specification indicating that heterochromatin is indeed an epigenetic state required for many other genome functions including chromosome segregation, gene regulation or maintenance of genome stability. Heterochromatin is a specific feature of eukaryotes and preferential localization at the nuclear periphery and to the nucleolus is observed in most organisms. This spatial organization of heterochromatin is maintained through the cell cycle although DNA is replicated, chromatin is condensed into chromosomes, and the nuclear envelope is disrupted and reformed. Whether spatial positioning participates in heterochromatin function and how plant cells are able to establish and maintain such a genome organization across the cell cycle is still largely unknown in *Arabidopsis*. Mechanisms leading to the appropriate positioning of heterochromatin within the nucleus will be discussed in the light of data coming from other species. A better understanding of

Annual Plant Reviews Volume 46: Plant Nuclear Structure, Genome Architecture and Gene Regulation, First Edition. Edited by David E. Evans, Katja Graumann and John A. Bryant.

heterochromatin organization and functions will be an important field of investigation for plants such as maize and wheat with considerably larger genome sizes than *Arabidopsis*.

Keywords: heterochromatin; epigenetics; chromocenters; chromatin organization; chromosome; chromosome territory; histone

6.1 Heterochromatin structure

In order to better understand how heterochromatin is assembled and what functions are assumed by it, it is first necessary to describe some structural aspects of heterochromatin. Emil Heitz (1928) historically identified heterochromatin as the nuclear material that remains highly condensed within the interphase nucleus. He named these regions 'heterochromatin' to distinguish them from the regions showing variable staining and condensation, which he called 'euchromatin'. The boundaries of heterochromatin as specified by such cytological analyses are not very well defined and may vary in diverse tissues or with different analytical techniques. Subsequently, heterochromatin was subdivided into two classes, constitutive and facultative heterochromatin, depending on its persistent presence throughout the cell cycle (Brown, 1966). Formation of facultative heterochromatin at specific genomic loci is fundamentally important in defining cellular properties such as differentiation, stress response and reaction to developmental, physiological, or environmental stimuli.

At the same time, renaturation kinetics of DNA, also known as Cot analysis, gave striking results when applied to genomic DNA from eukaryotes (Britten and Kohne, 1968). The eukaryotic genome includes three distinct fractions according to their rates of renaturation because repeated sequences renature faster than single DNA sequences (Figure 6.1a). The repeated nature of DNA turns out to be a very interesting criterion for better defining heterochromatin, as most of heterochromatic sequences fall into the highly repeated (HR) fraction whereas euchromatin sequences are included in the single or low (SL) fraction. Since these early cytological and kinetic characterizations, euchromatin and heterochromatin are subsequently distinguished according to many other criteria such as their chromosome distribution, DNA sequence composition, protein binding or epigenetic features as will be detailed below.

6.1.1 Heterochromatic sequences

In small genome species such as Arabidopsis (150Mbp), heterochromatin represents 10–15% of the whole genome (Arabidopsis Genome Initiative, 2000). The Arabidopsis genome is organized in five linear chromosomes with dedicated domains made of heterochromatin (Figures 6.1b and 6.1c). The

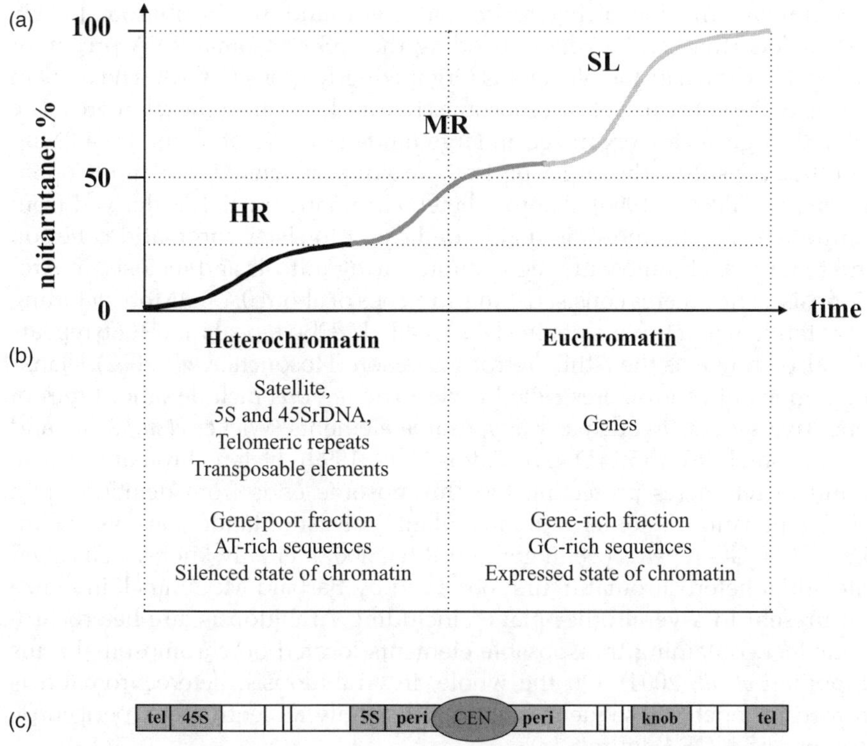

Figure 6.1 Distinctive structural features of the two states of chromatin. a) The renaturation curve of double-strand DNA highlights the three main sequence components of the eukaryotic genome in respect to their repetitiveness. The curve expressed as a percentage of renaturation (vertical axis) over time (horizontal axis) consists of highly repetitive (HR), moderately repetitive (MR), and single/low-copy (SL) components characterized by fast, intermediate and slow reassociation, respectively. b) Simple criteria to distinguish the two states of chromatin. Heterochromatin is the most repeated fraction of the genome. Constitutive and facultative heterochromatin are included respectively within the HR and MR fractions. Heterochromatin includes highly repeated sequences known as AT rich and gene poor and is thus considered to be a silenced state of chromatin. Euchromatin is mostly included within the SL fraction but can in some cases be part of the MR fraction. Euchromatin is the expressed and gene-rich fraction of the genome; it contains sequences homologous to Expressed Sequence Tags (EST) known to be more GC-rich. It is somehow more difficult to describe the MR fraction which can include both euchromatic and heterochromatic sequences. c) Chromosome schematic of a hypothetical eukaryotic chromosome where heterochromatic sub-domains are depicted. Telomeres (tel), 45S and 5S rDNA (45S and 5S), pericentric repeats (peri), centromere (CEN) and knob are indicated as grey boxes and interrupted by euchromatic segments (in white) which may include dispersed repeated sequences such as transposable elements (vertical bars)

predominant fraction of heterochromatin is found at 45S ribosomal[1] (45S rDNA) loci (that is, the genes encoding the 45S ribosomal RNA precursor molecule) known as the Nucleolus Organizing Regions (NOR), which are at the top of the short arm of chromosomes two and four, near the telomeres. The 45S rRNA genes are organized in large tandem arrays of about 3.5–4.0Mbp and thus constitute the larger part of Arabidopsis heterochromatin (Copenhaver and Pikaard, 1996). Another heterochromatic fraction is derived from centromeres, a chromosome region dedicated to sister chromatid cohesion and to normal chromosome segregation during mitosis and meiosis. In Arabidopsis, centromeres consist of long stretches of about 0.4–1.4Mpb including short tandem repetitive 'satellite' DNA, called 180bp repeats and 106B repeats and also known as the Athila retrotransposon (Hosouchi *et al.*, 2002). Flanking regions of centromeres called pericentric regions include other types of repetitive sequences such as transposable elements (Wicker *et al.*, 2007) and 5S ribosomal DNA (5S rDNA) (Tutois *et al.*, 1999). Heterochromatin is also found in telomeres protecting the chromosome ends from deletion upon DNA replication. Telomeres in most plants consist of tandem arrays of simple TTTAGGG repeats (Zellinger and Riha, 2007). Finally knobs, also called interstitial heterochromatin, first observed by Barbara McClintock in maize but present in several other plants, including Arabidopsis, are heterochromatic loci containing transposable elements located on chromosomal arms (Lippman *et al.*, 2004). On the whole, in Arabidopsis, heterochromatin is formed at repetitive sequences clustered mainly at centromere, pericentric regions and NOR (Figure 6.1c).

Transposable elements, rDNA and satellites are common components of heterochromatin in other plant species. They can vary in term of sequences but their properties remain the same. Indeed, most plant centromeres include satellite DNA elements and a survey of these sequences can be found in Wang *et al.* (2009). Heterochromatin content and heterochromatin organization can be very different among eukaryotes and this is especially true for plants where a three-log range of variation in genome size is observed. Over the past decades, although polyploidy and gene duplication have been described as important mechanisms occurring in plants, it turns out that an increase in heterochromatic sequences was the largest contributor to genome size expansion (Gaut and Ross-Ibarra, 2008). Ma and collaborators carried out a comparative analysis of plant centromere organizations including Arabidopsis, *Oriza sativa* and *Zea mays* with a genome size of 150Mb, 370Mb and 2700Mb, respectively (Ma *et al.*, 2007). Centromere sizes among these species do not explain the variation in genome sizes. Alternatively, transposition of mobile DNA on chromosome arms was shown as the main explanation of genome size expansion (Gaut and Ross-Ibarra, 2008). Thus, in many genomes other than Arabidopsis, heterochromatic sequences are not only concentrated to centromeres and rDNA but can also be found in large numbers on chromosome

[1]That is, the genes encoding the 45S ribosomal RNA precursor molecule.

arms. This is now well described in cereals such as maize and wheat in which 70–80% of the genome is made of transposable elements (Choulet *et al.*, 2010). Using Arabidopsis as a plant model system to study heterochromatin should therefore be carefully considered since it has a very compacted genome in which heterochromatin is mostly centric and pericentric compared to most plant species.

6.1.2 Epigenetic marks

Heterochromatin forms at repeated sequences including rDNA, transposable elements and tandem repeats. These repeated sequences are tightly associated with epigenetic marks which define the heterochromatin state. Main epigenetic modifications such as DNA methylation, histone modifications, non-coding RNA and histone variants are among the major players in the propagation and modulation of the heterochromatic status and will be described in the following section.

6.1.2.1 DNA methylation

In plants, DNA methylation occurs at cytosine residues in three different contexts, CG, CHG and CHH, where H is either A, C or T. DNA methylation depends on a large number of proteins but with central roles for METHYLTRANSFERASE1 (MET1) for CG, SU(VAR)3-9 HOMOLOG4 / KRYPTONITE (SUVH4/KYP) a member of the Su(var)3-9 class of histone methyltransferases, CHROMOMETHYLASE3 (CMT3) for CHG and CHH and DOMAIN REARRANGED METHYLTRANSFERASE2 (DRM2) for CHH. The function of all three methylases is in some way overlapping as DRM2 is a *de novo* DNA methylase for CG, CHG and CHH whereas MET1 and CMT3 respectively maintain methylation at CG and CHG (reviewed in Feng *et al.*, 2010). Many other factors are involved in this process including DECREASE in DNA METHYLATION1 (DDM1) required for both methylation and histone tail modification at H3K9 (see below). It is worth noting that *met1* and *ddm1* mutants were indeed identified in screens for loss of centromeric-repeat methylation (Vongs *et al.*, 1993; Kankel *et al.*, 2003). Further investigations of those two key players on transposable elements such as the retrotransposon *ÉVADÉ* (EVD), and transposons from the CACTA family (named CACTA because of the occurrence of CATCA sequences at the edges of the elements) or from MUtator-Like Element (MULE) class, which are usually silent, clearly show that they can be reactivated in *met1* and *ddm1* mutants (Mirouze *et al.*, 2009; Tsukahara *et al.*, 2009). On a genome-wide level, 24% of CG, 6.7% of CHG and 1.7% of CHH are methylated. CG, CHG and CHH methylation are enriched in heterochromatic sequences however, with a marked increase for CG methylation in the gene coding sequences (Cokus *et al.*, 2008). Within the gene coding sequences, up to 30% of CG are methylated while less than 1% of the CHG and CHH are methylated (Widman *et al.*, 2009). Conversely heterochromatic sequences are enriched in CHH as a consequence of methylated cytosine deamination in thymine, one of the most frequent point mutations

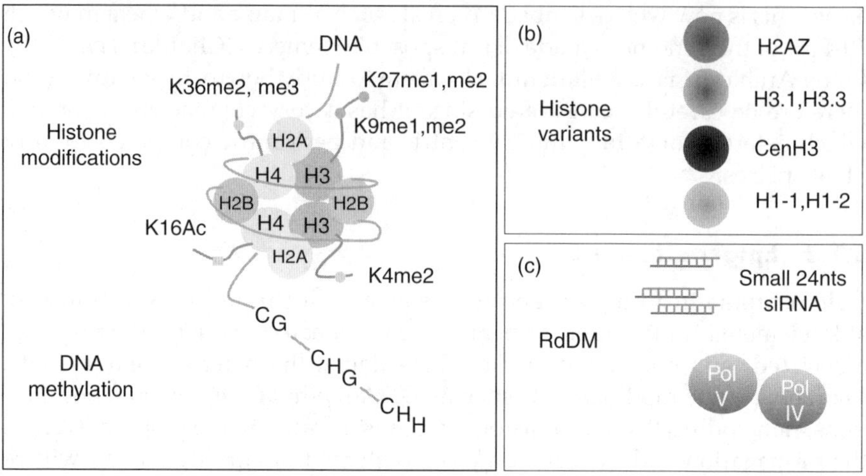

Figure 6.2 Heterochromatin features. a) The nucleosome is the basic structure of chromatin and includes histone octamer H2A, H2B, H3 and H4 as well as DNA. Lysine residues (K) lying within the histone tail at the N-terminal end of histone H3 and H4 can be modified either by methylation (me) or acetylation (Ac). Main repressive marks are indicated in red in Plate 6.2 (H3K9me1, me2 and H3K27me1, me2) whereas activating marks are indicated in green (H3K4me2, H4K36me2, me3 and H4K16Ac). DNA can also be subjected to DNA methylation at CG, CHG and CHH residues. Heterochromatin is enriched in methylated cytosine especially in CHH context, contains histone repressive marks and low levels of histone activating marks. b) Histone variants can substitute for canonical histones. H1-1, H1-2, H2AZ, H3.1, H3.3 and centromeric H3 (homologous to animal CenPA) are discussed in the frame of this review. H2AZ and H3.1 are enriched in actively transcribed regions while H3.3 and centromeric H3 are mostly found within heterochromatic regions. H1 variants are involved in chromosome segregation in plants. c) RNA Directed RNA Methylation (RdDM) occurs at heterochromatic loci and produces small non-coding RNA of 24nts called small interfering RNA (siRNA). RdDM involves plant specific polymerases called Polymerases IV and V (Pol IV and V) and is involved in the establishment of heterochromatin. (*For colour details please see colour plate section.*)

within eukaryotic genomes (Widman *et al.*, 2009). The overall consequence is that heterochromatic sequences are usually AT-rich sequences which are heavily methylated at CHH (and to a lesser extends at CHG) (Figure 6.1b and 6.2a).

6.1.2.2 Histone code

Core histones H2A, H2B, H3 and H4 are implicated in the formation of the nucleosome core including 147bp of DNA. Histone H1 are linker histones involved in the formation of more condensed chromatin fibres limiting the access for regulatory proteins to nucleosomal components (Happel and Doenecke, 2009). Histones have important implications in gene regulation through modifications at the C-terminal histone tails through methylation, acetylation, phosphorylation or ubiquitination; these modifications are referred to as the histone code (reviewed in Grant-Downton and Dickinson, 2005).

6.1.2.3 Histone-repressive marks

Heterochromatin-specific histone methylation marks are mainly mono- and dimethyl H3K9 (H3K9me1 and me2) and mono- and dimethyl H3K27 (H3K27me1 and me2) (Figure 6.2a). Specific patterns of histone marks at heterochromatic loci are believed to promote a close chromatin configuration not favourable for gene expression. Basically, two major histone marks are usually followed as heterochromatic marks. These are H3K9me2 implemented by KYP, and H3K27me1 established by histone methyltransferases *ARABIDOPSIS* TRITHORAX-RELATED PROTEIN5 and 6 (ATRX5, ATRX6). H3K9me2 and H3K27me1 are usually found in constitutive heterochromatin but H3K9me2 is dependent upon DNA methylation while H3K27me1 is not, probably reflecting two independent mechanisms occurring at heterochromatic loci (Liu *et al.*, 2010).

6.1.2.4 Histone-activating marks

On the other hand, heterochromatin H3K36me, H3K4me2 and H4K16 acetylation are poorly represented in heterochromatin and mostly found in euchromatic gene-rich regions (Figure 6.2a). H3K36me and H3K4me2 are established respectively by the histone methyltransferases ABSENT SMALL OR HOMEOTIC DISCS1 (ASH1) and *ARABIDOPSIS* TRITHORAX1 (ATX1) acting mostly at euchromatic loci (reviewed in Liu *et al.*, 2010). Interestingly, histone methyltransferases are unable to methylate target lysine residues that are acetylated, and therefore histone deacetylases (HDAC) are required to allow methylation at those regions (Noma *et al.*, 2001). To date 16 Arabidopsis HDACs have been identified and among them HDA6, a homolog of RPD3 in yeast, has a central role in heterochromatin (Pandey *et al.*, 2002). The *Hda6* mutation causes enrichment for euchromatic epigenetic marks such as H3K4 methylation, leading to a decondensed state of chromatin, visible at the cytological level at rDNA loci (Probst *et al.*, 2004). Furthermore, up-regulated loci in *hda6* overlap with those in *met1*, and the *hda6* mutation causes the complete loss of DNA methylation on some HDA6 target loci. These results suggest that HDA6 and MET1 DNA targets are overlapping. Further, this indicates that HDA6 is required together with MET1 to establish and/or maintain heterochromatin (To *et al.*, 2011).

6.1.2.5 Histone variants

Apart from C-terminal core histone modifications, H2A, H2B and H3 histone variants also modulate gene expression (Figure 6.2b). Studies of H3 variants have recently started in plants. Histone H3 variants include H3.1, H3.3 and CENP-A, a centromeric H3 variant, and are encoded by 15 *HISTONE THREE RELATED* (*HTR*) genes (Ingouff and Berger, 2009). These variants could contribute to the epigenetic memory of chromatin states in somatic cells.

Centromeres are epigenetically specified by incorporation of CENH3 (CENP-A in humans; CenH3 is encoded by the *HTR12* gene in Arabidopsis), replacing conventional H3 in centromeric nucleosomes (Henikoff and Dalal, 2005; Ravi and Chan, 2010). Histone variants H2A and H2B were reviewed in Wyrick and Parra (2009) but, to date, very little is known in plants. H2A variant H2A.Z deposition and DNA methylation seem to be mutually exclusive both in actively transcribed genes and in methylated transposons. H2A.Z marks actively transcribed genes and is found at promoter sequences (Zilberman *et al.*, 2008). This may be of particular interest as the yeast H2A histone variant HTZ1 (homologous to plant H2AZ) might tether active promoters to NUP2, a component of the Nuclear Pore Complex (NPC). Recently, it has also been proposed that nucleoplasmic NPC components such as NUP98 and NUP50 can activate expression of internally localized genes, whereas DNA interaction with NUP88 leads to gene silencing (see also Meier this issue; reviewed in Arib and Akhtar, 2011). Future investigations of plant NPC functions will then be of particular interest for gene regulation and heterochromatin.

6.1.2.6 Non-coding RNA

The structural features of heterochromatin led to a simple model where the gene-poor heterochromatin is the condensed state of chromatin and is poorly transcribed while euchromatin is less compacted allowing active transcription of the gene-rich regions. The epigenetic era drastically modified this oversimplified view. Indeed, heterochromatin is transcribed by plant-specific Polymerase II variant enzymes known as Polymerase IV and Polymerase V (Lahmy *et al.*, 2010) and produces large non-coding RNAs (ncRNA) processed through the RNA-directed DNA methylation pathway (RdDM) (Figure 6.2c). This involves ARGONAUTE (AGO) and DICER-LIKE (DCL) families of silencing factors and yields short RNA molecules, 24 nucleotides (24nts) in length, called short-interfering RNA (siRNA) homologous to heterochromatic sequences. The RdDM pathway is a complex network including many proteins and a more detailed description can be found elsewhere (Law and Jacobsen, 2010). SiRNAs are responsible for repression of repeated sequence expression both at the transcriptional and post-transcriptional levels. From an evolutionary point of view, RdDM is believed to be a defence mechanism evolving to protect eukaryotic genomes against mutagenesis by endogenous transposable elements and against exogenous viruses (see discussion in Cavalier-Smith, 2010). As a consequence of the production of 24nts RNA molecules, chromatin is modified both at the DNA level through DNA methylation and at the histone level through post-translational modifications; among them acetylation and methylation are the most studied (see Sections 6.1.2.3 and 6.1.2.4 above; Liu *et al.*, 2010).

6.1.3 Non-histone protein binding

In the early years of the use of Arabidopsis as a plant model species, attempts were made to apply data collected from Drosophila and yeast. The best example related to heterochromatin was the search for Suppressor and Enhancer of variegation (the so called *Su(var)* and *E(var)* genes) involved in the Position Effect Variegation (PEV) in Drosophila. In PEV, when a gene is brought close to heterochromatin, this led to its stochastic repression (variegation). In the course of the *Su(var)* and *E(var)* screens, it was demonstrated that PEV was induced by non-histone proteins specific to heterochromatin, which have the property of spreading from heterochromatic to euchromatic loci inducing a silent state of chromatin. Among them, SU(VAR)3-9, a histone methyltransferase establishing methylation at H3K9 and HETEROCHROMATIN PROTEIN 1 (HP1) a highly conserved protein found in yeast, human and plants, were discovered. Drosophila HP1 interacts with H3K9me and is a main actor in heterochromatin assembly through interaction with other heterochromatic factors. Once HP1 binds to a given locus, it can induce the spread of heterochromatin in adjacent regions unless a blocking element (called 'boundary elements', see Section 6.2.3.1) blocks heterochromatin propagation (reviewed in Grewal and Jia, 2007). Indeed, plant homologues for SU(VAR) were identified and as an example, SUVH4/KYP (see Section 6.1.2.1) is the functional homologue of SU(VAR)3-9 (Fischer *et al.*, 2006). Arabidopsis HP1, called LIKE HETEROCHROMATIN PROTEIN1 (LHP1) was also discovered but very unexpectedly binds euchromatic loci and recognizes H3K27me3, a function assumed in Drosophila by another class of protein called the Polycomb group proteins. LHP1 should then not be considered as the functional homologue of Drosophila HP1 (Gaudin *et al.*, 2001; Turck *et al.*, 2007). Another striking specificity of plant heterochromatin came from the characterization of a duplicated segment of the long arm of chromosome four translocated onto the short arm of chromosome four. This new segment is called hk4S (Lippman *et al.*, 2004). Sequence comparison of the two duplicated segments indicated that hk4S includes the same numbers of genes (33 genes) as the original chromosomal segment but was invaded by 34 retrotransposons and 40 transposons. As expected for heterochomatic sequences, transposable elements at hk4S have high DNA methylation and increased H3K9 methylation while genes are almost unaffected and fully expressed. This was very unexpected and very different from Drosophila in which genes would have been repressed by the vicinity of heterochromatic sequences. In plants, spreading of heterochomatin does not seem to occur. Instead, using the *ddm1* and *met1* mutants, Lippman *et al.* (2004) proposed that heterochromatic epigenetic marks are set up by siRNA instead of SU(VAR) in Arabidopsis, in a more accurate and sequence specific manner than the SU(VAR)-mediated process in Drosophila.

6.1.4 Heterochromatin is an epigenetic state

An in depth review of heterochromatin can be found in Grewal and Jia (2007) but in summary four main epigenetic features are associated with a repressed chromatin state; DNA methylation (in all three sequence contexts), 24nts siRNA (RdDM), histone methylation at H3K9me2 and H3K27me1 and histone hypoacetylation at H4K16. The exact function of histone variants are as yet largely unknown except for the centromeric H3 variant that will be discussed below. To date there is not a clear understanding of the various steps leading to heterochromatin assembly and this is further complicated by the fact that many processes are interconnected. As an example, DNA methylation, histone modifications and RdDM are tightly linked processes. Indeed if *met1* context is maintained over several generations, RdDM becomes misdirected. The data suggest that CG methylation directs non-CG and H3K9 methylation and is essential for long-term inheritance of epigenetic information (Mathieu *et al.*, 2007).

6.2 Heterochromatin organization

In the previous section, the main properties of heterochromatin were described in terms of sequences and chromatin features. These structural features clearly highlight the fact that heterochromatin distribution is not random but rather clustered at centromeres, pericentromeres, telomeres and knobs although repeated sequences can also be found within euchromatic chromosome arms especially when the genome size increases such as in maize or wheat (Figure 6.1a). However, this linear distribution of heterochromatin along the chromosome does not account for its real organization within the nucleus. Strikingly, the very first investigations of chromosome organization within interphase nuclei in Arabidopsis indicated that chromosomes are not arranged in a linear way but rather are distributed within the nucleus with a special location of heterochromatin near the nuclear periphery and around the nucleolus (Fransz *et al.*, 2002). These localizations turned out to reflect specific functional aspects required for proper genome expression. Since then, many nuclear sub-compartments have been described and this nuclear organization of chromatin and sub-compartments are collectively referred to as the 'Nuclear Architecture' (Lanctôt *et al.*, 2007). This section will review our knowledge regarding nuclear architecture focusing on heterochromatin and will give possible mechanisms to explain the positioning of heterochromatin within the nucleus.

6.2.1 Heterochromatin and nuclear architecture

6.2.1.1 Chromosome territories in Arabidopsis

Early studies of chromosome spatial organization came from the comparative analysis of active (Xa) and inactive (Xi) X chromosome in humans (Eils *et al.*, 1996). The results suggested that Xa exhibited a larger surface than Xi territories, which should be in a more condensed heterochromatic domain. According to this model, chromosomes would then occupy discrete nuclear domains known as Chromosome Territories (CTs). CTs are an evolutionary conserved structural feature of chromosome organization among plants and animals and are believed to have functional significance for appropriate genome expression. Chromosome painting was performed on the five Arabidopsis chromosomes using 669 Bacterial Artificial Chromosomes (Figure 6.3a). Chromosome Territories were then defined in Arabidopsis interphase nuclei in which each CT corresponds to about 25Mbp of DNA, a number very similar to the physical chromosome size ranging from 17 to 25Mbp. Chromosome Territories occur in differentiated and actively dividing cells such as meristematic cells, in various tissues including roots and leaves and in cells of different ploidy levels (Pecinka *et al.*, 2004). Positioning of CTs is random, meaning that there is no specific pattern arrangement in respect to the Nuclear Envelope (NE) or between various CTs except for NOR which are clustered at the nucleolus (Berr and Schubert, 2007; Schubert and Shaw, 2011). The nucleolus was found to be localized in most cases at or near the centre of the nucleus. In addition, the nucleolus represents a significant volume excluding the remaining chromatin at a more peripheral position within the nucleus and thus influences the organization of CTs. In addition, CTs are also constrained by nuclear shape and nuclear size (De Nooijer *et al.*, 2009).

6.2.1.2 Chromocentres and the rosette-loop model of chromatin organization

Each interphase chromosome of Arabidopsis consists of a single heterochromatic domain, or chromocentre (CC), from which euchromatic loops of about 0.2–2 Mbp emanate (Van Driel and Fransz, 2004). This has been called the 'Rosette-loop model' of heterochromatin (Figure 6.3b). One possible mechanism of loop assembly may rely on pericentric heterochromatin, which includes ancient transposable elements used as a nucleation site to recruit homologous sequences scattered through the chromosome arms (Fransz and De Jong, 2011). The small genome size of *Arabidopsis* provides a unique model species where most of heterochromatin is organized in chromocentres that can be easily followed by light microscopy as dense chromatin foci (Figure 6.3c). Usually, six to ten CCs can be scored in somatic cells because two or more CCs can fuse together (Fransz *et al.*, 2002; Fang and Spector, 2005; Berr and Schubert, 2007). Chromocentres contain epigenetic markers for silent chromatin, such as DNA methylation and repressive histone marks. In contrast, euchromatic loops are enriched in gene-coding regions and contain acetylated histones as well as H3K4me (Fransz *et al.*, 2002). Together, a given CC and associated euchromatic loops form a chromosome territory.

Euchromatin loops can extend from the heterochromatin domains and form either a single rosette structure or multiple rosettes per chromosome, depending on the genome size. In Arabidopsis, centromeres are randomly distributed in peripheral positions and near the nucleolus (Fransz *et al.*, 2002). Using centromeric H3 variant CENH3 fused to a fluorescent protein, Fang and Spector (2005) clearly showed that centromeres as part of the CC localize predominantly at the nuclear periphery. Further information on CC assembly was gained by studying heterochromatin in protoplasts. Indeed CCs are virtually absent from protoplasts except for heterochromatic foci formed at 45S rDNA. Reduction of pericentric heterochromatin is not accompanied by a global loss of histone H3K9me2, indicating that the H3K9me2 histone mark is not sufficient on its own to induce compaction of heterochromatin. Following heterochromatin during a longer period of protoplast culture, reformation of CCs could then been observed. The repetitiveness of the sequences explains the kinetics of CC reformation starting with 45S rRNA genes (sub-telomeric NOR), followed by the 180 bp repeats (centromeric repeats) and the 5S rDNA genes (pericentric repeats) (Tessadori *et al.*, 2007). Spatial positioning of Arabidopsis centromeres was similar in several diploid cell types including guard cells, small leaf epidermal pavement cells, root meristematic cells and sepal and petal epidermal pavement cells (Fang and Spector, 2005). However, CC conformation changes have been documented as for instance in endosperm where an Endosperm-Specific Interspersed (ESI) heterochromatin has been described (Baroux *et al.*, 2007). In this tissue, the Rosette model still stands but with a more relaxed conformation. Another situation in which CCs are altered is the use of epigenetic modifier mutations such as *ddm1* and *met1*. In both mutants hypomethylation is observed as well as a reduction of CC size (Soppe *et al.*, 2002). Chromocentres can still be detected in the null *met1-3*

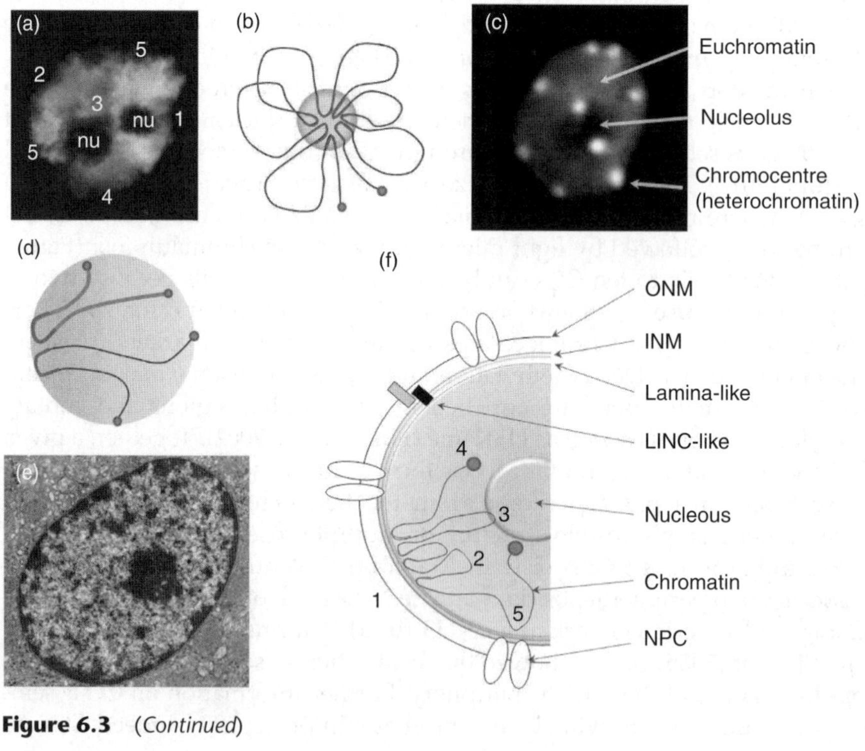

Figure 6.3 (*Continued*)

mutant, where both H3K9m2 and CG methylation are lacking indicating that those epigenetic marks are not essential for initial heterochromatin assembly (Tariq *et al.*, 2003). Using one specific LacO transgenic line called EL702C and containing three copies of the LacO arrays in two separate loci located on chromosome three, Pecinka *et al.* (2005) found that LacO tandem repeats are often co-localized with chromocentres. This co-localization has also been observed for other tandem arrays such as those of transgenic multicopies

---

Figure 6.3 Chromatin organization in the nucleus. a) Chromosome Territories (CTs) were identified by FISH experiments using BAC probes against the five chromosomes from Arabidopsis with five differential labellings. Chromosome painting was then performed on a tetraploid nucleus including two nucleoli (nu) at interphase stage and revealed discrete CTs corresponding to the five chromosomes (1–5). With kind permission from Springer Science and Business media: Pecinka, A., Schubert, V., Meister, A., Kreth, A.G., Klatte, M., Lysak, M.A., Fuchs, J., Schubert, I. (2004) Chromosome territory arrangement and homologous pairing in nuclei of *Arabidopsis thaliana* are predominantly random except for NOR-bearing chromosomes. *Chromosoma* **113**, 258–269. b) The rosette-loop chromosome conformation as defined for Arabidopsis, from Fransz *et al.* (2002). Heterochromatic loci (in red in Plate 6.3) are grouped together at chromocentres and allow euchromatin to form loops. The chromocentre is indicated as a large yellow circle and corresponds to dense foci observed in 1C. Telomeres are indicated as blue circles. c) *Arabidopsis thaliana* nuclei at interphase from cotyledon stage stained with DAPI (kindly provided by Dr Sylvette Tourmente, http://www.gred-clermont.fr). Heterochromatin is more intensely stained due to its large amount of AT-rich sequences and appears as dense foci. Each spot represents a chromocentre although some chromocentres can merge and appear as a single spot. The nucleolus is seen as a non-stained region in the centre of the nucleus. Two chromocentres localize in the vicinity of the nucleolus whereas the six remaining chromocentres are located close to the nuclear periphery. d) The Rabl chromosome conformation is found in various genomes such as yeast (small genome) or wheat (large genome). Telomeres (blue circle at chromosome extremities in Plate 6.3) and centromeres (indicated in red) are located at opposite sites at the nuclear periphery. e) Electron microscopy image a of nucleus from a parietal stomach cell from *Rattus norvegicus*, showing the nuclear membrane, light euchromatin and dark heterochromatin, some of which is attached to the central nucleolus (Photo credit: image from Dr. Jastrow's electron microscopic atlas on the WWW http://www.drjastrow.de). f) Schematic of the chromatin organization in respect to the Nuclear Envelope (NE). The NE is made of an inner nuclear membrane (INM), an outer nuclear membrane (ONM) and contains numerous NPC. A putative plant LINC-like complex is believed to connect the nucleoskeleton and the cytoskeleton. A lamina-like structure has been described in plants although its components remain to be described. Chromatin is represented as a blue line forming rosette-like structures. Heterochromatin (in red) is clustered at the NE (1) whereas euchromatic loops in blue (2) are at more internal locations within the nucleus. Various heterochromatic sequences, such as MARs and TFIIIC binding sites may participate in the recruitment of chromatin to the nuclear periphery. Loci encoding transfer RNA or ribosomal RNA (3) and telomeres (4) are brought into, or close to, the nucleolus (large grey circle). Data from other organisms suggest that other loci (5) may be recruited to the NPC where they may be actively transcribed. Panels D, E and F give insights about chromosome organization within the nucleus. (*For colour details please see colour plate section.*)

of the *HYGROMYCIN PHOSPHOTRANSFERASE* (*HPT*) locus (Probst *et al.*, 2003). In EL702C, LacO arrays are marked with H3K9me2 but a mutation at *SUVH4/KYP* does not alter its co-localization with the chromocentre (Jovtchev *et al.*, 2011).

Altogether, these studies indicate that H3K9me2 is not responsible for CCs formation. Instead, the repeated nature of heterochromatic sequences, histone variants or siRNA could play an important role in heterochromatin assembly while histone marks and DNA methylation would help to maintain and reinforce heterochromatin compaction. Interestingly, Cavalier-Smith (2010) proposed that heterochromatin first evolved from the characteristic folding of repeated sequences at centromeres. This process would have been subsequently reinforced by the emergence of a specific centromeric histone. siRNA would then be viewed as a mediator of the recognition between heterochromatic sequences and heterochromatin establishing proteins.

6.2.1.3 Chromatin organization in large genome species

Chromocentres are not specific to Arabidopsis, although in this species they are easily visualized. Of 67 plants species studied, most display chromocentres in various numbers (Ceccarelli *et al.*, 1998). Organisms with medium-sized genomes (500–3000Mbp), such as tomato contain many heterochromatin domains along their chromosomes. In interphase nuclei, chromosomes form decondensed loops in a rosette-like structure around one or a few number of chromocentres along the chromosome. In mouse, chromocentres are more internally positioned than in Arabidopsis (Fedorova and Zink, 2008). In organisms with very large genomes (more than 3000Mpb), such as barley (5000Mbp) and wheat (16000Mbp), in which a large proportion of the genome consists of transposable elements, chromosomes are more condensed. Therefore, the chromocentre model does not really apply in these species (or at least there is a high number of very small chromocentres) and a high number of chromatin loops occur over the complete length of the chromosome (reviewed in Van Driel and Fransz, 2004). In some species or tissues, chromosomes are in 'Rabl' conformation – i.e. telomeres at one side of the nucleus and centromeres at the other side (Rabl, 1885). The Rabl configuration seems to keep the anaphase configuration of chromosomes in a 'frozen' state until the next cell cycle (Figure 6.3d). This organization of the chromosomes clearly supports the existence of links between centromeres and telomeres and the NE. These links will be discussed in the following section. In wheat and its close relatives, the interphase chromosomes adopt a highly regular Rabl configuration with centromeres and telomeres located at opposite poles of the nuclei (Heselop-Harrison *et al.*, 1990). Peter Shaw's group investigated if heterochromatic marks such as methylation and acetylation could affect the Rabl organization of chromosomes. For that purpose they used 5-azacytidine, which reduces DNA methylation, and Trichostatine A, an inhibitor of histone deacetylase (HDAC), and in both cases, the Rabl configuration was not altered, indicating that the links between centromeres, telomeres and the NE are rather strong (Santos *et al.*, 2002). It is quite difficult to draw a clear picture of why a given species adopts the Rabl conformation, as great variability is observed between species and among tissues or developmental stages. A correlation may exist between chromosome length and organisation where species with large chromosomes adopt the Rabl confomation. Exceptions for this are, however, yeast and Drosophila, which have small chromosomes in the Rabl conformation; Arabidopsis does not.

6.2.2 Recruitment of heterochromatin at the nuclear periphery

Using electron microscopy, heterochromatin has been defined as a dense structure concentrated close to the NE in most eukaryotic nuclei including plant and animals as illustrated in Figure 6.3e. In Arabidopsis, chromocentres clearly co-localize with the NE whereas some are found at the vicinity of the nucleolus (Figure 6.3b). Although preferential localization to the nucleolus has been ascribed to the presence of the NOR at the tips of chromosome two and four (reviewed in Fransz and De Jong, 2011), little is known about the preferential localization of the remaining chromocentres at the NE. Links between heterochromatin and NE should then exist and the NE has a special place within this review as it is believed to participate in chromatin anchorage in plants. A more exhaustive description of NE composition and structure is given in Chapter 2 and the NPC is described in Chapter 3. This section will focus on chromatin-NE interactions. Over the past decades, many interactions involving NE and DNA/chromatin have been described, especially in animals, but with similarities in plants. These data may provide insights into possible mechanisms for maintaining heterochromatin close to the NE in plants.

6.2.2.1 The central role of lamins in animals

From a cytological point of view, heterochromatin is in the immediate vicinity of the internal face of the NE (the Inner Nuclear Membrane, INM) and in contact with the lamina, a thin meshwork made of lamins A, B and C (Figure 6.3f). Lamins belong to the large coiled-coil family of proteins and form a scaffold, which connects the various NPCs and lends mechanical strength to the nucleus. In humans, the premature aging disease Hutchinson–Gilford progeria syndrome (HGPS) is caused by a mutant lamin A (LAΔ50). In LAΔ50, nuclei are abnormally shaped and display a loss of heterochromatin. In humans, H3K27me3 and H3K9me3 identify facultative heterochromatin, whereas H4K20me3 marks constitutive heterochromatin. Consistently with the loss of heterochromatin, constitutive and facultative heterochromatic marks are altered in LAΔ50 (Shumaker *et al.*, 2006). *In vitro* lamin B has some affinity to H2A but not to the other core histones (see Mattout *et al.*, 2007 and references herein) and interactions between lamins and chromatin may involve other protein factors (see below).

6.2.2.2 The inner nuclear membrane and heterochromatin

At least 80 unique proteins have been found localized to the NE in animal cells (Tzur *et al.*, 2006), but few of them have been well characterized. These include KLARSICHT/ANC-1/SYNE HOMOLOGY (KASH), LAMIN B RECEPTOR (LBR), LEM proteins (LAMINA-ASSOCIATED POLYPEPTIDE-EMERIN-MAN1), SAD-1/UNC-84 (SUN) and NESPRIN (see Chapter 2 in this volume). LBR is an integral NE protein interacting both with lamins and heterochromatin. It has a complex function and structure that have been recently reviewed in Olins *et al.* (2010). *In vitro* it binds directly to DNA within the linker region between nucleosomes and interacts with a large number of partners including HP1 and Methyl Binding Protein 2 (MeCP2) (Olins *et al.*, 2010). HP1 has a central role in the assembly of heterochromatin in other organisms such as yeast and Drosophila where it recognizes histone H3K9 methylation while MeCP2 recognizes methylated cytosine at the DNA level. Strikingly, the LBR loss of function led to HP1α mislocalization. Whether this is a direct or indirect effect remains to be determined. As previously stated (see Section 6.2.1.3), chromosome organization in mouse chromocentre-like structures recalls the rosette-loop model adopted by the Arabidopsis chromosomes. Indeed, the LBR knock-out mouse displays a reduced number of chromocentre-like structures, which are more diffused suggesting a more relaxed state of heterochromatin (Cohen *et al.*, 2008).

Another interesting NE component was shown to directly interact with centromeric heterochromatin in fission yeast Schizosaccharomyces pombe. In this species, centromeres are connected to the primary microtubule organizing centre (MTOC; or spindle pole body (SPB), equivalent to the animal centrosome) through interactions between a SUN domain protein called SPINDLE POLE BODY-ASSOCIATED PROTEIN1 (SAD-1) and two KASH domain proteins KARYOGAMY MEIOTIC SEGREGATION PROTEIN1 and 2 (KMS1 and KMS2). Connections are stabilized by the INM protein ISOMALTASE 1 (IMA1) (reviewed in Mekhail and Moazed, 2010). SUN and KASH proteins are part of the LInker of Nucleoskeleton and Cytoskeleton (LINC) complex connecting the cytoskeleton and the nucleoskeleton (see Chapter 2; see also Figure 6.3f in the current chapter). Recently, the first KASH protein has been discovered in plants and has been shown to interact with SUN domain proteins (Zhou *et al.*, 2012). Further investigations will be needed to establish if SUN domain proteins or other LINC-like complexes are used in Arabidopsis to link the NE and the chromocentres, offering one possible mechanism of chromocentre positioning in Arabidopsis.

6.2.2.3 Heterochromatin positioning in plants

In plants, no functional lamina homologue has been described so far, although heterochromatin still localizes peripherally. Electron microscopy performed on tobacco BY-2 cells identified a filamentous structure closely attached to the INM and linked to NPCs. The organization and dimensions of these filaments are very similar to those observed for the lamina in *Xenopus laevis* oocytes used as a control in the experiments (Fiserova *et al.*, 2009). These data suggest the existence of a lamina-like structure in plants but are not a definite proof of its existence. As lamins belong to the coiled-coil family, several researchers then investigated this large gene family in plants. In tomato, the Meier group identified MATRIX ATTACHMENT REGION (MAR) BINDING FILAMENT-LIKE PROTEIN1 (MFP1, Gindullis and Meier, 1999a), MFP1 ASSOCIATED FACTOR1 (MAF1, Gindullis *et al.* 1999b) a serine threonine-rich protein interacting with MFP1 and the FILAMENT-LIKE PLANT proteins (FPP family, Gindullis *et al.*, 2002) also interacting with MAF1. MAF1 was subsequently shown to be the first WPP-domain protein conserved in Arabidopsis. However, WPP-domain proteins were recently shown to be part of the plant NPC and not directly connected with heterochromatin (Meier *et al.*, 2010). A similar approach was developed in carrot, in which a coiled-coil protein called NUCLEAR MATRIX CONTTTUENT PROTEIN1 (NMCP1) was identified as a plant-specific insoluble nuclear protein localized at the nuclear periphery. Using NMCP1 as a starting point for a reverse genetics screen, Dittmer *et al.* (2007) identified a new family of proteins called LITTLE NUCLEI 1-4 also known as CROWDED NUCLEI1-4 (LINC1-4, CRWN1-4) in Arabidopsis. Disruption of either the *LINC1* or *LINC2* gene induces a dwarf phenotype and causes a reduction in nuclear size associated with an increased DNA compaction. Moreover, combining the *linc1* and *linc2* mutations reduced by two-fold the number of chromocentres (8.6 compared to 4.6 chromocentres per nucleus in wild type and double mutant respectively). This suggests that barriers against association between distinct chromocentres are lowered, resulting in the fusion of some of the chromocentres. Altogether, the *linc1 linc2* double mutant displays a reduced nuclear size, an increased DNA compaction and a reduction of the number of chromocentres. One possible explanation could be that the nuclear shape alters chromatin organization or *vice versa*. Although the function and structure of the LINC proteins may recall those of animal lamins, further work on LINC proteins is needed and will determine whether they influence NE structure or modify links between chromatin and NE.

6.2.3 Higher order of chromatin organization

Spatial positioning of heterochromatin in rosette-loops at the chromocentre suggests that heterochromatin can bring distant loci closer together. For Arabidopsis, one possible mechanism has already been proposed in Section 6.2.1.2 in which transposable elements on chromosome arms are recruited to the chromocentre (Fransz and De Jong, 2011). In other species, dedicated sequences and proteins that can fold chromatin into loops were identified with similar properties. In Drosophila, boundary elements like CCCTC-BINDING FACTOR (CTCF), SUPPRESSOR OF HAIRY-WING (Su(Hw)), ZESTE WHITE 5 (ZW5), BOUNDARY ELEMENT ASSOCIATED FACTOR (BEAF) or CENTROSOMAL PROTEIN 190 (CP190) are involved in chromatin organization (Bushey *et al.*, 2008) and some of them have been implicated in chromatin loop formation (Blanton *et al.*, 2003). Some of these proteins were hypothesized to interact with the nuclear lamina, recruiting the boundary elements in a way very similar to that described for heterochromatic sequences of the rosette-loop model (Gerasimova *et al.*, 2000). Boundary elements and their binding partners are thus among the possible mechanisms responsible for chromatin loop formation and defining chromosome territories in eukaryotes.

6.2.3.1 Boundary elements

6.2.3.1.1 Drosophila and animal CTCF In order to recover an enriched fraction of NE-associated chromatin, lamin was used in human and drosophila cell cultures as a fusion with Dam methylase from *Escherichia coli*, the so-called 'DamID' method (Van Steensel *et al.*, 2001). Dam methylase targets adenines while the lamin counterpart tethers the fusion protein to the NE and binds specific DNA sequences. This led to the discovery of Lamin-Associated Domains (LADs) of about 0.5 to 1Mpb each (Pickersgill *et al.*, 2006; Guelen *et al.*, 2008). In Drosophila Kc cells, these domains contain essentially silent genes characterized by a low level of H3K4 methylation and low H4K16 acetylation (see Figure 6.2a). Careful analyses revealed that CTCF-binding sites are at the edges of LADs (Pickersgill *et al.*, 2006). CTCF is a zinc finger protein binding CCCTC motifs; it is well known as both an insulator (enhancer-blocking property) and a boundary element (blocking repressive effects mediated by heterochromatin) (reviewed in Gaszner and Felsenfeld, 2006). To date no CTCF homologues have been identified in plants

6.2.3.1.2 Transcription factor IIIC How do organisms that lack CTCF homologues accomplish the same goal? In yeast, the transcription factor IIIC (TFIIIC) could have a similar function. TFIIIC is an associated factor of RNA Polymerase III (Pol III), which has been linked to insulator activity. TFIIIC binds B-Box elements found in Pol III promoters such as in tDNA, rDNA, U6snDNA or SINE transposable elements all transcribed at the nucleolus. TFIIIC recognizes the 274 tRNA gene clusters distributed throughout the 16 chromosomes and could contribute significantly to organizing the yeast genome by tethering tRNA gene clusters to the nucleolus. This can be viewed as an alternative mechanism to establish or maintain loops, which bring loci from distinct chromosomal positions closer together. As an example, in fission yeast, transition from CENP-A centric heterochromatin to outside repeats coincides with the presence of two to four tRNA genes, which may define boundaries of centromeres (reviewed in White and Allshire, 2008).

Genome-wide studies identified a second type of TFIIIC binding site not associated with RNA Pol III binding. While RNA Pol III-associated TFIIIC loci are recruited at the nucleolus, such sites are most often anchored to the NE and defined as 'Chromosome Organizing Clamp' (COC) (Noma *et al.*, 2006). The recent discovery of hundreds (possibly even thousands) of COC sites in the human genome points towards an important, conserved function for these sites in organizing eukaryotic genomes (Moqtaderi *et al.*, 2010; Oler *et al.*, 2010).

6.2.3.2 Condensin and cohesin

Condensins and cohesins, two chromosome scaffold proteins, are members of the Structural Maintenance of Chromosomes (SMC) family of proteins. They act respectively on chromosome compaction prior to and during mitosis and cohesion between the sister chromatids upon replication during the S phase until mitosis (reviewed in Wood *et al.*, 2010). Arabidopsis cohesins (SMC1 and 3) and condensins (SMC2 and 4) have been identified and mutants in SMC1, 2 and 3 display a similar 'Titan' (TTN) phenotype: giant endosperm nuclei and arrested embryos with a few small cells (Liu *et al.*, 2002). One of the functional roles of TTN is to provide normal microtubule function during seed development, which may explain the enlargement of the nuclei in this tissue. Strikingly, DMS3, another Arabidopsis SMC-like protein, was shown to be involved in the RdDM connecting proteins with nuclear scaffold function and epigenetic regulation (Kanno *et al.*, 2008).

In animals, SMC complexes are bound to chromatin all through the cell cycle and recent investigations have highlighted implications of this binding in some aspects of genome organization. First, cohesin was shown to co-localize with CTCF binding sites. It is even hypothesized that the boundary function first described for CTCF may be achieved by cohesin. Cohesin may be recruited at CTCF binding sites and subsequently bring together unlinked loci to form DNA loops. The current hypothesis would then be that CTCF defines binding sites for cohesin, which in turn induces DNA topology responsible for insulator and/or boundary effects (Wendt *et al.*, 2008). Second, genome-wide studies identified condensin-binding sites at tRNA loci. A mutation in the condensin sub-units induces a loss of preferential localization of tRNA loci to the nucleolus. As for cohesin, it is suggested that condensin facilitates tRNA clustering to the nucleolus by participating in long range interactions between distant chromosome sites (D'Ambrosio *et al.*, 2008).

6.2.3.3 Matrix Attachment Regions

Matrix Attachment Regions (MARs) are A-T rich repeated DNA sequences found in animals and plants. At least 12 proteins were described as binding at MARs in various organisms, including lamins, histone H1, topoisomerase II, as well as plant-specific factors already discussed such as MFP1 and MAF1, which are components of the NE (Wang *et al.*, 2010). In *Mus musculus*, another MAR binding factor called SPECIAL AT-RICH BINDING PROTEIN1 (SATB1) is responsible for repression of numerous genes mediated by deacetylation at histone H3K9 meaning that MARs can also be defined at the epigenetics level (Cai *et al.*, 2003). MARs are attached to the nuclear scaffold and are believed to organize metaphase chromosomes into rosette-like structures forming loops. It is worth noting that transposable elements can be recovered, to a certain extent, as MARs (Tikhonov *et al.*, 2001) according to their possible function in the rosette-loop model in Arabidopsis. The attachment sites of the loops are hypersensitive to DNaseI treatment and co-localize with Topoisomerase II binding sites. Using these two criteria, MARs were identified in plants in which they form loops of various sizes ranging from 25 kb in Arabidopsis to 45 kb in maize (Paul and Ferl, 1998). *In silico* analyses predicted smaller loops around 6–7 kpb as 21705 MAR sequences were detected across the Arabidopsis genome by Rudd *et al.* (2004). Among them, 10% lie within genes where they may influence gene expression and confer tissue, organ, and developmental specificity of gene expression in Arabidopsis (Tetko *et al.*, 2006). Finally, in vertebrates, the β-globin insulator element that is binding CTCF can also be recovered as a MAR (Yusufzai and Felsenfeld, 2004). The authors clearly ruled out the fact that all the sequences isolated as MARs are also insulator elements (there can be as many as 30 000 to 80 000 MARs in the human genome). Alternatively, CTCF may have some yet unknown function in MARs. The exact significance of MARs is still elusive and their effective relationships with the rosette-loop model in Arabidopsis remains to be defined.

6.2.3.4 Future prospects in plants

In summary, interactions of heterochromatin to form chromatin loops may rely on putative structural elements such as transposable elements, boundary elements, MARs, or on protein complexes such SMC or TFIIIC complexes or by directly interacting with the NE or through a lamina-like structure (Figure 6.3f). In plants, no boundary or insulating elements similar to CTCF or other *Drosophila* insulators have been identified. Boundary elements may not be needed and alternatively, as proposed by Lippman *et al.* (2004), siRNA-dependent mechanism may be sufficient to fulfil the function of a boundary element by defining well delimited chromatin domains.

6.3 Functional significance of heterochromatin positioning

The structure of heterochromatin led us to discover that it is composed of repeated sequences and constitutes a specific chromatin state distinct from euchromatin. The study of its organization has revealed that heterochromatin is grouped in specific regions of the chromosome and that this organization may participate in the positioning of heterochromatin at the periphery of the nucleus and euchromatin at a more internal part of the nucleus. If this positioning is a biological reality, it is probably relevant for some functional properties of the cell. In the following section, three main roles in cell functions will be reviewed: chromosome segregation, transcriptional regulation and genome protection against instability. Whether these functions are dependent upon the peripheral positioning of heterochromatin will be discussed.

6.3.1 Centric heterochromatin directs chromosome segregation

Clearly, centromere positioning at the nuclear periphery is an essential process to achieve chromosome segregation. Thus mechanisms dragging chromosomes to each side of the cell can yield clues about one of the possible mechanisms for positioning chromatin at the nuclear periphery. At the sequence level, centromere identity relies on particular chromatin specificities made of dedicated repeated sequences (satellite DNA in plants) and involving small non-coding RNA and specific histone variants such as CenH3/CENP-A and H3.1.

During mitosis and meiosis, centromeric heterochromatin is attached to spindle microtubules through the kinetochore, a proteinaceous structure connecting the centromere and the spindle. CENP-A together with CENP-C are required to build up a fully functioning kinetochore and thus have a central role in the centromeric function in chromosome segregation (Verdaasdonk and Bloom, 2011). In Arabidopsis, the 180bp satellite is believed to define a key centromeric feature of all centromeres. Indeed 180bp arrays bind the centromeric histone variant CenH3 but in extended chromatin fibres studied by FISH, only 10–12% of the repeats bind CenH3. These were compacted chromatin regions forming knobs and were hypothesized to be sites of kinetochore formation (Shibata and Murata, 2004). What makes some of the 180bp competent to bind CenH3 and to form kinetochores remains to be elucidated. CenH3 mutants have been recently obtained and this will provide the opportunity to investigate the functions of the centromeric histone variant in chromocentre positioning (Lermontova *et al.*, 2011).

Once kinetochores are formed, they are connected to spindle microtubules and allow chromosome segregation. It is worth noting that, in plants, spindle attachment does not rely on specialized organelles (centrosomes in animals). Instead, plants are believed to follow a specific mechanism of spindle formation known as 'spindle self-organization' relying on the Ran pathway (reviewed in Zhang and Dawe, 2011). Surprisingly, in this model, Histone H1 acts as a microtubule organizing factor at the nuclear periphery and indeed Histone H1 was shown in tobacco BY2 cells to be localized at the nuclear periphery (Nakayama *et al.*, 2008). The identified histone H1 in this experiment was shown to be similar to one of the two main histone variants from tobacco BY2 cells. The authors observed that histone H1 and DNA do not co-localize suggesting that the recognized histone H1 variant may, in this case, have a distinct function from the classical linker histone association.

In Arabidopsis, H1 variants were characterized and shown to be encoded by three genes. *H1-1* and *H1-2* share extensive homology with each other whereas *H1-3* is a more divergent variant related to a drought-inducible class of gene (Wierzbicki and Jerzmanowski, 2005). RNAi plants with ~90% decrease in the overall level of histone H1 display developmental phenotypes, with stochastic alterations of DNA methylation at various loci including 180bp, 5S rDNA, transposable element and FLOWERING WAGENINGEN (FWA). It is currently difficult to reconcile the two reported functions of histone H1, namely linker histone involved in chromatin compaction (reviewed in Jerzmanowski, 2007) and its microtubule organizing function (Wierzbicki and Jerzmanowski, 2005). This may be solved in the future by carefully looking at each H1 variant to address its specific functions and localization within the nucleus.

In most plant species, centromeres are close to the NE whether or not they display the Rabl configuration. This close proximity to the NE favours rapid interaction with microtubules during NE breakdown. Further investigations are needed to define clearly the function of histone variants and ncRNA or unknown factors in centromere positioning. This may be of importance for deciphering one of the possible mechanisms to achieve heterochromatin positioning.

6.3.2 Spatial positioning of heterochromatin affects transcriptional activity

As previously discussed, chromosomes are spatially organized in interphase nuclei and heterochromatin tends to be situated next to the NE and the nucleolus (see Figure 6.3). As heterochromatin represents a repressive state of chromatin, it has long been suggested that transcription may be repressed at the NE and this has been well described for telomeric silencing in yeast (Akhtar and Gasser, 2007). Yeast and mammals were largely used to investigate the effect of heterochromatin on gene expression but also the effect of gene tethering to the nuclear periphery or to the nucleolus (Towbin *et al.*, 2009; Sáez-Vásquez and Gadal, 2010). However this turned out to be a difficult task, and the effect on transcription of the perinuclear positioning of chromatin is indeed highly dependent upon the chosen locus, how it is recruited to the nuclear periphery and how it is positioned relative to various NE components such as the NPC. What about plants? One example in plants came from the study of *FWA* gene (Soppe *et al.*, 2000). In its wild-type configuration, *FWA* is methylated at the DNA level whereas mutants *fwa-1* and *fwa-2* display a complete loss of DNA methylation associated with a late-flowering phenotype. *FWA* contains in its promoter region some sequence repeats that were mapped as the key determinant of this epigenetic regulation. It is interesting to note that *FWA* is among the 33 genes of the hk4S translocation (see Section 6.1.3). As described above, most genes included within hk4S are not altered by the spreading of repressive marks from the nearby transposable elements as only very slight differences can be recorded in their DNA methylation profiles. However, using the *ddm1* mutant background, Lippman *et al.* (2004) identified a derepression of FWA at the transcriptional level, indicating that genes can be modulated by DNA repeats such as transposable elements but only when they are located close to their transcription units. Is positioning of *FWA* responsible for this derepression? The position (inside or outside a given chromosome territory) of the FWA gene was then determined when *FWA* was in the active (*ddm1* background) and in the silent state (*wild type* background). Unfortunately, no significant difference in the nuclear position of the FWA locus could be identified (Pecinka *et al.*, 2004). Of course this is only one example and, as stated before, the effect of spatial positioning may not apply to every gene or may be dependent upon chromosomal position or sequence environment.

In an attempt to get a more general tool to investigate the function of spatial positioning in gene activity, Eric Lam's group engineered a set of 277 transgenic lines containing a lac operator (LacO) and expressing Luciferase under the CaMV 35S promoter (Rosin *et al.*, 2008). Each locus has been characterized for its chromosomal position and its level of gene expression through the LUC activity. LacO arrays can then be targeted using fluorescent protein fused to the Lac repressor (LacI) in order to track chromosome positioning (a given LacO transgene) in living plants. A similar approach has also been developed by the Matzke group using the tet operator (Matzke *et al.*, 2005). Using *ddm1*, *met1* and 5-azacytidine, they showed that changing the epigenetic status can, for some transgenes, decrease the Luciferase activity and modify the nuclear position of the LacO transgene within the nucleus (Rosin *et al.*, 2008). However, this is not a general rule and the effect differs depending on the chromosomal position of the transgene array (reviewed in Lam *et al.*, 2009).

Gene regulation by spatial positioning is a new field of investigation in plants. Tools to study transgene tethering to sub-nuclear localizations are becoming available and future investigations will be needed to establish whether the nuclear periphery induces a repressive or activation effect in respect to gene expression.

6.3.3 Heterochromatin positioning protects against genome instability

In most organisms, pericentric heterochromatin contains inactive transposable elements that are used to silence the homologous euchromatic copies. Transposable elements are then mainly in a repressed state, controlled by the RdDM pathway in plants. One clear consequence of this genomic organization is that pericentric heterochromatic repeats protect the genome against transposable element mobilization, which would otherwise lead to genome mutagenesis. The repressed state can be released in some epigenetic mutants such as *ddm1* or *met1*, in which transposable element mobilization can then be observed (Mirouze *et al.*, 2009; Tsukahara *et al.*, 2009). In *ddm1*, there is a clear correlation between heterochromatin decondensation illustrated by a reduced number of chromocentres and derepression of heterochromatic sequences.

One striking feature of heterochromatin not yet investigated in plants is that of rDNA as highlighted by recent yeast studies. This is an essential gene encoding ribosomal RNA but it also belongs to heterochromatic sequences due to its tandem repeat organization (see Section 6.1 and Figure 6.1). rDNA is found in all organisms from prokaryotes to eukaryotes where it is organized in tandem arrays, meaning that a gene amplification system maintains rDNA cluster(s) in a tandemly repeated organization. Each species displays a given number of copies: seven in *Escherichia coli*, 150 in *Saccharomyces cerevisea*, 350 in *Homo sapiens*, at least 1000 in Arabidopsis and up to 12 000 in *Zea mays*. rDNA arrays are an evolutionary conserved process to produce a large amount of rRNA without increasing the transcription rate of a given gene but surprisingly not all the copies within an array are transcribed. Why are untranscribed units kept within the rDNA array? Unexpectedly, deletion of untranscribed rDNA copies induces an increased sensitivity of DNA to mutagenic agents. Silent extra rDNA copies were shown to facilitate condensin association and sister-chromatid cohesion, thereby facilitating recombination repair between sister-chromatids. Losing the extra copy thus reduces the capacity of DNA repair using the sister-chromatid. In Arabidopsis, chromosome pairing between sister-chromatids occurs at transgene repeats such as LacO or HPT. Homologous pairing is correlated with DNA hypermethylation at the array (Watanabe *et al.*, 2005). Yeast genetics yielded one more role of condensin, which not only functions in DNA compaction but is also required for the attachment of rDNA arrays from the sister-chromatids to each other with important implication in genome stability. This remains to be studied in plants.

6.4 Perspectives

The definition of heterochromatin relies on various criteria. In this review, three main criteria were used: structure, organization and biological functions (Table 6.1). Despite this, it is difficult to establish a general definition of heterochromatin that would suit all situations encountered. Some scientists will use the repeated nature of DNA sequences, its position on the chromosome, a combination of epigenetic marks or a sub-nuclear localization of chromatin. However, none of them are sufficient by themselves to define the heterochromatin state of chromatin. As an example, repeated gene families such as the *NUCLEOTIDE-BINDING SITE–LEUCINE-RICH REPEAT* (NBS-LRR) disease resistance genes (150 genes in *Arabidopsis*, 400 in rice – McHale *et al.*, 2006) are organized in clusters but cannot be considered as heterochromatic loci; methylation at H3K9 can also be found at euchromatic loci and cannot be strictly assigned to heterochromatin; and recruitment at the nuclear periphery does not always lead to repression as actively transcribed genes may be recruited to NPCs. When considering well established heterochromatic sequences such as the pericentric 5S rDNA repeats at chromosome five, some repeats are clearly expressed whereas others remain repressed. Active and repressed 5S genes are associated respectively with active and repressive epigenetic marks although all these genes are within the same repeated array (Mathieu *et al.*, 2003). Finally, it remains to be elucidated how to define heterochromatin in species with large genomes, such as wheat, in which a large number of transposable elements are scattered along the

Table 6.1 Criteria used in the definition of heterochromatin

Criteria	Heterochromatin Features	Heterochromatin Characteristics	Short Description	References
Structure	DNA sequences	Gene poor	Heterochromatin is made of repeated sequences	Cokus *et al.* (2008)
		Contains satellites	180bp repeats in Arabidopsis	Hosuchi *et al.* (2002)
		Contains transposable elements	Mostly pericentric in Arabidopsis	Arabidopsis Genome Initiative (2000)
		May form knobs	hk4S knob in Arabidopsis	Lippman *et al.* (2004)
		Includes 5S and 45S rDNA	Subtelomeric 45S and pericentric 5SrDNA	Copenhaver & Pikaard (1996); Tutois *et al.* (1999)
	Epigenetic marks	High DNA methylation	CG, CHG and CHH methylation at cytosine residues	Cokus *et al.* (2008)
		Contains repressive histones marks	Enriched H3K9me2, H3K27mel	Liu *et al.* (2010)
		Depleted in activating histone marks	Decreased H3K4me2, H4K16Ac	Probst *et al.* (2004)
		Contains specific Histone variants	H3.1, CenH3 Centromeric variant (CenPA homolog)	Ingouff et Berger (2009)
		Is targeted by non-coding RNA	Heterochromatin targeting by RNA Directed DNA Methylation (RdDM)	Law and jacobsen (2010)
	Non-histone protein binding	Binds specific Su(var) proteins	Arabidopsis LHP1 has functions distinct from animal HP1	Gaudin *et al.* (2001)

(continued)

Table 6.1 (*Continued*)

Criteria	Heterochromatin Features	Heterochromatin Characteristics	Short Description	References
Organization	Chromosome distribution	Located at centromere, telomere, pericentromeres, knob	See DNA sequences	Arabidopsis Genome Initiative (2000)
	Nuclear architecture	Involved in Chromosome Territories	Arabidopsis Chromosomes are organized in CTs	Pecinka *et al.* (2005)
		Forms chromocentre vs Rabl conformations	Arabidopsis Chromosomes form chromocentres	De Jong and Fransz (2011)
	Nuclear positioning at NE	Enriched at nuclear periphery	Repressive context for transcription	Fang and Spector (2005); Andrey *et al.* (2010)
		Under represented at NPC	Activating context for transcription in yeast	Arib and Akhtar (2011)
	Higher order of chromatin	May contain boundary elements, MARs	To be investigated in Arabidopsis	–
Functions	Chromosome segregation	Forms centromere	Satellite, histone variants and epigenetic state are involved	Zhang and Dawe (2011)
	Transcription regulation	Is a silenced state of chromatin	Epigenetic repressive marks, chromatin compaction	Lam *et al.* (2009)
	Protection against genome instability	Repressed transposition and recombination	Transposable elements, rDNA (in yeast)	Mirouze *et al.* (2002); Watanabe (2005)

chromosome arms. It is therefore important to direct future research towards better defining heterochromatin. For instance, in Drosophila, sub-classes in relation to the various criteria set out above have been established (Filion *et al.*, 2010).

While heterochromatin components are well documented, the chronological events leading to its assembly are still largely unknown. Clearly, several studies suggest that histone methylation at H3K9 is not needed to build heterochromatin (Tariq *et al.*, 2003; Tessadori *et al.*, 2007; Jovtchev *et al.*, 2011). Cavalier-Smith (2010) proposed that centric heterochromatin evolved first, the repeated nature being the primary determinant of heterochromatin subsequently stabilized by centrosomal histone variants CENP-A. According to this hypothesis, plasmid partitioning in prokaryotes involves three simple components: a repeated DNA sequence (centromeric repeats in eukaryotes), a centromere binding protein (histone variants in eukaryotes), and an associated motor protein (microtubules in eukaryotes) to separate the segregating plasmids (Wilson and Dawson, 2011). In plants, it will be very relevant to study the role of the various histone variants to determine whether they are needed in the initial steps of heterochromatin assembly.

Nuclear architecture is a recent field of investigation in plants and is hypothesized to participate in the regulation of genome expression. Heterochromatin may be an important component in this mechanism by anchoring chromatin loops, a mechanism that nicely explains the formation of chromocentres in Arabidopsis (Fransz and De Jong, 2011). Chromocentre positioning falls into two main classes as some chromocentres cluster at the nucleolus while others are in close proximity to the NE. Nucleolus association can be explained by the fact that 45S rDNA repeats have to be actively transcribed within the nucleolus and, as a more general rule, genes transcribed by RNA Pol III and Pol I may be recruited to the nucleolus (Sáez-Vásquez and Gadal, 2010). However, preferential association of chromocentres at the nuclear periphery remains elusive. This may be reminiscent of chromosome segregation or other functions discussed in Section 6.3. Further, mechanisms responsible for chromocentre positioning still remain to be discovered in plants. Collectively, authors suggest that the nucleolus, because of its central position, excludes the chromocentres from internal localization except those bearing NOR. Chromosome territories were shown to be randomly distributed but form distinct areas within the nucleus suggesting a mutual exclusion between different territories (Pecinka *et al.*, 2004; Berr and Schubert, 2007; De Nooijer *et al.*, 2009; Andrey *et al.*, 2010). Data coming from other species suggest specific interactions between NE components and heterochromatin, providing an alternative hypothesis that could explain heterochromatin positioning. Many issues are yet unsolved: lamin and LBR are not yet described in plants and KASH proteins have only recently been discovered. The possible implication of SUN and LINC proteins in NE-heterochromatin interactions also remain to be established.

In most species heterochromatin is the main form of chromatin. Its functional significance is constantly growing as our understanding of the cell process is increasing. Research into its assembly, maintenance and positioning will give us a better view of the processes involved in the regulation of genome expression.

Acknowledgements

The authors thank Aline Probst and David Evans for helpful discussions and helpful comments on the manuscript.

References

Akhtar, A. and Gasser, S.M. (2007) The nuclear envelope and transcriptional control. *Nature Reviews Genetics* **8**, 507–517.

Andrey, P., Kiêu, K., Kress, C. *et al.* (2010) Statistical analysis of 3D images detects regular spatial distributions of centromeres and chromocenters in animal and plant nuclei. *PLOS Computational Biology* **6**, e1000853.

The *Arabidopsis* Genome Initiative (2000) Analysis of the genome sequence of the flowering plant *Arabidopsis thaliana*. *Nature* **408**, 796–815.

Arib, G. and Akhtar, A. (2011) Multiple facets of nuclear periphery in gene expression control. *Current Opinion in Cell Biology* **23**, 346–353.

Baroux, C., Pecinka, A., Fuchs, J. *et al.* (2007) The triploid endosperm genome of *Arabidopsis* adopts a peculiar, parental-dosage-dependent chromatin organization. *Plant Cell* **19**, 1782–1794.

Berr, A. and Schubert, I. (2007) Interphase chromosome arrangement in *Arabidopsis thaliana* is similar in differentiated and meristematic tissues and shows a transient mirror symmetry after nuclear division. *Genetics* **176**, 853–863.

Blanton, J., Gaszner, M. and Schedl, P. (2003) Protein:protein interactions and the pairing of boundary elements in vivo. *Genes and Development* **17**, 664–675.

Britten, R.J. and Kohne, D.E. (1968) Repeated sequences in DNA. Hundreds of thousands of copies of DNA sequences have been incorporated into the genomes of higher organisms. *Science* **161**, 529–540.

Brown, S.W. (1966) Heterochromatin. *Science* **151**, 417–425.

Bushey, A.M., Dorman, E.R. and Corces, V.G. (2008) Chromatin insulators: regulatory mechanisms and epigenetic inheritance. *Molecular Cell* **32**, 1–9.

Cai, S., Han, H.-J. and Kohwi-Shigematsu, T. (2003) Tissue-specific nuclear architecture and gene expression regulated by SATB1. *Nature Genetics* **34**, 42–51.

Cavalier-Smith, T. (2010) Origin of the cell nucleus, mitosis and sex: roles of intracellular coevolution. *Biology Direct* **5** (7).

Ceccarelli, M., Morosi, L. and Cionini, P.G. (1998) Chromocenter association in plant cell nuclei: determinants, functional significance, and evolutionary implications. *Genome* **41**, 96–103.

Choulet, F., Wicker, T., Rustenholz, C. *et al.* (2010) Megabase level sequencing reveals contrasted organization and evolution patterns of the wheat gene and transposable element spaces. *The Plant Cell Online* **22**, 1686–1701.

Cohen, T.V., Klarmann, K.D., Sakchaisri, K. *et al.* (2008) The lamin B receptor under transcriptional control of C/EBPε is required for morphological but not functional maturation of neutrophils. *Human Molecular Genetics* **17**, 2921–2933.

Cokus, S.J., Feng, S., Zhang, X. *et al.* (2008) Shotgun bisulphite sequencing of the *Arabidopsis* genome reveals DNA methylation patterning. *Nature* **452**, 215–219.

Copenhaver, G.P. and Pikaard, C.S. (1996) RFLP and physical mapping with an rDNA-specific endonuclease reveals that nucleolus organizer regions of *Arabidopsis thaliana* adjoin the telomeres on chromosomes 2 and 4. *The Plant Journal* **9**, 259–272.

D'Ambrosio, C., Schmidt, C.K., Katou, Y., *et al.* (2008) Identification of cis-acting sites for condensin loading onto budding yeast chromosomes. *Genes and Development* **22**, 2215–2227.

De Nooijer, S., Wellink, J., Mulder, B. and Bisseling, T. (2009) Non-specific interactions are sufficient to explain the position of heterochromatic chromocenters and nucleoli in interphase nuclei. *Nucleic Acids Research* **37**, 3558–3568.

Dittmer, T.A., Stacey, N.J., Sugimoto-Shirasu, K. and Richards, E.J. (2007) LITTLE NUCLEI genes affecting nuclear morphology in *Arabidopsis thaliana*. *Plant Cell* **19**, 2793–2803.

Eils, R., Dietzel, S., Bertin, E. *et al.* (1996) Three-dimensional reconstruction of painted human interphase chromosomes: active and inactive X chromosome territories have similar volumes but differ in shape and surface structure. *Journal of Cell Biology* **135**, 1427–1440.

Fang, Y. and Spector, D.L. (2005) Centromere positioning and dynamics in living *Arabidopsis* plants. *Molecular Biology of the Cell* **16**, 5710–5718.

Fedorova, E. and Zink, D. (2008) Nuclear architecture and gene regulation. *Biochimica et Biophysica Acta (BBA) – Molecular Cell Research* **1783**, 2174–2184.

Feng, S., Jacobsen, S.E. and Reik, W. (2010) Epigenetic reprogramming in plant and animal development. *Science* **330**, 622–627.

Filion, G.J., van Bemmel, J.G., Braunschweig, U. *et al.* (2010) Systematic protein location mapping reveals five principal chromatin types in drosophila cells. *Cell* **143**, 212–224.

Fischer, A., Hofmann, I., Naumann, K. and Reuter, G. (2006) Heterochromatin proteins and the control of heterochromatic gene silencing in *Arabidopsis*. *Journal of Plant Physiology* **163**, 358–368.

Fiserova, J., Kiseleva, E. and Goldberg, M.W. (2009) Nuclear envelope and nuclear pore complex structure and organization in tobacco BY-2 cells. *The Plant Journal* **59**, 243–255.

Fransz, P., De Jong, J.H., Lysak, M. *et al.* (2002) Interphase chromosomes in *Arabidopsis* are organized as well defined chromocenters from which euchromatin loops emanate. *Proceedings of the National Academy of Sciences of the USA* **99**, 14584–14589.

Fransz, P. and de Jong, H. (2011) From nucleosome to chromosome: a dynamic organization of genetic information. *The Plant Journal* **66**, 4–17.

Gaszner, M. and Felsenfeld, G. (2006) Insulators: exploiting transcriptional and epigenetic mechanisms. *Nature Reviews Genetics* **7**, 703–713.

Gaudin, V., Libault, M., Pouteau, S. *et al.* (2001) Mutations in LIKE HETEROCHROMATIN PROTEIN 1 affect flowering time and plant architecture in *Arabidopsis*. *Development* **128**, 4847–4858.

Gaut, B.S. and Ross-Ibarra, J. (2008) Selection on major components of angiosperm genomes. *Science* **320**, 484–486.

Gerasimova, T.I., Byrd, K. and Corces, V.G. (2000) A chromatin insulator determines the nuclear localization of DNA. *Molecular Cell* **6**, 1025–1035.

Gindullis, F. and Meier, I. (1999a) Matrix attachment region binding protein MFP1 is localized in discrete domains at the nuclear envelope. *The Plant Cell Online* **11**, 1117–1128.

Gindullis, F., Peffer, N.J. and Meier, I. (1999b) MAF1, a novel plant protein interacting with matrix attachment region binding protein MFP1, is located at the nuclear envelope. *The Plant Cell Online* **11**, 1755–1768.

Gindullis, F., Rose, A., Patel, S. and Meier, I. (2002) Four signature motifs define the first class of structurally related large coiled-coil proteins in plants. *BMC Genomics* **3**, 9.

Grant-Dowton, R.T. and Dickinson, H.G. (2005) Epigenetics and its implications for plant biology. 1. The epigenetic network in plants. *Annals of Botany* **96**, 1143–1164.

Grewal, S.I.S. and Jia, S. (2007) Heterochromatin revisited. *Nature Reviews Genetics* **8**, 35–46.

Guelen, L., Pagie, L., Brasset, E. *et al.* (2008) Domain organization of human chromosomes revealed by mapping of nuclear lamina interactions. *Nature* **453**, 948–951.

Happel, N. and Doenecke, D. (2009) Histone H1 and its isoforms: contribution to chromatin structure and function. *Gene* **431**, 1–12.

Heitz, E. (1928) Das Heterochromatin der Moose. *Jahrbücher für Wissenschaftliche Botanik* **69**, 762–818.

Henikoff, S. and Dalal, Y. (2005) Centromeric chromatin: what makes it unique? *Current Opinion in Genetics and Development* **15**, 177–184.

Heslop-Harrison, J. and Bennett, M. (1990) Nuclear architecture in plants. *Trends in Genetics* **6**, 401–405.

Hosouchi, T., Kumekawa, N., Tsuruoka, H. and Kotani, H. (2002) Physical map-based sizes of the centromeric regions of *Arabidopsis thaliana* chromosomes 1, 2, and 3. *DNA Research* **9**, 117–121.

Ingouff, M. and Berger, F. (2009) Histone3 variants in plants. *Chromosoma* **119**, 27–33.

Jerzmanowski, A. (2007) SWI/SNF chromatin remodeling and linker histones in plants. *Biochimica et Biophysica Acta* **1769**, 330–345.

Jovtchev, G., Borisova, B.E., Kuhlmann, M. *et al.* (2011) Pairing of lacO tandem repeats in *Arabidopsis thaliana* nuclei requires the presence of hypermethylated, large arrays at two chromosomal positions, but does not depend on H3-lysine-9-dimethylation. *Chromosoma* **120**, 609–619.

Kankel, M.W., Ramsey, D.E., Stokes, T.L. *et al.* (2003) *Arabidopsis* MET1 Cytosine methyltransferase mutants. *Genetics* **163**, 1109–1122.

Kanno, T., Bucher, E., Daxinger, L. *et al.* (2008) A structural-maintenance-of-chromosomes hinge domain-containing protein is required for RNA-directed DNA methylation. *Nature Genetics* **40**, 670–675.

Lahmy, S., Bies-Etheve, N. and Lagrange, T. (2010) Plant-specific multisubunit RNA polymerase in gene silencing. *Epigenetics* **5**, 4–8.

Lam, E., Luo, C. and Watanabe, N. (2009) Charting functional and physical properties of chromatin in living cells, *Current Opinion in Genetics and Development* **19**, 135–141.

Lanctôt, C., Cheutin, T., Cremer, M. *et al.* (2007) Dynamic genome architecture in the nuclear space: regulation of gene expression in three dimensions, *Nature Reviews Genetics* **8**, 104–115.

Law, J.A. and Jacobsen, S.E. (2010) Establishing, maintaining and modifying DNA methylation patterns in plants and animals. *Nature Reviews Genetics* **11**, 204–220.

Lermontova, I., Koroleva, O., Rutten, T. *et al.* (2011) Knockdown of CENH3 in *Arabidopsis* reduces mitotic divisions and causes sterility by disturbed meiotic chromosome segregation. *The Plant Journal* **68**, 40–50.

Lippman, Z., Gendrel, A.-V., Black, M. *et al.* (2004) Role of transposable elements in heterochromatin and epigenetic control. *Nature* **430**, 471–476.

Liu, C., McElver, J., Tzafrir, I. *et al.* (2002) Condensin and cohesin knockouts in *Arabidopsis* exhibit a titan seed phenotype. *The Plant Journal* **29**, 405–415.

Liu, C., Lu, F., Cui, X. and Cao, X. (2010) Histone methylation in higher plants. *Annual Review of Plant Biology* **61**, 395–420.

Ma, J., Wing, R.A., Bennetzen, J.L. and Jackson, S.A. (2007) Plant centromere organization: a dynamic structure with conserved functions. *Trends in Genetics* **23**, 134–139.

Mathieu, O., Jasencakova, Z., Vaillant, I. *et al.* (2003) Changes in 5S rDNA chromatin organization and transcription during heterochromatin establishment in *Arabidopsis*. *Plant Cell* **15**, 2929–2939.

Mathieu, O., Reinders, J., Caikovski, M. *et al.* (2007) Transgenerational stability of the *Arabidopsis* epigenome is coordinated by CG Methylation. *Cell* **130**, 851–862.

Mattout, A., Goldberg, M., Tzur, Y. *et al.* (2007) Specific and conserved sequences in D. melanogaster and C. elegans lamins and histone H2A mediate the attachment of lamins to chromosomes. *Journal of Cell Science* **120**, 77–85.

Matzke, A.J.M., Huettel, B., van der Winden, J. and Matzke, M. (2005) Use of two-color fluorescence-tagged transgenes to study interphase chromosomes in living plants. *Plant Physiology* **139**, 1586–1596.

McHale, L., Tan, X., Koehl, P. and Michelmore, R.W. (2006) Plant NBS-LRR proteins: adaptable guards. *Genome Biology* **7**, 212-1–212-11

Meier, I., Zhou, X., Brkljacić, J. *et al.* (2010) Targeting proteins to the plant nuclear envelope. *Biochemical Society Transactions* **38**, 733.

Mekhail, K. and Moazed, D. (2010) The nuclear envelope in genome organization, expression and stability. *Nature Reviews Molecular Cell Biology* **11**, 317–328.

Mirouze, M., Reinders, J., Bucher, E. *et al.* (2009) Selective epigenetic control of retrotransposition in *Arabidopsis*. *Nature* **461**, 427–430.

Moqtaderi, Z., Wang, J., Raha, D. *et al.* (2010) Genomic binding profiles of functionally distinct RNA polymerase III transcription complexes in human cells. *Nature Structural and Molecular Biology* **17**, 635–640.

Nakayama, T., Ishii, T., Hotta, T. and Mizuno, K. (2008) Radial microtubule organization by histone H1 on nuclei of cultured tobacco BY-2 cells. *Journal of Biological Chemistry* **283**, 16632–16640.

Noma, K, Allis, C.D, Grewal, S.I. (2001) Transitions in distinct histone H3 methylation patterns at the heterochromatin domain boundaries. *Science* **293**, 1150–1155.

Noma, K., Cam, H.P., Maraia, R.J. and Grewal, S.I.S. (2006) A role for TFIIIC transcription factor complex in genome organization. *Cell* **125**, 859–872.

Oler, A.J., Alla, R.K., Roberts, D.N. *et al.* (2010) Human RNA polymerase III transcriptomes and relationships to Pol II promoter chromatin and enhancer-binding factors. *Nature Structural and Molecular Biology* **17**, 620–628.

Olins, A.L., Rhodes, G., Welch, D.B.M. *et al.* (2010) Lamin B receptor. *Nucleus* **1**, 53–70.

Pandey, R., Müller, A., Napoli, C.A. *et al.* (2002) Analysis of histone acetyltransferase and histone deacetylase families of *Arabidopsis thaliana* suggests functional diversification of chromatin modification among multicellular eukaryotes. *Nucleic Acids Research* **30**, 5036–5055.

Paul, A.-L. and Ferl, R.J. (1998) Higher order chromatin structures in maize and *Arabidopsis*. *The Plant Cell Online* **10**, 1349–1360.

Pecinka, A., Schubert, V., Meister, A. *et al.* (2004) Chromosome territory arrangement and homologous pairing in nuclei of *Arabidopsis thaliana* are predominantly random except for NOR-bearing chromosomes. *Chromosoma* **113**, 258–269.

Pecinka, A., Kato, N., Meister, A. *et al.* (2005) Tandem repetitive transgenes and fluorescent chromatin tags alter local interphase chromosome arrangement in *Arabidopsis thaliana*. *Journal of Cell Science* **118**, 3751–3758.

Pickersgill, H., Kalverda, B., de Wit, E. *et al.* (2006) Characterization of the Drosophila melanogaster genome at the nuclear lamina. *Nature Genetics* **38**, 1005–1014.

Probst, A.V., Fagard, M., Proux, F. *et al.* (2004) *Arabidopsis* histone deacetylase HDA6 is required for maintenance of transcriptional gene silencing and determines nuclear organization of rDNA repeats. *Plant Cell* **16**, 1021–1034.

Probst, A.V., Fransz, P.F., Paszkowski, J. and Mittelsten Scheid, O. (2003) Two means of transcriptional reactivation within heterochromatin. *The Plant Journal* **33**, 743–749.

Rabl, C. (1885) Uber Zelltheilung. *Morphol Jahrbuch* 214–330.

Ravi, M. and Chan, S.W.L. (2010) Haploid plants produced by centromere-mediated genome elimination. *Nature* **464**, 615–618.

Rosin, F.M., Watanabe, N., Cacas, J.-L. *et al.* (2008) Genome-wide transposon tagging reveals location–dependent effects on transcription and chromatin organization in *Arabidopsis*. *The Plant Journal* **55**, 514–525.

Rudd, S., Frisch, M., Grote, K. *et al.* (2004) Genome-wide in silico mapping of scaffold/matrix attachment regions in *Arabidopsis* suggests correlation of intragenic scaffold/matrix attachment regions with gene expression. *Plant Physiology* **135**, 715–722.

Sáez-Vásquez, J. and Gadal, O. (2010) Genome organization and function: a view from yeast and *Arabidopsis*. *Molecular Plant* **3**, 678–690.

Santos, A.P., Abranches, R., Stoger, E. *et al.* (2002) The architecture of interphase chromosomes and gene positioning are altered by changes in DNA methylation and histone acetylation, *Journal of Cell Science* **115**, 4597–4605.

Schubert, I. and Shaw, P. (2011) Organization and dynamics of plant interphase chromosomes. *Trends in Plant Science* **16**, 273–281.

Shibata, F. and Murata, M. (2004) Differential localization of the centromere-specific proteins in the major centromeric satellite of *Arabidopsis thaliana*. *Journal of Cell Science* **117**, 2963–2970.

Shumaker, D.K., Dechat, T., Kohlmaier, A. *et al.* (2006) Mutant nuclear lamin A leads to progressive alterations of epigenetic control in premature aging. *Proceedings of the National Academy of Sciences* **103**, 8703–8708.

Soppe, W.J., Jacobsen, S.E., Alonso-Blanco, C. *et al.* (2000) The late flowering phenotype of fwa mutants is caused by gain-of-function epigenetic alleles of a homeodomain gene. *Molecular Cell* **6**, 791–802.

Soppe, W.J.J., Jasencakova, Z., Houben, A. *et al.* (2002) DNA methylation controls histone H3 lysine 9 methylation and heterochromatin assembly in *Arabidopsis*. *EMBO Journal* **21**, 6549–6559.

Tariq, M., Saze, H., Probst, A.V. *et al.* (2003) Erasure of CpG methylation in *Arabidopsis* alters patterns of histone H3 methylation in heterochromatin. *Proceedings of the National Academy of Sciences* **100**, 8823–8827.

Tessadori, F., Chupeau, M.-C., Chupeau, Y. *et al.* (2007) Large-scale dissociation and sequential reassembly of pericentric heterochromatin in dedifferentiated *Arabidopsis* cells. *Journal of Cell Science* **120**, 1200–1208.

Tetko, I.V., Haberer, G., Rudd, S. *et al.* (2006) Spatiotemporal expression control correlates with intragenic scaffold matrix attachment regions (S/MARs) in *Arabidopsis thaliana*. *PLOS Computational Biology* **2**, e21.

Tikhonov, A.P., Lavie, L., Tatout, C. *et al.* (2001) Target sites for SINE integration in Brassica genomes display nuclear matrix binding activity. *Chromosome Research* **9**, 325–337.

To, T.K., Kim, J.-M., Matsui, A. *et al.* (2011) *Arabidopsis* HDA6 regulates locus-directed heterochromatin silencing in cooperation with MET1. *PLOS Genetics* **7**, e1002055.

Towbin, B.D., Meister, P. and Gasser, S.M. (2009) The nuclear envelope – a scaffold for silencing? *Current Opinion in Genetics and Development* **19**, 180–186.

Tsukahara, S., Kobayashi, A., Kawabe, A. *et al.* (2009) Bursts of retrotransposition reproduced in *Arabidopsis*. *Nature* **461**, 423–426.

Turck, F., Roudier, F., Farrona, S. *et al.* (2007) *Arabidopsis* TFL2/LHP1 Specifically Associates with Genes Marked by Trimethylation of Histone H3 Lysine 27. *PLOS Genetics* **3**, e86.

Tutois, S., Cloix, C., Cuvillier, C. *et al.* (1999) Structural analysis and physical mapping of a pericentromeric region of chromosome 5 of *Arabidopsis thaliana*. *Chromosome Research* **7**, 143–156.

Tzur, Y.B., Wilson, K.L. and Gruenbaum, Y. (2006) SUN-domain proteins: 'Velcro' that links the nucleoskeleton to the cytoskeleton. *Nature Reviews Molecular Cell Biology* **7**, 782–788.

van Driel, R. and Fransz, P. (2004) Nuclear architecture and genome functioning in plants and animals: what can we learn from both? *Experimental Cell Research* **296**, 86–90.

van Steensel, B., Delrow, J. and Henikoff, S. (2001) Chromatin profiling using targeted DNA adenine methyltransferase. *Nature Genetics* **27**, 304–308.

Verdaasdonk, J.S. and Bloom, K. (2011) Centromeres: unique chromatin structures that drive chromosome segregation. *Nature Reviews Molecular Cell Biology* **12**, 320–332.

Vongs, A., Kakutani, T., Martienssen, R.A. and Richards, E.J. (1993) *Arabidopsis thaliana* DNA methylation mutants. *Science* **260**, 1926–1928.

Wang, G., Zhang, X. and Jin, W. (2009) An overview of plant centromeres. *Journal of Genetics and Genomics* **36**, 529–537.

Wang, T.-Y., Han, Z.-M., Chai, Y.-R. and Zhang, J.-H. (2010) A mini review of MAR-binding proteins. *Molecular Biology Reports* **37**, 3553–3560.

Watanabe, K., Pecinka, A., Meister, A. *et al.* (2005) DNA hypomethylation reduces homologous pairing of inserted tandem repeat arrays in somatic nuclei of *Arabidopsis thaliana*. *The Plant Journal* **44**, 531–540.

Wendt, K.S., Yoshida, K., Itoh, T., Bando, M. *et al.* (2008) Cohesin mediates transcriptional insulation by CCCTC-binding factor. *Nature* **451**, 796–801.

White, S.A. and Allshire, R.C. (2008) RNAi-mediated chromatin silencing in fission yeast. *Current Topics in Microbiology and Immunology* **320**, 157–183.

Wicker, T., Sabot, F., Hua-Van, A. *et al.* (2007) A unified classification system for eukaryotic transposable elements. *Nature Reviews Genetics* **8**, 973–982.

Widman, N., Jacobsen, S.E. and Pellegrini, M. (2009) Determining the conservation of DNA methylation in *Arabidopsis*. *Epigenetics* **4**, 119–124.

Wierzbicki, A.T. and Jerzmanowski, A. (2005) Suppression of histone H1 genes in *Arabidopsis* results in heritable developmental defects and stochastic changes in DNA methylation. *Genetics* **169**, 997–1008.

Wilson, K.L. and Dawson, S.C. (2011) Functional evolution of nuclear structure. *The Journal of Cell Biology* **195**, 171–181.

Wood, A.J., Severson, A.F. and Meyer, B.J. (2010) Condensin and cohesin complexity: the expanding repertoire of functions. *Nature Reviews Genetics* **11**, 391–404.

Wyrick, J.J. and Parra, M.A. (2009) The role of histone H2A and H2B post-translational modifications in transcription: A genomic perspective. *Biochimica et Biophysica Acta* **1789**, 37–44.

Yusufzai, T.M. and Felsenfeld, G. (2004) The 5′-HS4 chicken β-globin insulator is a CTCF-dependent nuclear matrix-associated element. *Proceedings of the National Academy of Sciences of the United States of America* **101**, 8620–8624.

Zellinger, B. and Riha, K. (2007) Composition of plant telomeres. *Biochimica et Biophysica Acta* **1769**, 399–409.

Zhang, H. and Dawe, R.K. (2011) Mechanisms of plant spindle formation. *Chromosome Research* **19**, 335–344.

Zhou, X., Graumann, K., Evans, D.E. and Meier, I. (2012) Novel plant SUN-KASH bridges are involved in RanGAP anchoring and nuclear shape determination. *The Journal of Cell Biology* **196**, 203–211.

Zilberman, D., Coleman-Derr, D., Ballinger, T. and Henikoff, S. (2008) Histone H2A.Z and DNA methylation are mutually antagonistic chromatin marks. *Nature* **456**, 125–129.

Annual Plant Reviews (2013) **46**, 191–228
doi: 10.1002/9781118472507.ch7

http://onlinelibrary.wiley.com

Chapter 7

TELOMERES IN PLANT MEIOSIS: THEIR STRUCTURE, DYNAMICS AND FUNCTION

Nicola Y. Roberts[1], Kim Osman[1], F. Chris H. Franklin[1], Monica Pradillo[2], Javier Varas[2], Juan L. Santos[2] and Susan J. Armstrong[1]

[1] *School of Biosciences, University of Birmingham, B15 2TT, UK*
[2] *Universidad Complutense, Madrid, Spain*

Abstract: Although the primary role of the telomeres is to protect the chromosome ends from being recognized and processed as DNA double-strand breaks, evidence is emerging that they have a pivotal role in early events in the movement and synapsis of homologous chromosomes in the meiotic pathway. Attention has been paid to the bouquet, a nearly universal event, during which the telomeres cluster on the nuclear envelope (NE) in early prophase I. It has been suggested that their close proximity promotes homologous pairing. We have previously shown in wild-type *Arabidopsis thaliana* that the telomeres are organized around the nucleolus in somatic cells and during the early stages of meiosis. While still associated with the nucleolus, homologous telomeres undergo pairing at the transition from G2 to leptotene at around the same time as assembly of the axial elements. We do not observe a classical bouquet, but as the homologues synapse during zygotene, the paired telomeres occasionally reveal a loose clustering on the NE, which may represent a transient bouquet. As Arabidopsis homologous telomere pairing precedes transient bouquet formation, we have suggested that close juxtaposition of the homologues by virtue of the tethering of the paired telomeres to the NE may facilitate subsequent chromosome alignment and synapsis. Identifying proteins that link the telomeres and the NE has been stimulated by observations in *Schizosaccharomyces pombe*, where meiotic telomere clustering at the spindle polar body (SPB) involves Sad1, a SPB protein that is indirectly connected to a telomere binding protein, Rap1. Sad1 and the related protein UNC-84 from *Caenorhabditis elegans* contain a so-called SUN domain consisting of conserved C-terminal protein

Annual Plant Reviews Volume 46: Plant Nuclear Structure, Genome Architecture and Gene Regulation, First Edition. Edited by David E. Evans, Katja Graumann and John A. Bryant.

regions a few hundred amino acids long. SUN domains are usually found following a transmembrane domain and a less conserved region of amino acids. Similar proteins have subsequently been identified in other yeasts and mammalian species. In the mouse and C. *elegans*, they appear to be required for telomere attachment to the NE and for moving the chromosomes via the telomeres in meiosis. Several components of the NE of plants have only recently been identified. This review focuses on the structure of the telomeres in Arabidopsis and their behaviour in the meiotic pathway. We also discuss recent observations linking a role for the meiotic telomeres and their association with the NE in meiotic prophase I.

Keywords: Arabidopsis; telomeres; meiosis; bouquet; telomere binding proteins; dynamics

7.1 Introduction

Telomeres are specialized nucleoprotein structures that protect the ends of linear eukaryotic chromosomes from degradation and inappropriate recognition and processing as damaged DNA. They are also involved in distributing and organizing chromosomes within somatic and meiotic nuclei. Although their precise role(s) in meiosis is not yet fully understood, telomere-mediated meiotic chromosome movement (Chikashige *et al.*, 1994) and attachment to the NE during prophase I (Moens, 1969) are thought to promote the establishment of homologous chromosome pairing and subsequent synapsis and recombination.

7.1.1 The meiotic pathway

Meiosis is a highly conserved process that occurs in all sexually reproducing organisms, resulting in the production of haploid gametes or spores. During the meiotic pathway, a single round of DNA replication is followed by two rounds of chromosome segregation, which results in a halving of the chromosome number (Roeder, 1997). The diploid state is then restored during sexual reproduction by fusion of the male and female gametes (Figure 7.1).

In most eukaryotes, homologous recombination (HR) is necessary to establish physical connections between homologous chromosomes in order for them to undergo accurate segregation during the first meiotic division (reviewed in Osman *et al.*, 2011). Homologous recombination is initiated during leptotene by the programmed formation of DSBs, which is mediated by the Spo11 protein (Keeney, 2001). Subsequently, homologous chromosomes begin to pair (align) along their length and become connected by a proteinaceous structure called the synaptonemal complex (SC). Formation of the SC is initiated during zygotene, often from the telomeric ends and progresses through to pachytene where homologues can be observed to be fully synapsed (López *et al.*, 2008). The processes of pairing and synapsis have been shown to require progression through the early stages of HR, with the final stages of recombination taking place during pachytene. The

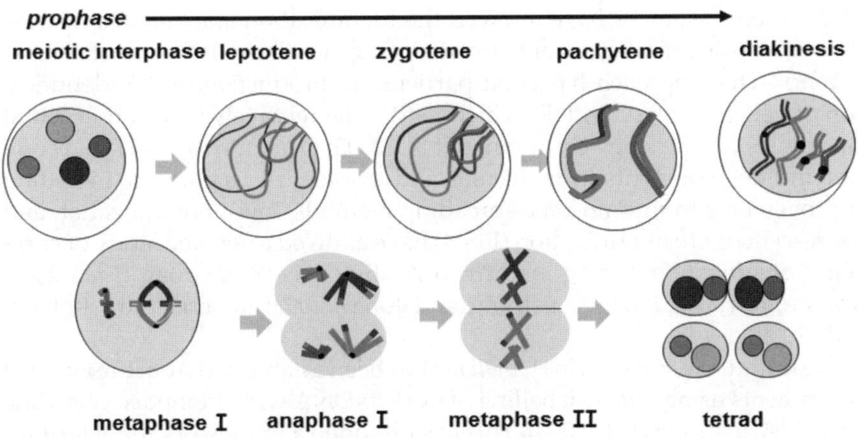

Figure 7.1 The behaviour of two pairs of chromosomes (red and blue in Plate 7.1) during meiosis. (*For colour details please see colour plate section.*)

SC begins to disassemble during diplotene and by diakinesis the products of recombination can be observed cytologically. At this stage homologues are held together at sites of reciprocal crossovers (chiasmata) between non-sister chromatids, which involve the exchange of genetic material between homologues (Jones, 1984). These connections, together with sister chromatid cohesion and the attachment of the spindle apparatus to the kinetochores, are crucial for ensuring the faithful segregation of homologues at anaphase I. Segregation of the homologues at anaphase I and the sister chromatids at anaphase II is permitted by the tightly regulated cleavage of meiotic cohesin proteins (reviewed in Ishiguro and Watanabe, 2007). The second division is less complex and occurs much more quickly than the first, resembling haploid mitosis with sister chromatids separating to each pole.

7.1.2 *Arabidopsis thaliana* as a model for meiosis

A. thaliana (2n = 10) is now well established as an excellent plant model for the study of meiosis, combining a wealth of genetic resources with well-developed molecular and cytological techniques. The significance of Arabidopsis as a model for meiosis is underlined by its relatively close evolutionary relationship with the Brassica crop species (Arabidopsis Genome Initiative, 2000). An understanding of the mechanisms of meiotic recombination in Arabidopsis may help us to develop crop lines with novel traits. For example, knowledge of the factors controlling CO distribution may allow the manipulation of recombination events towards those regions where they would otherwise rarely occur, thereby producing recombinant plants with novel combinations of traits. Arabidopsis possesses a fully sequenced genome (Arabidopsis Genome Initiative, 2000), a range of online databases (for example, TAIR) and several T-DNA insertion mutant collections (Alonso *et al.*,

2003). These resources have allowed the identification and characterization of Arabidopsis orthologues of many meiotic genes through the use of reverse genetics. This approach has been particularly productive in the identification of genes involved in HR, which tend to be relatively well conserved at the species level (reviewed in Jones *et al.*, 2003; Osman *et al.*, 2011). Accompanying developments in molecular and cytological techniques, including improvements to chromosome spreading methods, immunolocalization and fluorescent *in situ* hybridization (FISH) have allowed a detailed study of chromatin organization during meiosis in Arabidopsis (Armstrong *et al.*, 2001; Armstrong *et al.*, 2002; Armstrong and Jones, 2003; Roberts 2009; Roberts *et al.*, 2009).

The duration of meiosis in *A. thaliana* has been established from time course experiments using pulsed labelling of nuclear DNA with bromodeoxyuridine (BrdU) (Armstrong *et al.*, 2003). This has provided a framework for determining the relative timing and duration of key molecular events in meiosis in relation to cytologically defined landmarks. The unique cytologically distinct stages of prophase I (Figure 7.1) have the longest duration in the meiotic pathway, occupying 30 hours of a total duration of approximately 33 hours (Armstrong *et al.*, 2003). The onset of leptotene was defined by reference to the loading of the chromosome axis-associated protein ASY1 (Asynaptic protein 1) (Caryl *et al.*, 2000; Armstrong *et al.*, 2002) as a continuous signal and this permitted the detection of a definite G2 stage which had a maximum duration of 9 hours. The meiotic S-phase has a similar duration to that of G2 and is at least three times as long as S-phase in the somatic cycle cell (Armstrong *et al.*, 2003). This feature of the meiotic S phase has been noted previously in a range of organisms (Bennett *et al.*, 1971; Bennett, 1984) and may be linked to the need for essential preconditions for chromosome pairing (alignment), recombination and synapsis to be established during meiotic interphase. Telomere biology in plants, including Arabidopsis, is now a well established area of research (reviewed in Riha and Shippen, 2003; Watson and Riha, 2010). A number of mutant lines where the progression of meiosis is delayed or the maintenance of telomeres is impaired are now available and the presence of unique sub-telomeric sequences on each chromosome has allowed the study of individual chromosomes using probes specific to each chromosome arm (Riha and Shippen, 2003). In this review, we describe the structure of the telomeres in *Arabidopsis thaliana* and their behaviour during meiosis. We also discuss recent evidence linking the role of telomeres in plants and their association with the NE during meiotic prophase I in the context of chromosome alignment and synapsis.

7.2 The telomeres and associated proteins

Telomeres are protein-DNA complexes that cap the ends of linear chromosomes to protect them from chromosome instability. This protection requires a minimum number of tandem repeats of a short sequence of DNA to allow the binding of a specialized protein complex known as 'shelterin'. With the exception of *Drosophila melanogaster* (Biessmann and Mason, 1990), DNA sequences at the telomeres are highly conserved between organisms (Blackburn, 2001). In plants they are usually formed by TTTAGGG repeats, although some exceptions have been described (Cuñado *et al.*, 2001; de la Herrán *et al.*, 2005 and references therein). The length of arrays of telomeric repeats is species specific, ranging from 2–5 kb in *Arabidopsis thaliana* (Richards and Ausubel, 1988) to up to 60–160 kb in tobacco (Riha and Shippen, 2003). The number of copies of the repeat also differs between chromosomes and between chromosome arms of the karyotype (Schwarzacher and Heslop-Harrison, 1991) and is variable from cell to cell and tissue to tissue (Kilian *et al.*, 1995).

Telomeres are thought to exist in two conformations. The first is an open conformation, which is thought to be necessary for telomerase, the enzyme that synthesizes telomeric sequences (Blackburn, 2001), to access the DNA. The second is a closed conformation (T-loop), thought to be a protected state. The T-loop is formed by the 3′ G-rich single-stranded overhang folding back and invading the telomeric duplex DNA to form a D-loop. Electron microscopy has revealed the existence of the T-loop conformation in the common garden pea, with loops of up to 75 kb being observed (Cesare *et al.*, 2003). Telomeres cannot be completely copied by the normal DNA polymerases of the cell as these require short RNA primers for initiation. Since chromosomes are linear, when the final RNA molecule that primes lagging strand synthesis is removed from the extreme 5′ end of the daughter strand it leaves a gap. This creates a single stranded 3′ overhang, which is usually rich in guanine bases. Due to this end replication problem, telomeric DNA can be progressively lost during successive rounds of DNA replication (Riha and Shippen, 2003; Watson and Riha, 2010). Maintenance of the telomere structure is essential for function. Where telomeres have been degraded past a critical length, chromosome ends lose their protection and are processed as DSBs. This leads to nucleolytic degradation and end-to-end joining of non-homologous chromosomes, resulting in chromosome fusions and genomic instability (Heacock *et al.*, 2004). Telomerase compensates for telomere attrition through the *de novo* addition of DNA repeats onto the chromosome ends by the activity of the highly conserved reverse transcriptase catalytic subunit (TERT) using an associated RNA component (TERC) as a template. For examples, see Tetrahymena (Greider and Blackburn, 1985) and Arabidopsis (Fitzgerald *et al.*, 1999; McKnight and Shippen, 2004). Telomerase can also attach telomeric sequences to new chromosome ends, providing a mechanism to stabilize and repair broken chromosomes (Wang *et al.*, 1992). Telomerase activity in tissues containing dividing cells and telomere length stability during development have been reported in *Silene latifolia* (Riha *et al.*, 1998). Here, the highest telomerase activity was found in germinating seedlings and root tips, whereas a hundredfold decrease in telomerase activity in leaves and no activity in quiescent seeds were observed. Telomerase was also found in mature pollen grains. These findings suggest the precise control of telomere length during plant ontogenesis. However, the telomere length regulation mechanism could be unbalanced during *in vitro* de-differentiation and a conspicuous increase in the telomerase activity in calli of tobacco compared to the source leaves has been reported (Fajkus *et al.*, 1998).

Fitzgerald *et al.* (1999) have shown that Arabidopsis plants that are homozygous for a T-DNA insertion in the telomerase gene lack telomerase activity. This leads to a progressive shortening of the telomeres, from one generation to the next, of approximately 500 bp (Fitzgerald *et al.*, 1999). Plants deficient for telomerase may survive for up to ten generations. The growth and development of *tert* plants has been shown to be indistinguishable from wild type for the first five generations. Subsequent to this, progressively more severe abnormalities in growth and development become apparent. Firstly, defects in leaf or shoot meristem morphology occur. In late-generation mutants, plants exhibit severe defects in reproductive organs, including anthers, and at the terminal phenotype plants arrest in vegetative development (Riha *et al.*, 2001). A critical size threshold for *A. thaliana* telomeres has been identified, below which telomeres readily undergo fusion events. Heacock *et al.* (2007) found that the appearance of telomere fusions corresponded to the presence of at least one telomere of 1 kb or shorter. Numerous anaphase bridges have been observed in late generation *tert* mutants, which correlate with the onset of the developmental defects described (Heacock *et al.*, 2004). Breakage of anaphase bridges has been shown to cause genome rearrangements and aneuploidy (Siroky *et al.*, 2003). In humans and mammals telomerase is switched off in differentiated cells, leading to shortened telomeres and consequently triggering chromosomal rearrangements, which may be involved in cancers or cell death (apoptosis) (Blasco, 2007).

Intriguingly, a novel mechanism of chromosome end protection, which does not rely on single-stranded G-overhangs, has recently been discovered in flowering plants (Kazda *et al.*, 2012). Thus, around 50% of telomeres in *Arabidopsis thaliana* appear to be blunt ended or to contain short G-overhangs of 1-3 nucleotides and are stably-retained in post-mitotic tissue. The integrity of the blunt-ended telomeres is dependent on the Ku70/Ku80 heterodimer but not on another telomere-capping protein, STN1 (see Section 7.2.1). Following Ku-depletion, telomeres remain functional and undergo resection by exonuclease I, which promotes intrachromatid recombination, possibly facilitating the formation of an alternative protective structure. Besides Arabidopsis, Kazda *et al.* (2012) were able to detect blunt-ended telomeres in other Angiosperm plants but not in moss, yeasts and human. This study challenges the view that telomeres require ssDNA overhangs to form functional capping structures and indicates that there can be flexibility in the mechanisms used to protect the ends of chromosomes.

7.2.1 Telomere binding proteins

Mammalian telomeres are bound by a complex, termed 'shelterin', comprising six proteins: TRF1, TRF2, TIN2, POT1, TPP1 and RAP1) (de Lange 2005).

Shelterin has fundamental roles in the protection of chromosomes and in the regulation of telomerase activity at chromosome ends. It is thought to alter the structure of telomeres through its DNA remodelling activity, for example by the formation of T-loops (de Lange, 2005). Some of its component proteins, such as hTRF1 and hTRF2 (Human Telomere Repeat Binding Factor 1 and 2), bind to double-stranded DNA. hTRF1 has been observed to be a negative regulator of telomere length, with overexpression causing shortening of telomeres. hTRF2 has been shown to be involved in the formation and protection of T-loops and hTRF2 deficient telomeres are recognized as DSBs (de Lange, 2005). These proteins bind to DNA via a Myb domain (Kuchar, 2006). Some proteins, for example POT1 (Protection Of Telomeres 1), bind to the single-stranded DNA of the G rich 3′ overhang and they have been shown to bind DNA via an N-terminal oligonucleotide binding fold (OB fold). hPOT1 has a negative regulatory function, preventing telomerase from lengthening the chromosome (Shakirov *et al.*, 2005).

In addition to proteins which bind directly to telomeric DNA, several proteins within the complex are associated via protein-protein interactions. Thus, TIN2 binds TRF1 and TRF2 and recruits TPP1 and POT1. RAP1 (Repressor Activated Protein 1) then forms a complex with TRF2 (de Lange, 2005). It has recently been reported that RAP1 is dispensable for telomere capping but prevents telomere recombination and fragility (Martínez *et al.*, 2010; Sfeir *et al.*, 2010). Furthermore, bouquet formation and telomere attachment to the NE is RAP-1 independent in mammalian meiosis (Scherthan *et al.*, 2011).

S. cerevisiae telomeres are protected by a trimeric complex of proteins, CST, composed of Cdc13, Stn1 and Ten1 (Lustig, 2001; Pennock *et al.*, 2001; Bertuch and Lundblad, 2006; De Lange *et al.*, 2006). The complex has single-stranded DNA binding activity; Cdc13 and Stn1 contain at least one OB-fold and in the case of Cdc13 this is used to bind to the G-rich 3′ overhang (Mitton-Fry *et al.*, 2002; Gao *et al.*, 2007). Association of Stn1 and Ten1 with the overhang is predominantly through interaction with Cdc13. The CST complex has multiple roles in both telomere length regulation and chromosome end protection.

7.2.2 Arabidopsis telomere binding proteins

Telomere binding proteins have been identified in *A. thaliana*. Telomeric double-stranded DNA binding proteins may be defined according to whether they possess a Myb domain in the C-terminal or N-terminal (reviewed in Kuchar, 2006). Proteins containing an N-terminal Myb domain, known as SMH (Single Myb Histone) family proteins, include AtTRB1-3. Arabidopsis telomeric binding proteins containing a C-terminal Myb domain include AtTRP1 and AtTBP1 (Chen *et al.*, 2001; Hwang *et al.*, 2001). AtTBP1 was later shown to be necessary for telomere length homeostasis (Hwang and Cho, 2007). Two further C-terminal Myb-domain proteins, AtTBP2 and AtTRP2, were identified based on sequence homology to AtTBP1 (Hwang *et al.*, 2005). Both proteins were found to contain a putative nuclear localization signal and to be capable of sequence-specific binding to duplex telomeric plant DNA *in vitro*, where they induced a degree of DNA-bending comparable to that in hTRF1.

Based on homology to the human TRF1 and TRF2 proteins, which possess C-terminal Myb domains, Karamysheva *et al.* (2004) defined a total of 12 TRF-like (TRFL) proteins in Arabidopsis, including AtTRP1 and AtTBP1. These form two distinct families. TRFL family 1 proteins (TBP1, TRP1, TRFL1, TRFL2, TRFL4, and TRFL9) possess a highly conserved region C-terminal to the Myb domain, the so-called Myb-extension (Myb-ext) domain, which is absent from TRFL family 2 proteins (TRFL3, TRFL5-8 and TRFL10). The Myb-ext domain confers the ability for *in vitro* binding of plant telomeric DNA on TRFL family 1 proteins. Although there was no evidence of direct telomeric binding of TRFL family 2 proteins, an *in vivo* interaction via protein-protein interaction could not be ruled out for this group (Karamysheva *et al.*, 2004). It was noted that four of the TRFL genes are located in regions of the Arabidopsis genome that are known to be duplicated, with *AtTRP1* and *AtTRFL1* sharing 59% identity (70% similarity) and *AtTRFL3* and *AtTRFL6* exhibiting 52% identity (62% similarity). In view of this it is perhaps not surprising that a number of the TRFL proteins were found to be functionally redundant with single T-DNA insertion lines exhibiting no disruption in telomere length or genetic stability and no plant growth or developmental defects (Karamysheva *et al.*, 2004).

A study by Kuchar and Fajkus (2004) set out to ascertain how known or candidate telomere binding proteins interact in Arabidopsis. AtTRP1, AtTBP1, AtTRB1, AtKu70, and AtPOT1 were analysed using the yeast two-hybrid system. AtTRB1 was shown to interact with two similar proteins, AtTBP2, and AtTBP3, which had previously been observed to interact with each other (Schrumpfova *et al.*, 2004). AtTRB1 was found to interact with AtPOT1, similar to the interaction of human TRF1 with POT1. This study also observed an association between AtKu70 and AtTRP1. As it is known that in humans TRF2 interacts with Ku70, this may imply that the interaction is of functional importance (Kuchar and Fajkus, 2004). It was also postulated that AtTRP1 might be an orthologue of human TRF2 due to this interaction, and the fact that each contains a Myb domain in the C-terminal region. To establish whether these proteins are genuine TBPs, it will be important to demonstrate their interaction with plant telomeres *in vivo*.

In Arabidopsis the most well characterized telomere binding proteins are the POT1 proteins. To date, three orthologues of the yeast and human POT1 proteins have been identified in Arabidopsis, namely POT1a, POT1b, and POT1c (Shakirov *et al.*, 2005 and Rossignol *et al.*, 2007). It was thought that POT1c may have formed by a partial duplication of POT1a (Rossignol *et al.*, 2007). AtPOT1a and AtPOT1b have been shown to contain the structurally conserved OB fold domain, which is present in the POT1 proteins of other organisms. The proteins were identified based on sequence similarity with the *S. pombe* Pot1 protein sequence in a BLAST analysis (http://www.ncbi.nlm.nih.gov) of the Arabidopsis genome. It was observed that both genes were ubiquitously expressed using reverse transcription-PCR analysis and noted that the proteins had a weak, but specific, affinity for ssDNA of plants (Shakirov *et al.*, 2005). Another study by Tani and Murata (2005) observed three variants of AtPOT1a and two of AtPOT1b, suggesting that they are subject to alternative splicing. Antibodies raised against the N-terminus of AtPOT1a were shown to bind to three polypeptides, suggesting that each of the splice variants is translated. To assess the functions of these proteins, Shakirov *et al.* (2005) studied the effect of overexpressing truncations of the AtPOT1a and AtPOT1b genes. It was determined that AtPOT1a functions in telomere length regulation. Overexpression of the C-terminus of AtPOT1a caused telomeres to shorten. The average length of the telomeres after overexpression was 1-1.5 kb shorter than wild type, which suggested that AtPOT1a functioned as a positive regulator of telomere length. AtPOT1b was shown to be functionally divergent from AtPOT1a, with a role in telomere protection and genome stability. Overexpression of the N-terminus of AtPOT1b was shown to cause genome instability, sterility and growth defects, resembling late generation telomerase deficient mutants (Shakirov *et al.*, 2005). It was proposed that the known functions of single-stranded telomere-binding proteins have diverged to form two different proteins in Arabidopsis, as they were found to be functionally non-redundant.

A more recent study has further elucidated the role of AtPOT1a. Rossignol *et al.* (2007) showed that AtPOT1a interacts with the N-terminus of the telomerase TERT sub-unit by yeast two-hybrid analysis and, using GFP tags, demonstrated that these proteins both have a nuclear organization. A role for POT1a in the regulation of telomerase activity through direct interaction with the N-terminal region of TERT was proposed. In addition to this, AtPOT1a was shown to interact with a protein kinase, CIPK21, which, it was suggested, could be involved in DNA damage signalling. These interactors were shown to be specific to POT1a, supporting the findings of Shakirov *et al.* (2005) for divergent roles of the two proteins. Another study illustrated that mutant plants null for AtPOT1a appear morphologically similar to wild type and do not exhibit reduced fertility or growth defects for the first six generations (Surovtseva *et al.*, 2007). This supports previous findings that AtPOT1a is not required for chromosome end protection and is required for the maintenance of telomere length. Terminal restriction fragment analysis was used to measure the length of telomeres, which were found to be significantly reduced in POT1a null mutants and to shorten at the same rate as TERT mutants (Surovtseva *et al.*, 2007). AtPOT1a and TERT were shown by genetic analysis to act in the same pathway, as the *in vitro* levels of telomerase activity were significantly reduced in AtPOT1a null plants (Surovtseva *et al.*, 2007). The level of AtPOT1a associating with the telomeres was found to be cell cycle dependent. Binding of AtPOT1a to telomeres *in vivo* was shown, by chromatin immunoprecipitation, to be enhanced during S-phase when telomerase is active at the telomeres. Shakirov *et al.* (2009) went further by analysing POT1 proteins of two additional Brassicaceae species, *Arabidopsis lyrata* and *Brassica oleracea*. Recombinant POT1 proteins failed to bind single-stranded telomeric DNA *in vitro* or *in vivo* in all the species studied. Several single-stranded telomeric DNA binding activities were found but were associated with proteins, which had a much lower molecular weight than that predicted for the POT1 proteins in *B. oleracea*. It was proposed that the POT1 proteins had functionally diverged from those of other eukaryotes and that they did not represent the major single-stranded telomeric DNA binding proteins in *A. thaliana*. POT1a has recently been shown to be a component of telomerase in Arabidopsis, supporting this view (Cifuentes-Rojas *et al.*, 2011).

The protein domains involved in interactions between AtTRB1 and AtPOT1b, as well as domains engaged in the formation of homomeric and heteromeric complexes of AtTRB proteins, have been characterized by Schrumpfová *et al.* (2008). These results, together with the existence of at least two functionally divergent AtPOT1 proteins in *A. thaliana,* have led these authors to suggest that plant shelterins display specific characteristics not found in animal and fungal shelterins.

While the majority of organisms contain one POT1 protein, mouse telomeres have been found to associate with two POT1 paralogues, mPOT1a and mPOT1b. The mouse proteins share a higher degree of similarity than the Arabidopsis proteins; 72% and 49% respectively. The mouse proteins also exhibit functional divergence. Conditional deletions have shown that mPOT1a is important for preventing recognition of the telomere as damaged DNA and mPOT1b controls the length of telomeric DNA. However, some functional redundancy has been found, as both proteins appear to be important for telomere protection (Hockemeyer *et al.*, 2006). Antibodies generated against mPOT1a and mPOT1b have been shown to co-localize with TRF1, demonstrating that they are located at the telomeres of somatic chromosomes (Hockemeyer *et al.*, 2006).

Arabidopsis POT1 proteins exhibit a different distribution pattern to that described in mouse. Using antibodies to AtPOT1a (kindly provided by Dr D. Shippen) and AtPOT1b (Roberts, 2009), combined BrdU pulse labelling and immunolocalization indicated that AtPOT1a and AtPOT1b localize to somatic and meiotic cells only during S-phase/early G2. During this time the proteins are co-localized with telomeres in the nucleolar region. As pollen mother cells progress through G2 and into meiotic prophase I, telomeres are no longer associated with the nucleolus and AtPOT1a and AtPOT1b staining is lost (Osman and Armstrong, unpublished). This localization pattern is consistent with work carried out by the Shippen laboratory, where AtPOT1a association with telomeres was found to peak in S-phase in suspension culture cells (Surovtseva *et al.*, 2007).

Arabidopsis orthologues of components of the *S. cerevisiae* heterotrimeric CST complex have also been described (Song *et al.*, 2008; Surovtseva *et al.*, 2009). AtSTN1 has been shown to be a single-copy gene with a conserved OB fold (Song *et al.*, 2008). Plants deficient for STN1 showed developmental defects and a reduction in fertility in the first generation. They exhibited extensive telomere erosion together with chromosome end fusions, which occurred at a much greater rate than in plants deficient for TERT. YFP-tagged AtSTN1 was shown to co-localize with the telomere repeat probe in somatic cells. It was proposed that AtSTN1 was indispensable for chromosome end protection (Song *et al.*, 2008). In our laboratory, we have examined *Atstn1-1* mutant lines (kindly provided by Dr D. Shippen) for defects in the meiotic pathway. We have failed to find any meiotic abnormalities in the first generation, where telomere pairing and dynamics were indistinguishable from those of wild type. In the second generation, we found that the majority of meiocytes resembled wild type but a small proportion showed defects, for example, extensive telomeric single-stranded regions during pachytene and chromosome bridges at metaphase II (Osman and Armstrong, unpublished). This is consistent with observations in somatic cells (Song *et al.*, 2008). A functional orthologue of the budding yeast *CDC13* gene has also recently been identified in *A. thaliana*, named Conserved Telomere Maintenance Component 1 (*CTC1*) (Surovtseva *et al.*, 2009). AtCTC1 was found to contain two C-terminal OB folds and to interact with AtSTN1. Immunolocalization of AtCTC1 indicated that, like AtSTN1, it associates with telomeres *in vivo*. Loss of AtCTC1 was shown to cause extensive telomere erosion, resulting in developmental defects, reduced fertility and genome instability. An increase in the length of G-overhangs, and chromosome end-to-end fusions were also observed. Orthologues of *AtSTN1* and *AtCTC1* have also been identified in humans, suggesting the widespread existence of a CST-like complex in multicellular organisms (Surovtseva *et al.*, 2009). This challenged the previously held idea that shelterin and CST have evolved as two distinct telomere capping complexes in vertebrates and budding yeast respectively (reviewed in Linger and Price, 2009). It is now known that these and other telomeric complexes interact (reviewed in Pinto *et al.*, 2011). Telomerase has been shown to interact with shelterin, CST, DNA-dependent protein kinase (DNA-PK) and MRN and there is evidence to suggest that TERT may switch between different complexes (reviewed in Pinto *et al.*, 2011).

7.2.3 DNA repair proteins

Components of the mammalian shelterin complex may interact with several proteins involved in DNA repair, for example RAD50/MRE11, Ku70/Ku80 and a number of RECQ helicases (de Lange, 2005; Ghosh *et al.*, 2011). Homologues of DNA repair proteins have also been discovered in Arabidopsis. AtKu70 and AtKu80 were identified based on their amino-acid sequence homology to human proteins, sharing 28.6% and 22.5% identity with hKu70 and hKu80 respectively (Tamura *et al.*, 2002). Using reverse transcription-PCR both *AtKu70* and *AtKu80* were found to be expressed in all tissues studied, although the levels of expression were low. Yeast-two-hybrid analysis showed that these proteins form a heterodimer as in human cells. Expression of both genes was shown to be induced by treatment with DNA damaging agents, bleomycin and methylmethane sulfonate, which induced DSBs (Tamura *et al.*, 2002). Plants mutant in either of these genes show an increased sensitivity to DNA damage and have elongated telomeres (Riha and Shippen, 2003). These studies provide evidence that these genes are homologous to those in humans and exhibit an evolutionary conserved function.

AtRAD50 and *AtMRE11* have also been characterized in Arabidopsis (Gallego *et al.*, 2001; Bundock and Hooykaas, 2002). Plants carrying mutations in the *AtRad50* gene exhibit shortening of their telomeres and are sterile (Gallego and White, 2001). This study also noted an increase in cell death in cultures of homozygous mutants and it appeared that surviving cells possessed longer telomeres, which, it was suggested, might be indicative of a RAD50-independent mechanism for maintaining the telomeres. Mutants have been previously found to be sensitive to DNA damaging agents showing that the protein is involved in DNA repair at the telomeres (Gallego and White, 2001). *AtMRE11* has also been shown to be involved in telomere maintenance. T-DNA insertion mutants of this gene were observed to be sensitive to DNA damaging agents and telomeres were found to be elongated (Bundock and Hooykaas, 2002). Recently, Amiard *et al.* (2010) have shown that *Atrad50* and *Atmre11* mutants spontaneously activate a DNA damage response dependent on the protein kinase, ATR. These results provide evidence for a role for ATR in avoiding S-phase DNA damage. A dual role for ATM kinase, firstly in regulating telomere size by promoting elongation of short telomeres and, secondly, by preventing the accumulation of cells bearing large telomere deletions, has been demonstrated (Vespa *et al.*, 2007). Both ATR and ATM kinases also prevent propagation of genome damage due to telomere dysfunction (Amiard *et al.*, 2011). Repair of DSBs by non-homologous recombination (NHR), which joins DNA ends with little or no dependence on DNA sequences, has been the focus of research in recent years (Kozak *et al.*, 2009; Jain and Cooper, 2010; Charbonel *et al.*, 2010, 2011).

In humans, several RECQ helicase proteins have been shown to function in telomere maintenance and are associated with rare genetic disorders. For example, RECQL4 was shown to localize to telomeres and bind the shelterin proteins TRF1 and TRF2 (Ghosh *et al.*, 2011). BLM and WRN helicases are thought to resolve secondary structures at the telomeres and interact with TRF1, TRF2 and POT1 proteins (Opresko *et al.*, 2005). Arabidopsis contains several *RecQ*-like genes (Hartung *et al.*, 2000). AtRECQA was implicated in meiosis during an analysis of the meiotic proteome (Sanchez-Moran *et al.*, 1982). This protein was proposed to be a functional homologue of hBLM (Hartung *et al.*, 2007). Immunolocalization of AtRECQ4A shows numerous small foci throughout the chromosomes during early prophase I with considerable enrichment at the telomeres in late zygotene/pachytene (Higgins *et al.*, 2011). Telomeric localization is particularly marked in ecotype Cape Verde Island, which has significantly longer telomeres than Columbia. Mutants deficient in AtRECQ4A exhibited chromatin bridges between telomeres of non-homologous chromosomes that were dependent on recombination. It was proposed that AtRECQ4A enables dissolution of these telomeric associations during meiosis (Higgins *et al.*, 2011). Plants also contain several other classes of helicases that are known from other species to be involved in genome stability, for example FANCM in Arabidopsis has homology with the human Fanconi anaemia proteins (Knoll and Puchta, 2011). AtFANCM has recently been shown to have a role in the control of meiotic crossovers (Crismani *et al.*, 2012; Knoll *et al.*, 2012).

7.3 The behaviour of the telomeres in meiosis

7.3.1 The bouquet

The dynamics of telomere behaviour in meiosis has been known for many years, particularly the arrangement of the telomeres in early prophase I where they form a characteristic organization, known as the bouquet, which is widely conserved amongst many sexually reproducing eukaryotes (Bass *et al.*, 2000 and references therein). It was named from its resemblance to a bouquet of cut flowers and was first observed in the late 19th century (Scherthan, 2001 and references therein). The bouquet consists of a cluster of telomeres in a polarized organization, attached to the inner NE, as shown in Figure 7.1. The bouquet is formed during the leptotene-zygotene transition of prophase I, and persists until pachytene (Bass, 2003). Observations on the distribution of somatic telomeres in rapidly dividing cells have been collected for over a century following the observations of Rabl (1885). He found, using salamander cells, that the centromeres occupied one hemisphere of the nucleus and the telomeres occupied another, and this is now widely known as the 'Rabl organization'. Although this type of organization was subsequently observed in many plants and animals, for example the Triticeae cereals with large chromosomes (Harper *et al.*, 2004), it has become apparent that this is not the rule for all species. Fluorescence *in situ* hybridization (FISH) with probes to the telomeric consensus sequence and species-specific pericentromeric heterochromatin has shown that the telomeres and centromeres appear to be distributed throughout the nucleus in some plant species, e.g. Sorghum and rice (Dong and Jiang, 1998).

The movement of the telomeres during meiosis is thought to involve the location of the telomeres at the inner surface of the nuclear envelope. The telomere attachment at the NE is adjacent to the MTOCs, the centrosome in animals and the spindle pole body (SPB) in fungi (reviewed in Scherthan, 2001). In plants there appears to be no identified MTOCs and consequently there is an absence of an obvious attachment site. On the other hand, Cowan *et al.* (2001) have reported that the telomeres cluster in the microtubule-poor region in rye, and previously Holm (1977) found that in *Lilium longiflorum* cytoplasmic microtubules are rarely found at the bouquet.

The bouquet is formed *de novo* and requires chromosomes to switch their orientation, i.e. from the somatic Rabl organization to the meiotic bouquet formation (Harper *et al.*, 2004). This is apparent in a mono-telocentric wheat line, where the centromeric end of the telocentric chromosome initially associates with the other centromeres but switches to associate with the bouquet (Corredor and Naranjo, 2007). Further, in early meiotic prophase I of hexaploid wheat, there is a transition at which the typical centromeric association is lost whereas the telomeres aggregate into the bouquet (Aragón-Alcaide *et al.*, 1997; Martinez–Pérez *et al.*, 1999). The telomeric cluster of the bouquet also varies between species, ranging from those where little inter–telomere distance is visible, such as rye and wheat, to others with loose telomere clustering, for example maize and lily (Cowan *et al.*, 2001). Duration of the bouquet has also been reported to vary between the sexes (Roig *et al.*, 2004).

The bouquet is known to form in two stages. Telomeres first become attached to the NE and subsequently move along the NE to form a cluster (Golubovskaya *et al.*, 2002). Formation of the bouquet has been shown to require the telomeric repeat sequence. Ring chromosomes without telomeric repeats have been shown not to attach to the NE, for example in *S. pombe* and the mouse (Naito *et al.*, 1998; Voet *et al.*, 2003). In contrast, maize telocentric and ring chromosomes containing telomeric repeats have been shown to participate in bouquet formation (Carlton and Cande, 2002).

7.3.2 A role for the bouquet

The accepted hypothesis to date has been that telomere clustering during the bouquet may promote pairing and synapsis of homologous chromosomes (Scherthan *et al.*, 1996; Bass *et al.*, 2000; Naranjo and Corredor, 2008). It is thought to aid pairing by bringing homologues into close association, thereby allowing strand invasion to occur (Harper *et al.*, 2004). Bass *et al.* (2000) observed that homologous chromosomes paired and synapsed during the bouquet stage, suggesting that bouquet formation is necessary for homologue pairing in maize. Studies in a number of species, including humans, have also observed that synapsis is initiated close to the telomeres (Barlow and Hulten, 1996; Corredor *et al.*, 2007; López *et al.*, 2008). However, there is also evidence that contradicts this hypothesis. For example, Storlazzi *et al.* (2003) studied meiosis in *Sordaria macrospora*, and showed that homologue pairing took place prior to formation of the bouquet, suggesting that it is not necessary for the homology search process. Further support for a role of the bouquet in pairing, has come from studies of mutants defective in bouquet formation. Pairing has been found to be delayed in mutant budding and fission yeasts, where clustering is not present (Trelles-Sticken *et al.*, 2005).

The first protein discovered to be necessary for bouquet formation was Ndj1 (Non-Disjunction 1). This was identified in a screen for mutants defective in bouquet formation (Conrad *et al.*, 1997). Trelles-Sticken *et al.* (2000) observed that in *ndj1*Δ meiocytes, telomeres did not cluster into a bouquet but instead remained dispersed throughout the nucleus. This was shown to result in a delay in pairing of 2–3 hours. Pairing of larger chromosomes was delayed less than smaller chromosomes, presumably due to there being a higher chance of homologues finding each other randomly. Based on these findings, Trelles-Sticken *et al.* (2000) proposed that the bouquet may not be required for the homology search, but may facilitate homologue pairing by bringing telomeres into close association. Wu and Burgess (2006), in agreement with this, observed that single end strand invasion levels of *ndj1*Δ were similar to those of wild-type meiocytes. Rockmill and Roeder (1996) demonstrated that in budding yeast the chromosome ends in which the sub-telomeres are non-homologous were able, to a limited extent, to contribute to homologous recombination. It is possible that factors other than telomere clustering may contribute to the homology search and presynaptic alignment. In *S. pombe*, Cooper *et al.* (1998) showed that, in the absence of the telomere-associated protein Taz1, the clustering of telomeres at the SPB was disturbed and meiotic recombination levels were subsequently reduced. Mice deficient for telomerase, in which telomere lengths are severely reduced, show defects in pairing and synapsis (Liu *et al.*, 2004). In maize, a bouquet mutant has been described, named *pam1*, in which telomeres associate with the NE, but fail to cluster into the bouquet (Golubovskaya *et al.*, 2002). This defect was shown to impair pairing and synapsis of homologues, which resulted in a concomitant delay in meiotic progression. However, some meiocytes were observed to complete meiosis, suggesting that bouquet formation is not required for meiosis (Golubovskaya *et al.*, 2002). The bouquet is also achieved in haploids of rye and a considerable amount of non-homologous synapsis is observed. Further, in this situation some reciprocal recombination occurs near the telomeric regions (Santos *et al.*, 1994 and references therein). In many of these organisms, pairing has been shown to require formation and repair of DSBs, which is likely to involve the homology-recognition process (Zickler, 2006). However, in budding yeast *spo11* and *rad50S* recombination mutants, it appears that a bouquet is still formed, indicating that this step is recombination and synapsis independent (Trelles-Sticken *et al.*, 1999).

Recent work by Ding *et al.* (2012) has provided evidence in *S. pombe* that following on from attachment and clustering of the telomeres to the nuclear envelope subsequent pairing involves recognition of complexes containing meiotic-specific non-coding RNA transcripts. A number of other roles for the bouquet have been suggested, which include regulation of recombination and interlock resolution (for discussion see Zickler and Kleckner, 1998; Koszul and Kleckner, 2009). It has been suggested from studies in *S. pombe* that telomere-mediated chromosome movements promote pairing of chromosomes (Chikashige *et al.*, 1994). Indeed, recent studies have shown that when chromosome motility in prophase I is perturbed, pairing is delayed and synapsis and recombination are impaired (Sheenan and Pawlowski, 2009; Brown *et al.*, 2011). For further discussion see comments below.

7.4 Telomere dynamics in *Arabidopsis thaliana* meiosis

In Arabidopsis, the arrangement of centromeres and telomeres during interphase and early prophase I of meiosis have been studied using FISH to spread pollen mother cells (PMCs). In meiotic interphase cells, we have found that the centromeres remain unpaired and are widely dispersed and peripherally located in the nucleus, whilst the telomeres maintain a persistent association with the nucleolus throughout meiotic interphase (Armstrong *et al.*, 2001). The arrangement is similar to that observed in the Arabidopsis-cycling somatic cells. In Arabidopsis, homologues are not preferentially associated with each other prior to the onset of meiosis. Using a pair of distal and interstitial BAC probes from chromosome 1 we have found that there is no preferential association between homologous chromosome BACs in the meiotic interphase (Roberts, 2009). During the G2-to-leptotene transition the telomeres, still associated with the nucleolus, progressively associate in pairs, as shown in Figure 7.2a. FISH studies with sub-telomeric probes have demonstrated that during early leptotene, pairing of telomeres involves homologous chromosomes (Figure 7.2b). However, it is still unknown how homologous chromosomes recognize one another in plants (Armstrong *et al.*, 2001). It is thought that as the telomeres are composed of the same consensus sequence, the chromosome specific sub-telomeric sequences may be involved in homologue recognition. As leptotene progresses, the paired telomeres dissociate from the nucleolus and become widely dispersed to the NE. During zygotene, telomeres form a loose cluster within one hemisphere of the nucleus, which may represent a relic bouquet (Armstrong *et al.*, 2001). Pairing takes place whilst telomeres are associated with the nucleolus and the characteristic bouquet arrangement is not formed in Arabidopsis. It has therefore been proposed that the nucleolus-associated clustering could be equivalent to the bouquet of other species (Armstrong *et al.*, 2001). Nevertheless, in spreads of PMCs the paired telomeres are found at the NE from zygotene until diakinesis when they are observed to be dissociated from the envelope. Sections of young buds demonstrate that the telomeres remain attached to the NE until diplotene; after this stage they dissociate from the NE, which is itself dismantled (Figure 7.3).

7.4.1 Meiosis in *A. thaliana* telomere-deficient lines

We have investigated the role of Arabidopsis telomeres during meiosis by studying a generation eight *tert* mutant with severely shortened telomeres. Previous studies of the Arabidopsis *tert* mutant have observed a number of developmental defects, including reduced fertility (Riha *et al.*, 2001). Analysis of late generation telomerase deficient oocytes in mice has revealed defects in meiotic synapsis and recombination (Liu *et al.*, 2004). It was found that the number of oocytes that formed an SC per ovarian section was severely reduced. In addition to this, the numbers of MLH1 foci (indicative of crossover sites) were significantly reduced in late generation telomerase-deficient mice (Liu *et al.*, 2004). FISH analysis using sub-telomeric chromosome-specific

(a)

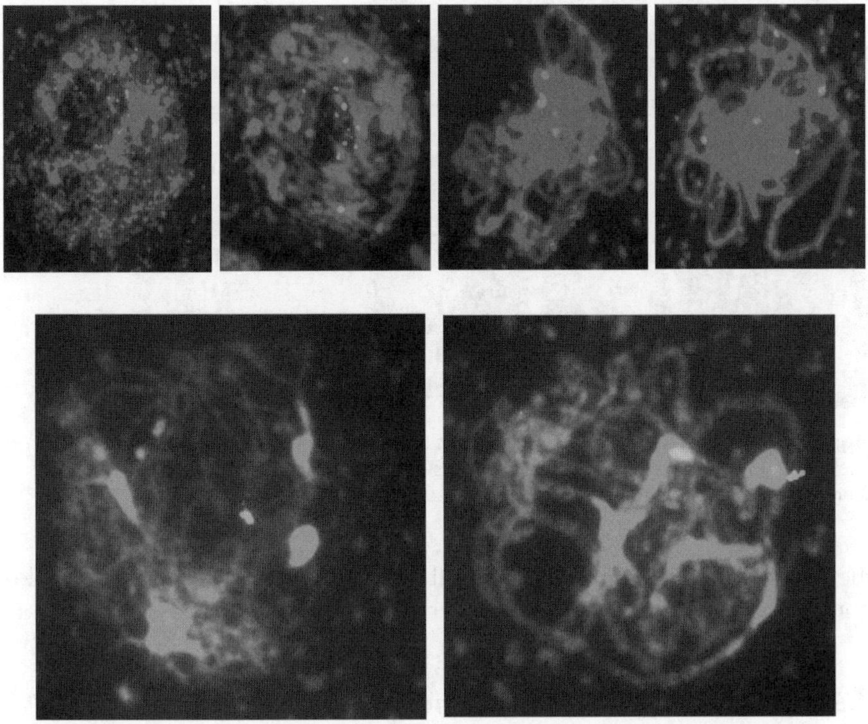

(b)

Figure 7.2 Panel a). FISH of WT Arabidopsis pollen mother cells probed with the telomere repeat probe (red), counterstained with DAPI, showing the dynamics and pairing of the telomeres. The first cell is in meiotic interphase, the telomeres are situated around the nucleolus and are unpaired (up to eighteen signals can be observed), in the second cell, at leptotene the telomeres start to pair in the nucleolar region (up to eleven signals observed). Following on from this paired telomeres (around eight signals) move to the nuclear periphery in the subsequent zyogotene and pachytene cells. Panel b). FISH with 2 BACs (subtelomeric probe detected with FITC, interstitial probe detected with Texas Red) of WT Arabidopsis pollen mother cells. In the first panel the cell is at leptotene the subterminal probe is paired (one signal) whereas the interstitial probe is unpaired at this stage, in the second cell at pachytene both probes are paired. (*For colour details please see colour plate section.*)

BACs showed that telomeres fail to pair in Arabidopsis generation eight *tert* mutants. Failure of telomere pairing did not result in failure of synapsis, as indicated by immunolocalization of the SC protein AtZYP1. Wild-type levels of chiasmata were also observed, indicating that recombination was not affected (Roberts, 2009). This shows that in Arabidopsis, telomere pairing is dispensable for pairing and synapsis of homologues. Whether the timing of meiotic progression is affected in this mutant remains to be elucidated. We postulate that the reduction in fertility in this mutant is due to defects in reproductive development rather than meiotic progression. It is possible that

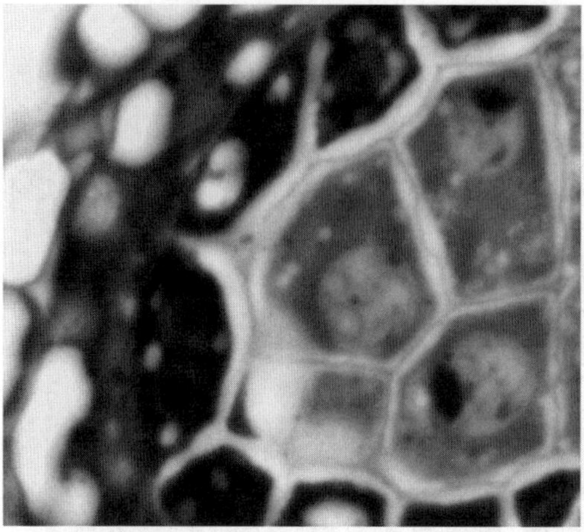

Figure 7.3 LM of part of a WT Arabidopsis anther showing several zygotene/pachytene cells. The telomeres in the nucleus close to the centre of the image can be seen to be associated with the NE.

the localization of telomeres around the nucleolus is sufficient for homologous chromosomes to find one another and pair in this species.

7.5 How are the telomeres moved in meiotic prophase I?

The mechanism of bouquet formation is not well understood. Recent studies have focused on the mechanism by which telomeres and chromosomes move. It has been speculated that, because in mammals and yeast, telomeres are either associated with the centrosome or spindle pole body respectively, cytoplasmic microtubules may be involved in telomere clustering (for example, see Scherthan *et al.*, 2011). Numerous investigations using drugs to disrupt the progression of meiosis have been used to study the mechanisms that are involved in movement of the telomeres around the cell during meiosis and, importantly, during bouquet formation and subsequent synapsis, in a range of species (Driscoll and Darvey, 1970; Salonen *et al.*, 1982; Corredor and Naranjo, 2007). The most widely studied of these is colchicine.

7.5.1 Colchicine disrupts meiotic progression

Colchicine is a microtubule depolymerizing drug, which is known to disrupt spindle formation. It has been known for many years to disrupt the progression of meiosis, blocking the later stages due to its effect on spindle formation and leading to mis-segregation of chromosomes at anaphase. This has been observed in a number of plant species, including Allium (Levan, 1939) and Lilium (Bennett *et al.*, 1979). Treatment with colchicine has also been shown to reduce the frequency of chiasma, as noted in Lilium (Bennett *et al.*, 1979),

Triticum (Driscoll *et al.*, 1967) and Triticale (Thomas and Kaltsikes, 1977) and to disrupt the formation of the SC in many organisms, including Allium (Loidl, 1988) and mouse (Tepperberg *et al.*, 1997). For a detailed discussion see Zickler and Kleckner (1998).

Evidence for an effect of colchicine on bouquet formation first came from experiments investigating the timing of the colchicine sensitive stage; the point of the pathway where meiosis is disrupted, which has been shown to differ between species. A number of studies have found the leptotene-zygotene transition to be the sensitive point. Bennett *et al.* (1979) found that, in Lilium, sensitivity to colchicine started at pre-meiotic interphase and lasted until mid-zygotene, with treatments after this stage having no consequence for meiosis. This is also the case in mice. Tepperberg *et al.* (1997) observed that the number of ^{3}H-thymidine-labelled spermatocytes with damage to the SC was highest after injection of colchicine at the leptotene-zygotene transition. A study of the hybrid hexaploid Triticale x tetraploid wheat also found that the colchicine sensitive stage was the bouquet at leptotene-zygotene (Thomas and Kaltsikes, 1977). Driscoll *et al.* (1967) observed that addition of colchicine prior to meiosis prevented the association of pre-meiotic homologues in hexaploid wheat. It was also found that after pairing of homologues had been achieved, colchicine no longer had an effect.

Evidence that the colchicine-sensitive stage was during the actual formation of the bouquet came from a later study, in which a wheat line carrying an isochromosome was treated with colchicine and shown to pair and form chiasma at the same frequency as an untreated control (Driscoll and Darvey, 1970). It was proposed that the movement of homologous chromosomes (bouquet formation) was sensitive to colchicine. An isochromosome consists of two homologous arms joined via a centromere such that movement of chromosomes is not required for the homologous sequences to come into close association. This is in contrast to normal homologous chromosomes, which only come into close association during the bouquet. As the homologues of the isochromosome were already closely associated prior to bouquet formation, colchicine had no effect (Driscoll and Darvey, 1970).

Cowan and Cande (2002) used cultured rye anthers to investigate the mechanism of bouquet formation. This study used a number of microtubule depolymerizing drugs to determine whether microtubules were necessary for telomere movement and bouquet formation. The authors found that depolymerization of cytoplasmic microtubules did not affect clustering of the telomeres. On the other hand, they demonstrated that colchicine and a related drug, podophyllotoxin, did prevent clustering of telomeres during meiosis. Colchicine was added to anthers in culture at pre-meiotic interphase/early leptotene for 10–16 hours. A FISH analysis was carried out with a probe specific for the telomere repeat sequence to investigate the distribution of telomeres. Colchicine prevented bouquet formation at a concentration, which had no effect on cytoplasmic microtubule organization. These results led to the proposal that in rye, the target of colchicine may be involved in the movement of telomeres during bouquet formation but that this did not include cytoplasmic microtubules. Cowan and Cande (2002) suggested that the target of colchicine may be a specialized non-microtubule tubulin or related protein because colchicine and podophyllotoxin bind to the same site of β-tubulin. Colchicine has also been shown to inhibit bouquet formation and result in impaired synapsis in wheat-rye addition lines (Corredor and Naranjo, 2007). Both of these studies suggest that bouquet formation in these cereals may play a role in homologous chromosome pairing (Cowan and Cande, 2002; Corredor and Naranjo, 2007).

In Arabidopsis we have found that pairing of telomeres and progression of meiosis is not affected by treatment with 100 μM colchicine unlike in cereals (Roberts, 2009). Treatment with 5 mM colchicine also failed to prevent progression of meiosis, although mis-segregation was observed at metaphase II indicating that the colchicine was adequately taken into the transpiration stream. A BrdU time-course showed that colchicine treated plants completed meiosis within a wild-type time frame. Combined BrdU and FISH analysis using a probe directed against the telomere repeat sequence showed no significant difference in the numbers of telomere probe foci between colchicine treated plants and a wild-type control. This difference in colchicine sensitivity could be due to the unique localization of telomeres in Arabidopsis. It is known that telomere pairing occurs while they are localized around the nucleolus and is independent of subsequent NE attachment (Armstrong *et al.*, 2001).

Microtubules have been shown to control telomere movements in *S. pombe*. In contrast to this, Trelles-Sticken *et al.* (2005) observed that microtubule-disrupting drugs had no effect on the movement of telomeres into the bouquet in budding yeast. In this study, telomeres were visualized using a Rap1-GFP construct and live cell imaging. To examine the role of microtubules in telomere movement, the drugs nocodazole and benomyl were simultaneously used in the culture medium. It was observed that astral and radial microtubules were broken down and that addition of these drugs at a number of different time points throughout meiosis did not block telomere clustering. Meiocytes deficient for the *rec8*Δ cohesin, which show a persistent bouquet arrangement, were also treated. In this mutant, telomere clustering did not reach the level achieved in untreated cells. It was also reported that disruption of astral microtubules led to a small reduction in telomere clustering in wild-type and *rec8*Δ cells and Trelles-Sticken *et al.* (2005) suggested that microtubules may provide a scaffold support for telomeres to undertake ordered movements.

7.5.2 The role of actin in telomere movement

Research has also focused on the role of actin in the movement of telomeres. To examine this, meiocytes were treated with the chemical Latrunculin B (LatB), which blocks the polymerization of G-actin. Using live-cell imaging of Rap1-GFP, Trelles-Sticken *et al.* (2005) observed that addition of LatB to a budding yeast wild-type meiotic culture prevented clustering of telomeres. They also found that after incubation of *rec8*Δ meiocytes with LatB the usual persistent clustering of telomeres was dispersed, providing evidence for a role of actin in the movement of telomeres. Disruption of telomere clustering resulted in a delay in pairing of homologous chromosomes, as has been previously observed for colchicine. It was proposed that actin was only important for the clustering of telomeres in meiosis, as movements of telomeres in vegetative nuclei were not affected. Telomere movement on the NE was prevented but disruption of actin did not affect telomere attachment to the membrane. This demonstrates that actin is not required for attachment. Whether cytoplasmic or nuclear actin is involved in clustering was not ascertained in this study (Trelles-Sticken *et al.*, 2005).

Further work in budding yeast has unveiled the nature of actin-mediated chromosome movements. Analysis of Zip1-GFP lines revealed that the chromosome at the leading edge of the motion moves dramatically and the remaining chromosomes then move in the direction of this lead chromosome (Koszul *et al.*, 2008). Rap1-GFP labelling demonstrated that telomeres were present at the leading end (Koszul *et al.*, 2008). These dramatic movements also cause protrusions in the NE and were shown to be associated with the onset of zygotene and continued throughout pachytene (Scherthan, 2007; Koszul *et al.*, 2008). Cytoskeletal, rather than nuclear, actin cables were found to be involved, with a few being intimately associated with the NE. These associated actin cables were mostly observed close to the nucleus/cell periphery and the SPB. It was proposed that this might explain the location of the bouquet in yeast (Koszul *et al.*, 2008).

The precise role of these mid-prophase I chromosome movements remains to be elucidated. It has been proposed that these movements may create stirring motions, which could bring homologous sequences into close proximity (Salonen *et al.*, 1982; Scherthan, 2007; Koszul *et al.*, 2008). However, Koszul *et al.* (2008) thought that this was unlikely and postulated that these movements may prevent chromosome entanglements. The movements observed were directed along actin cables and seemed unlikely to reflect a stirring motion. It appears that homology recognition begins prior to the onset of these chromosome movements. They may also serve to remove non-specific connections between non-homologous chromosomes. The introduction of LatB at around the time of DSB formation resulted in reduced crossover and non-crossover formation, which was thought to be due to the prevention of some DSBs finding the opposing chromosome axis (Koszul *et al.*, 2008). More recent work in budding yeast has shown that sustained and rapid movement of chromosomes is necessary for effective pairing and meiotic progression (Brown *et al.*, 2011).

Actin has also been shown to control chromosome movements in plants. A study in maize developed a method of live imaging of meiocytes inside intact anthers (Sheenan and Pawlowski, 2009). In zygotene, rapid chromosome movements of short segments of the chromatin led by telomeres attached to the NE were observed. This chromosome motility was coincident with deformations of the NE, as observed in budding yeast. Rotational movements of the entire chromatin were also observed at this stage. Actin was shown to be required for chromosome motility, as treatment with LatB prevented all chromosome movements during zygotene and pachytene stages. Colchicine was also shown to prevent chromosome movements, although whether this was due to its action on microtubules or a non-microtubule tubulin (Cowan and Cande, 2002) remains to be shown. The pattern of chromosome motility was observed to differ between zygotene and pachytene stages. Pachytene was associated with long sweeping motions of whole chromosome arms. It was thought that these movements were likely to be generated by a different mechanism to those in zygotene since mid-pachytene chromosomes are no longer attached to the NE. Sheenan and Pawlowski (2009) proposed that the chromosome movements in zygotene may serve to facilitate pairing of homologues by bringing homologous loci into a close association for recombination dependent homology recognition to take place. It was thought that the pachytene movements may play a different role, for example in removing interlocks. Recent studies have focused on how chromosome movements are regulated and controlled. Components of the NE that link telomeres to the force generating mechanism have been studied and are discussed below.

The mechanism for linking chromosome motility during meiosis to the cytoskeleton was first observed in *S. pombe*. Chikashige *et al.* (2006) carried out a microarray analysis to study meiosis-specific genes in *S. pombe*. Genes that bound to RNA from cells reacting to the mating pheromone, known to induce meiosis in yeast, were selected. Knockout mutants of the selected genes were then produced and one gene, named Bouquet 1 (*BQT1*), was found to prevent meiotic telomere clustering. A yeast two-hybrid screen then led to the discovery of an interacting protein called Bqt2. Both Bqt1 and Bqt2 were observed to localize to the SPB and yeast two-hybrid screens were used to find interacting proteins. Chikashige *et al.* (2006) found that Bqt1 bound to a component of the SPB called Sad1, and that Bqt1 and Bqt2 together bind to Rap1. It was proposed that these proteins may interact to form a link between the nuclear membrane and telomeres. Sad1 binds to the nuclear membrane and Bqt1, which subsequently recruits Bqt2. The complex then binds Rap1, which is known to localize to the telomeres. Bqt1 was shown to bind to a SUN domain of Sad1 and it was noted that proteins that tether other proteins to the nuclear membrane contain this characteristic SUN domain (Tomita and Cooper, 2006) (see Section 7.6). This protein complex has been shown to link telomeres to cytoplasmic microtubules via the SPB protein, Kms1 (Shimanuki *et al.*, 1997), in the outer nuclear membrane (ONM) (Chikashige *et al.*, 2006). Kms1 is analogous to the KASH-domain proteins of other organisms. This linkage leads to clustering of the telomeres into a bouquet formation at the SPB. The entire nucleus is subsequently moved by microtubules and their associated motor proteins, with telomeres at the leading edge; this characteristic rapid movement, from side to side of the cell, led by the telomeres, has been likened to the movement of a horsetail (also known as the horsetail nucleus) (Chikashige *et al.*, 2006).

7.6 Components of the nuclear envelope

The NE is a double membrane structure conserved in eukaryotes (Figure 7.4), separating the nuclear contents from the cytoplasm. It consists of an inner and outer nuclear membrane and the nucleus and cytoplasm are connected by nuclear pore complexes (NPCs). Several families of plant NE proteins have recently been identified which are important during meiosis (Evans *et al.* 2011 and references therein).

The SUN family of proteins were first discovered in *S. pombe* in the form of the Sad1 protein (Hagan and Yanagida, 1995). The term SUN (*S*ad-1 and *UN*C-84) was coined when a protein in *C. elegans,* called UNC-84, was found that had a region of around 120 amino acid residues, located in the C-terminal domain, with a high degree of similarity to the corresponding region of Sad1 (Malone *et al.*, 1999). The SUN-domain proteins are essential components of the nuclear envelope and play a conserved role in chromosome dynamics in eukaryotes. These proteins interact with various partners to form 'bridges' across the inner and outer nuclear membranes; known as the linker of nucleoskeleton and cytoskeleton (LINC) complex (Crisp *et al.*, 2006). The SUN-domain proteins display diverse roles including centrosome localization, germ-cell development and telomere positioning. In Arabidopsis, SUN proteins seem to be involved in the maintenance or formation of nuclear shapes as components of the nucleocytoskeleton complex (Oda and Fukuda, 2011).

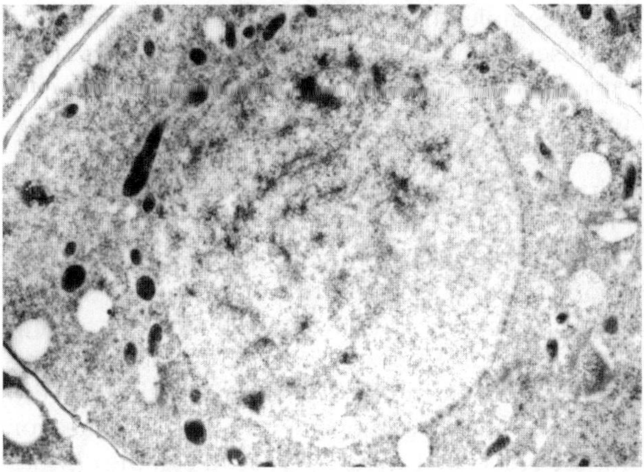

Figure 7.4 EM transmission micrograph. WT Arabidopsis zygotene cell showing the double layer of the NE, transversed by many NE pores.

In meiosis, SUN-domain proteins play an important role in linking telomeres to the force-generating mechanism in the cytoplasm. The proteins have a highly conserved structure, containing an N-terminal trans-membrane domain linked to a less well conserved coiled-coil domain that is linked to the SUN domain located at the C-terminal (Tzur *et al.*, 2006). They are positioned in the inner nuclear membrane (INM), with the N-terminal thought to localize to the nuclear face. This domain then interacts with nuclear lamin proteins of the INM in some cases. The SUN domain is localized in the lumen between the INM and outer nuclear membrane (ONM), where it connects to the cytoskeleton via a KASH-domain protein. KASH-domain proteins were named for their homology to regions of Klarsicht (Drosophila), ANC-1 (*C. elegans*), and Syne (mammals) proteins (Starr and Fischer, 2005). The weakly conserved domain is located at the C-terminus and is composed of a predicted trans-membrane domain followed by approximately 30 residues (Starr and Fischer, 2005). The localization of KASH-domain proteins in the ONM has been shown to be dependent on their interaction with SUN-domain proteins. The N-terminal of KASH-domain proteins is highly variable, with differing functions. In most cases they contain cytoskeletal linker motifs. Studies have shown that these proteins interact with cytoskeletal structures such as actin (Starr and Fischer, 2005; Tzur *et al.*, 2006).

SUN-domain proteins have been shown to be highly conserved between different organisms. The number of SUN-domain proteins has increased with progression through evolution. As such, *S. pombe* contains one SUN protein, *C. elegans* two, and humans at least five SUN proteins (Tzur *et al.*, 2006). SUN-domain proteins have been shown to link telomeres with the NE during meiosis in a number of organisms. Schmitt *et al.* (2007) discovered a rat SUN-domain protein, named SUN2, localized to the attachment sites of meiotic telomeres at the NE. Electron microscopy showed that SUN2 is part of a membrane-spanning complex that connects attached telomeres to cytoplasmic structures, proposed to be the actin cytoskeleton. This study demonstrated that the mechanism for telomere attachment to the NE is conserved between eukaryotes. Mice deficient for the NE protein, SUN1, show no telomere attachment to the membrane and no subsequent clustering. Ding *et al.* (2007) showed that pairing of homologous chromosomes, together with synapsis and recombination, was impaired in this mutant, and that mice were completely sterile. This supports the notion that attachment of telomeres to the nuclear envelope is important for meiotic progression. A meiosis-specific KASH-domain protein, KASH5, has been identified as a binding partner of SUN1 (Morimoto *et al.*, 2012). KASH5 localizes with SUN1 and telomeres during prophase I and has been shown to interact with dynactin, suggesting a possible mechanism for chromosome movement through connection to the cytoplasmic microtubule-associated dynein-dynactin complex (Morimoto *et al.*, 2012).

In mammals, a nuclear lamin protein has also been implicated in telomere dynamics at the NE during meiosis. Lamin C2, which is an alternatively spliced isoform of somatic A-type lamins, is expressed solely in meiotic cells (Smith and Benevente, 1992), where it localizes to telomere attachment sites at the NE (Alsheimer *et al.*, 1999). This N-terminally truncated variant possesses a significantly increased mobility compared to its somatic counterpart and it has been proposed to form dynamic platforms at the NE, allowing meiotic telomeres to slide at the plane of the NE (Jahn *et al.*, 2010).

In *C. elegans* the SUN-1 protein has also been implicated in facilitating homologue pairing and restricting synapsis to regions of homologous pairing (Penkner *et al.*, 2007; Sato *et al.*, 2009). Loss of SUN-1, or the KASH-domain protein ZYG-12, results in defects in synapsis. ZYG-12 has also been shown to play an essential role in nuclear positioning in the germ-line by mediating the recruitment of dynein to the NE to maintain microtubule organization (Zhou *et al.*, 2009). In *C. elegans*, telomeres do not play a key role in meiotic chromosome motility. Instead, chromosomes contain special homology recognition sites, known as pairing centres (PCs), which interact with the NE via association with one of a family of zinc-finger proteins (Phillips and Dernburg, 2006). During early prophase I, ZYG-12, which in interphase is distributed evenly around the nuclear surface, aggregates into patches at the nuclear periphery, where it appears to associate with SUN-1 at sites of PC attachment (Penkner *et al.*, 2007; Sato *et al.*, 2009). This remodelling of the NE, and subsequent homologous chromosome pairing and synapsis, is induced by recruitment of the Polo-like kinase, PLK-2, to the PCs, with a second, closely related kinase, PLK-1, sharing partial overlapping activities with PLK-2 (Harper *et al.*, 2011; Labella *et al.*, 2011). Meiosis-specific phosphorylation of SUN-1 at Ser12 is also dependent on recruitment of the PLKs to the PCs, although this residue may not be a direct PLK target (Harper *et al.*, 2011; Labella *et al.*, 2011).

Unlike in *S. pombe* and *C. elegans*, telomere movement in budding yeast has been shown to require the actin cytoskeleton (see Section 7.5.2; see also Scherthan, 2007; Kozul *et al.*, 2008). Chromosome movements in budding yeast are mediated by SUN KASH protein connections to meiotic telomeres, as in other organisms. The meiosis-specific telomere associated protein, Ndj1/Tam1 (Chua *et al.*, 1997; Conrad *et al.*, 1997), is thought to couple telomeres to the NE because, in an *ndj1*Δ mutant, prophase I chromosome movements are greatly reduced but NE deformations are still evident (Koszul *et al.*, 2008). Interaction of Ndj1 with the SUN-domain protein, Mps3, is critical for bouquet formation (Conrad *et al.*, 2007). A meiosis-specific protein, Csm4, has been shown to perform a role analogous to that of KASH-domain proteins, despite possessing no significant sequence similarity to known KASH domains. In a *csm4*Δ mutant, chromosome movements and NE deformations were eliminated, suggesting that this protein is involved in coupling the telomere/NE complexes to the force generating mechanism (Koszul *et al.*, 2008; Wanat *et al.*, 2008). In agreement with this, Conrad *et al.* (2008) showed that the *ndj1*Δ and *mps3*Δ mutants were defective for telomere attachment to the NE, whilst the *csm4*Δ mutant showed nuclei with peripherally located chromosomes, indicating that Csm4 is not required for anchoring telomeres to the NE

7.7 Components of the plant nuclear envelope

Putative SUN-domain proteins have been identified in several plant species, with most studied encoding at least two proteins (Graumann *et al.*, 2010). In Arabidopsis, recent attention has been paid to AtSUN1 and AtSUN2 as they share homology with other C-terminal SUN-domain proteins. Both AtSUN1 and AtSUN2 proteins share similar structural features with animal and fungal SUN-domain proteins, including a highly conserved C-terminal SUN domain, a functional coiled-coil domain, and an N-terminal transmembrane domain (Graumann *et al.*, 2010). Both proteins were also found to contain a bipartite nuclear localization signal (NLS), necessary for localization to the NE. AtSUN1 and AtSUN2 gene expression was observed in leaf and inflorescence tissue. Analysis of published microarray data showed AtSUN2 expression to be up-regulated in proliferating tissue. A more detailed study of AtSUN1 expression remains to be completed. Both proteins were shown to localize to the NE using fluorescently tagged proteins. Using fluorescence recovery after photobleaching (FRAP), Graumann *et al.* (2010) found that half of AtSUN1 and AtSUN2 fluorescently tagged proteins were immobile, suggesting that they are involved in strong binding interactions. The AtSUN1 and AtSUN2 proteins were also found to exist as both homomers and heteromers *in vivo* within the nuclear membrane, and as such tend not to be found as single discreet proteins. Graumann *et al.* (2010) proposed that the Arabidopsis SUN-domain proteins may function in a multimeric protein complex, as has been observed for mammalian SUN proteins (Wang *et al.*, 2006). Therefore, their ability to function may depend upon the availability and co-localization of other SUN proteins within the nuclear membrane.

A more recent study in maize has identified the existence of two classes of SUN-domain proteins in the plant kingdom (Murphy *et al.*, 2010) and five SUN-domain proteins have been shown to be encoded by the maize genome. ZmSUN1 and ZmSUN2 were named canonical C-terminal SUN domain (CCSD) proteins, due to their homology with previously identified mammalian SUN proteins (Murphy *et al.*, 2010). ZmSUN3, -4, and -5 were named plant-prevalent mid-SUN 3 transmembrane (PM3) proteins. This second group of SUN-domain proteins was shown to constitute a novel, conserved structural variant class. PM3 SUN proteins are characterized by the presence of three trans-membrane domains, a conserved domain of unknown function and a centrally positioned SUN domain (Murphy *et al.*, 2010). Arabidopsis encodes two PM3-class proteins, AtSUN3 and AtSUN4 (see Chapters 2 and 6). Mid-SUN-domain proteins have been identified in non-plant species, although their function remains unknown. To ascertain the role of SUN proteins in plants, a comparison of the maize gene expression levels was carried out. Expression levels of *ZmSUN1-4* genes were shown to be low in several tissues, including pollen. In contrast to this, *ZmSUN5* gene expression was observed to be distinctly higher in pollen than other tissues, which may suggest a pollen-specific role. However, whether the ZmSUN5 protein is involved in meiotic telomere function remains to be investigated. Immunolocalization of ZmSUN3 and -4 in post-meiotic pollen mother cells, showed that PM3 SUN-domain proteins localize to the NE, as has previously been observed for CCSD SUN proteins in Arabidopsis (Graumann *et al.*, 2010). Labelling of ZmSUN3 or -4 was not detected in meiotic prophase nuclei, which could suggest that these proteins are not involved in meiosis (Murphy *et al.*, 2010).

These studies provide support for the presence of a LINC-like complex that bridges the cytoskeleton and the nucleoskeleton in plants. It will be important to identify SUN domain interacting proteins to characterize the nature of these complexes. To date, no homologues of KASH domain or lamin proteins have been identified in plants. It will also be important to elucidate which of the SUN-domain proteins play a role in meiotic prophase I chromosome movements and whether disruption of telomere attachment to the force-generating mechanism in plants has a similar affect as in yeast and mammalian species.

7.8 Conclusions and future prospects

Telomeres clearly play an important role in the movement of chromosomes during meiosis. Telomeres have been thought to play a role in facilitating pairing, by bringing homologous chromosomes into a close association during the bouquet. Studies in which the bouquet is disrupted affect progression of meiosis (Trelles-Sticken *et al.*, 2000; Golubovskaya *et al.*, 2002). New developments in techniques, for example live imaging (Sheenan and Pawlowski, 2009), have led to a greater understanding of the role of chromosome movements during prophase I and how these are controlled. Mechanisms that control chromosome dynamics in plants are less well understood. Recent evidence shows that actin controls movements in maize (Sheenan and Pawlowski, 2009). Further improvements in live imaging – for example the ability to detect labelled telomeres – will further elucidate the nature of meiotic prophase chromosome movements in plants. Chromosome movements are mediated by attachment of telomeres to the NE, particularly during the bouquet. In yeast and mammalian species, proteins that link telomeres to the cytoskeleton have been identified, which are important in controlling chromosome movements. SUN-domain proteins have recently been identified in Arabidopsis (Graumann *et al.*, 2010) and maize (Murphy *et al.*, 2010). These proteins form part of the bridging complex between the telomeres and the force-generating mechanism of the cytoskeleton. This provides evidence for the presence of a similar mechanism of control in plants. A novel class of SUN proteins has been identified in plants and it will be important to identify which of the SUN-domain proteins in plants play a role in meiosis. It will be interesting to identify SUN interacting proteins in order to characterize this LINC-like complex in plants. Arabidopsis has proved to be a valuable model in the study of meiosis and will help us to unlock the link of the telomeres with the NE. It will be interesting to compare the differences in chromosome dynamics in Arabidopsis – a small genome plant with a unique arrangement of telomeres – with larger and more complex genome plants, like maize, which have a classical bouquet organization. Recent evidence suggests telomeres may facilitate pairing and synapsis, not by bringing homologues into close association during the bouquet, but by controlling chromosome movements during prophase I, by connecting chromosomes to the force-generating mechanism. It is thought that these chromosome movements may play a role in pairing and removal of interlocks and non-homologous interactions during prophase I.

Acknowledgements

Work in the Birmingham and Spanish laboratories is supported from the European Community's Seventh Framework Programme FP7/2007-2013 under grant agreement number KBBE-2009-222883.

We apologise to authors that we have failed to cite in this review.

References

Alsheimer, M., von Glasenapp, E., Hock, R. and Benavente, R. (1999) Architecture of the nuclear periphery of rat pachytene spermatocytes: distribution of nuclear envelope proteins in relation to synaptonemal complex attachment sites. *Molecular Biology of the Cell* **10**, 1235–1245.

Alsheimer, M. (2009) The dance floor of meiosis: evolutionary conservation of nuclear envelope attachment and dynamics of meiotic telomeres. *Genome Dynamics* **5**, 81–93.

Alonso, J.M., Stepanova, A.N., Leisse, T.J. *et al.* (2003) Genome-wide insertional mutagenensis of *Arabidopsis thaliana*. *Science* **301**, 653–657.

Amiard, S. Charbonnel, C. Allain, A. *et al.* (2010) Distinct roles of the ATR kinase and the Mre11-Rad50-Nbs1 complex in the maintenance of chromosomal stability in *Arabidopsis*. *The Plant Cell* **22**, 3020–3033.

Amiard, S., Depeiges, A., Allain, E. *et al.* (2011) *Arabidopsis* ATM and ATR kinases prevent propagation of genome damage caused by telomere dysfunction. *The Plant Cell* **23**, 4254–4265.

Arabidopsis Genome Initiative (2000) Analysis of the genome sequence of the flowering plant *Arabidopsis thaliana*. *Nature* **408**, 796–816.

Aragón-Alcaide, L., Reader, S., Beven, A. *et al.* (1997) Association of homologous chromosomes during floral development. *Current Biology* **7**, 905–908.

Armstrong, S.J., Franklin, F.C. and Jones, G.H. (2001) Nucleolus-associated telomere clustering and pairing precede meiotic chromosome synapsis in Arabidopsis thaliana. *Journal of Cell Science* **114** (23), 4207–4217.

Armstrong, S.J., Caryl, A., Jones, G.H. and Franklin, F.C.H. (2002) Asy1 a protein required for meiotic chromosome synapsis localizes to axis associated chromatin in Arabidopsis and Brassica. *Journal of Cell Science* **115**, 3645–3655.

Armstrong, S.J., Franklin, F.C.H. and Jones, G.H. (2003) A meiotic time-course for *Arabidopsis thaliana. Sexual Plant Reproduction* **16**, 141–149.

Armstrong, S.J. and Jones, G.H. (2003) Meiotic cytology and chromosome behaviour in wild-type Arabidopsis thaliana. *Journal of Experimental Botany* **54** (380), 1–10.

Barlow, A. and Hulten, M. (1996) Combined immunocytogenetic and molecular cytogenetic analysis of meiosis I human spermatocytes. *Chromosome Research.* **4**, 562–573.

Bass, H.W. (2003) Telomere dynamics unique to meiotic prophase: formation and significance of the bouquet. *Cellular and Molecular Life Sciences* **60** (11), 2319–2324.

Bass, H.W., Riera-Lizarazu, O., Ananiev, E.V. *et al.* (2000) Evidence for the coincident initiation of homolog pairing and synapsis during the telomere-clustering (bouquet) stage of meiotic prophase. *Journal of Cell Science* **113**, 1033–1042.

Bennett, M.D. (1984) Premeiotic events and meiotic chromosome pairing. *Symposia of the Society for Experimental Biology* **38**, 87–121.

Bennett, M.D., Chapman, V. and Riley, R. (1971) The duration of meiosis in pollen mother cells of wheat, rye and Triticale. *Proceedings of the Royal Society London Series B* **178**, 259–275.

Bennett, M.D. and Smith, J.B. (1979) The effect of colchicine on fibrillar material in wheat meiocytes. *Journal of Cell Science* **38**, 33–47.

Bennett, M.D., Toledo, L.A. and Stern, H. (1979) The effect of colchicine on meiosis in Lillium speciosum cv 'Rosemede'. *Chromosoma* **72**, 175–189.

Bertuch, A.A. and Lundblad, V. (2006) The maintenance and masking of chromosome termini. *Current Opinion in Cell Biology* **18**, 247–255.

Bianchi, A., Smith, S., Chong, L. *et al.* (1997) TRF1 is a dimer and bends telomeric DNA. *The EMBO Journal* **16**, 1785–1794.

Biessmann, H. and Mason, J.M. (1990) Chromosome ends in *Drosophila* without telomeric DNA sequences. *Proceedings of the National Academy of Sciences of the USA* **87**, 1758–1761.

Blackburn, E.H. (2001) Switching and signalling at the telomere. *Cell* **106** (6), 661–673.

Blasco, M.A. (2007) Telomere length, stem cells and aging. *Nature Chemical Biology* **3** (10), 640–649.

Brown, M.S., Zandero, S. and Alani, E. (2011) Sustained and rapid chromosome movements are critical for chromosome pairing and meiotic progression in budding yeast. *Genetics* **188**, 21–32.

Bundock, P. and Hooykaas, P. (2002) Severe developmental defects, hypersensitivity to DNA-damaging agents, and lengthened telomeres in Arabidopsis MRE11 mutants. *The Plant Cell* **14** (10), 2451–2462.

Carlton, P.M. and Cande, W.Z. (2002) Telomeres act autonomously in maize to organize the meiotic bouquet from a semipolarized chromosome orientation. *Journal of Cell Biology* **157** (2), 231–242.

Caryl, A.P., Armstrong, S.J., Jones, G.H. and Franklin, F.C.H. (2000) A homologue of the yeast HOP1 gene is inactivated in the Arabidopsis meiotic mutant asy1. *Chromosoma* **1009**, 62–71.

Celli, G.B. and De Lange, T. (2005) DNA processing is not required for ATM-mediated telomere damage response after TRF2 deletion. *Nature Cell Biology* **7**, 712–718.

Cesare, A.J., Quinney, N., Willcox, S. *et al.* (2003) Telomere looping in *P. sativum* (common garden pea). *The Plant Journal* **36**, 271–279.

Charbonnel, C., Allain, E., Gallego, M.E. and White, C.I. (2011) Kinetic analysis of DNA double-strand break repair pathways in Arabidopsis. *DNA Repair* **10**, 611–619.

Charbonnel, C., Gallego, M.E. and White, C.I. (2010) Xrcc1-dependent and Ku-dependent DNA double-strand break repair kinetics in Arabidopsis plants. *The Plant Journal* **64**, 280–290.

Chen, C.M., Wang, C.T. and Ho, C.H. (2001) A plant gene encoding a Myb-like protein that binds telomeric GGTTAG repeats in vitro. *The Journal of Biological Chemistry* **276** (19), 16511–16519.

Chikashige, Y., Ding, D.Q., Funabiki, H. *et al.* (1994) Telomere-led premeiotic chromosome movement in fission yeast. *Science* **264** (5156), 270–273.

Chikashige, Y., Tsutsumi, C., Yamane, M. *et al.* (2006) Meiotic proteins bqt1 and bqt2 tether telomeres to form the bouquet arrangement of chromosomes. *Cell* **125** (1), 59–69.

Chua, P.R. and Roeder, G.S. (1997) Tam1, a telomere-associated meiotic protein, functions in chromosome synapsis and crossover interference. *Genes and Development* **11**, 1786–1800.

Cifuentes-Rojas, C., Kannan, K., Tseng, L. and Shippen, D.E. (2011) Two RNA subunits and POT1a are components of Arabidopsis telomerase. *Proceedings of the National Academy of Sciences of the USA.* **108** (1), 73–78.

Conrad, M.N., Domínguez, A.M. and Dresser, M.E. (1997) Ndj1p, a meiotic telomere protein required for normal chromosome and segregation in yeast. *Science* **276**, 1252–1255.

Conrad, M.N., Lee, C.Y., Chao, G. *et al.* (2008) Rapid telomere movement in meiotic prophase is promoted by NDJ1, MPS3, and CSM4 and is modulated by recombination. *Cell* **133** (7), 1175–1187.

Conrad, M.N., Lee, C.Y., Wilkerson, J.L, and Dresser, M.E. (2007) MPS3 mediates meiotic bouquet formation in Saccharomyces cerevisiae. *Proceedings of the National Academy of Sciences of the USA* **104**, 8863–8868.

Corredor, E., Lukaszewski, A.J., Pachón, P. *et al.* (2007) Terminal regions of wheat chromosomes select their pairing partners in meiosis. *Genetics* **177**, 699–706.

Corredor, E. and Naranjo, T. (2007) Effect of colchicine and telocentric chromosome conformation on centromere and telomere dynamics at meiotic prophase I in wheat-rye additions. *Chromosome Research* **15** (2), 231–245.

Cooper, J.P., Watanabe, Y. and Nurse, P. (1998) Fission yeast Taz1 protein is required for meiotic telomere clustering and recombination. *Nature* **392** (6678), 828–831.

Cowan, C.R. and Cande, W.Z. (2002) Meiotic telomere clustering is inhibited by colchicine but does not require cytoplasmic microtubules. *Journal of Cell Science* **115**(19), 3747–3756.

Cowan, C.R., Carlton, P.M. and Cande, W.Z. (2001) The polar arrangement of telomeres in interphase and meiosis. Rabl organization and the bouquet. *Plant Physiology* **125**, 532–538.

Crismani, W., Girard, C., Froger, N. *et al.* (2012) FANCM limits meiotic crossovers. *Science* **336**, 1588–1590.

Crisp, M., Qian Liu., Q., Roux, K. *et al.* (2006) Coupling of the nucleus and the cytoplasm: role of the LINC complex. *Journal of Cell Biology* **172**, 41–53.

Cuñado, N., Sánchez-Morán, E., Barrios, J. and Santos, J.L. (2001) Searching for telomeric sequences in two *Allium* species. *Genome* **44**, 640–643.

de la Herrán, R., Cuñado, N., Navajas-Pérez, R. *et al.* (2005) The controversial telomeres of lily plants. *Cytogenetics Genome Research* **109**, 144–147.

De Lange, T. (2005) Shelterin: the protein complex that shapes and safeguards human telomeres. *Genes and Development* **19** (18), 2100–2110.

De Lange, T., Lundblad, V. and Blackburn, E.H. (eds) (2006) *Telomeres*, Cold Spring Harbor Laboratory Press, Cold Spring Harbor, NY, pp. 345–386.

Ding, D.-Q., Okamasa, K., Yamane, M. *et al.* (2012) Meiosis–specific noncoding of RNA mediates robust pairing of homologous chromosomes in meiosis. *Science* **366**, 732–736.

Ding, X., Xu, R. and Yu, J. *et al.* (2007) SUN1 is required for telomere attachment to nuclear envelope and gametogenesis in mice. *Developmental Cell* **12** (6), 863–872.

Dong, F. and Jiang, J. (1998) Non-Rabl patterns of centromere and telomere distribution in the interphase nuclei of plant cells. *Chromosome Research* **6**, 551–558.

Driscoll, C.J. and Darvey, N.L. (1970) Chromosome pairing: effect of colchicine on an isochromosome. *Science* **169** (942), 290–291.

Driscoll, C.J., Darvey, N.L. and Barber, H.N. (1967) Effect of colchicine on meiosis of hexaploid wheat. *Nature* **216** (18), 687–688.

Evans, D.E., Shvedunova, M. and Graumann, K. (2011) The nuclear envelope in the plant cell cycle: structure, function and regulation. *Biochemical Society Transactions* **38**, 741–746.

Fairall, L., Chapman, L., Moss, H. *et al.* (2001) Structure of the TRFH dimerization domain of the human telomeric proteins TRF1 and TRF2 *Molecular Cell* **8**, 351–361.

Fajkus, J., Fulneckova, J., Hulanova, M. *et al.* (1998) Plant cells express telomerase activity upon transfer to callus cultura, without extensively changing telomere lengths. *Molecular Genetics and Cell Biology* **260**, 470–474.

Fitzgerald, M.S., Riha, K., Gao, F. *et al.* (1999) Disruption of the telomerase catalytic subunit gene from Arabidopsis inactivates telomerase and leads to a slow loss of telomeric DNA. *Proceedings of the National Academy of Sciences of the USA* **96** (26), 14813–14818.

Gallego, M.E. and White, C.I. (2001) RAD50 function is essential for telomere maintenance in Arabidopsis. *Proceedings of the National Academy of Sciences* **98** (4), 1711–1716.

Gallego, M.E., Jeanneau, M., Granier, F. *et al.* (2001) Disruption of the Arabidopsis RAD50 gene leads to plant sterility and MMS sensitivity. *The Plant Journal* **25** (1), 31–41.

Gao, H., Cervantes, R.B., Mandell, E.K. *et al.* (2007) RPA-like proteins mediate yeast telomere function. *Nature Structural and Molecular Biology* **14**, 208–214.

Gerton, J.L. and Hawley, R.S. (2005) Homologous chromosome interactions in meiosis: diversity amidst conservation. *Nature Review Genetics* **6** (6), 477–487.

Ghosh, A.K., Rossi, M.L., Singh, D.K. *et al.* (2011) RECQL4, the Protein Mutated in Rothmund–Thomson Syndrome, Functions in Telomere Maintenance. *The Journal of Biological Chemistry*. Oct 28. Epub ahead of print.

Golubovskaya, I.N., Harper, L.C., Pawlowski, W.P. *et al.* (2002) The pam1 gene is required for meiotic bouquet formation and efficient homologous synapsis in maize (Zea mays L.). *Genetics* **162** (4), 1979–1793.

Grandin, N., Reed, S.I. and Charbonneau, M. (1997) Stn1, a new Saccharomyces cerevisiae protein, is implicated in telomere size regulation in association with Cdc13. *Genes and Development* **11**(4), 512–527.

Greider, C.W. and Blackburn, E.H. (1985) Identification of a specific telomere terminal transferase activity in Tetrahymena extracts. *Cell* **43** (2 Pt. 1), 405–413.

Graumann, K., Runions, J., and Evans, D. (2010) Characterisation of the Sun-domain proteins at the higher plant nuclear envelope. *The Plant Journal* **61**, 134–144.

Hagan, I. and Yanagida, M. (1995) The product of the spindle formation gene sad1+ associates with the fission yeast spindle pole body and is essential for viability. *Journal of Cell Biology* **129** (4), 1033–1047.

Harper, L., Golubovskaya, I. and Cande, W.Z. (2004) A bouquet of chromosomes. *Journal of Cell Science* **117** (18), 4025–4032.

Harper, N.C., Rillo, R., Jover-Gil, S. *et al.* (2011) Pairing centers recruit a Polo-kinase to orchestrate meiotic chromosome dynamics in *C. elegans*. *Developmental Cell* **21**, 934–947.

Hartung, F., Plchová, H. and Puchta, H. (2000) Molecular characterisation of RecQ homologues in *Arabidopsis thaliana*. *Nucleic Acids Research* **28** (21), 4275–4282.

Hartung, F., Suer, S. and Puchta, H. (2007) Two closely related RecQ helicases have antagonistic roles in homologous recombination and DNA repair in Arabidopsis thaliana. *Proceedings of the National Academy of Sciences of the USA* **104** (47), 18836–18841.

Heacock, M., Spangler, E., Riha, K. *et al.* (2004) Molecular analysis of telomere fusions in Arabidopsis: multiple pathways for chromosome end-joining. *The EMBO Journal* **23** (11), 2304–2313.

Heacock, M.L., Idol, R.A., Friesner, J.D. *et al.* (2007) Telomere dynamics and fusion of critically shortened telomeres in plants lacking DNA ligase IV. *Nucleic Acids Research* **35** (19), 6490–6500.

Higgins, J.D., Ferdous, M., Osman, K. and Franklin, F.C.H. (2011) The RecQ helicase AtRECQ4A is required to remove inter-chromosomal telomeric connections that arise during meiotic recombination in Arabidopsis. *The Plant Journal* **65** (3), 492–502.

Hockemeyer, D., Daniels, J.P., Takai, H. and de Lange, T. (2006) Recent expansion of the telomeric complex in rodents: Two distinct POT1 proteins protect mouse telomeres. *Cell* **126** (1), 63–77.

Holm, P.B. (1977) Three-dimensional reconstruction of chromosome pairing during zygotene stage of meiosis in *Lilium longiflorum* (Thunb.) *Carlsberg Research Communications* **42**, 103–151.

Hwang, M.G. and Cho, M.H. (2007) Arabidopsis thaliana telomeric DNA-binding protein 1 is required for telomere length homeostasis and its Myb-extension domain stabilizes plant telomeric DNA binding. *Nucleic Acids Research* **35** (4), 1333–1342.

Hwang, M.G., Chung, I.K., Kang, B.G. and Cho, M.H. (2001) Sequence-specific binding property of Arabidopsis thaliana telomeric DNA binding protein 1 (AtTBP1) *FEBS Letters* **503** (1), 35–40.

Hwang, M.G., Kim, K., Lee, W.K. and Cho, M.H. (2005) AtTBP2 and AtTRP2 in Arabidopsis encode proteins that bind plant telomeric DNA and induce DNA bending in vitro. *Molecular Genetics and Genomics* **273** (1), 66–75.

Ishiguro, K. and Watanabe, Y. (2007) Chromosome cohesion in mitosis and meiosis. *Journal of Cell Science* **120**, 367–368.

Jahn, D., Schramm, S., Benavente, R. and Alsheimer, M. (2010) Dynamic properties of meiosis-specific lamin C2 and its impact on nuclear envelope integrity. *Nucleus* **1**, 273–283.

Jain, D. and Cooper, J.P. (2010) Telomeric strategies: means to an end. *Annual Review of Genetics* **44**, 243–269.

Jones, G.H. (1984) The control of chiasma distribution. *Symposia of the Society for Experimental Biology* **38**, 293–320.

Jones, G.H., Armstrong, S.J., Caryl, A.P. and Franklin, F.C.H. (2003) Meiotic chromosome synapsis and recombination in *Arabidopsis thaliana*; an integration of cytological and molecular approaches. *Chromosome Research* **11** (3), 205–215.

Karamysheva, Z.N., Surovtseva, Y.V., Vespa, L. *et al.* (2004) A C-terminal Myb extension domain defines a novel family of double-strand telomeric DNA-binding proteins in Arabidopsis. *The Journal of Biological Chemistry* **279** (46), 47799–47807.

Kazda, A., Zellinger, B., Rossler, M. *et al.* (2012) Chromosome end protection by blunt ended telomeres. *Genes and Development* **26**, 1703– 1713.

Keeney, S. (2001) Mechanism and control of meiotic recombination initiation. *Current Topics in Developmental Biology* **52**, 1–53.

Kilian, A., Stiff, C. and Kleinhofs, A. (1995) Barley telomeres shorten during differentiation but grow in callus culture. *Proceedings of the National Academy of Sciences of the USA* **92**, 9555–9559.

Knoll, A. and Puchta, H. (2011) The role of DNA helicases and their interaction partners in genome stability and meiotic recombination in plants. *Journal of Experimental Botany* **62**, 1565–1579.

Knoll, A., Higgins, J.D., Seeliger, K. *et al.* (2012) The Fanconi anemia ortholog FANCM ensures ordered homologous recombination in both somatic and meiotic cells in Arabidopsis. *Plant Cell* **24**, 1448–1464.

Koszul, R. and Kleckner, N. (2009) Dynamic chromosome movements during meiosis: a way to eliminate unwanted connections? *Trends in Cell Biology* **19**, 716–724.

Kozak, J., West, C., White, C. *et al.* (2009) Rapid repair of DNA double strand breaks in *Arabidopsis thaliana* is dependent proteins involved in chromosome structure maintenance. *DNA Repair* **8**, 413–441.

Koszul, R., Kim, K.P., Prentiss, M. *et al.* (2008) Meiotic chromosomes move by linkage to dynamic actin cables with transduction of force through the nuclear envelope. *Cell* **133** (7), 1188–1201.

Kuchar, M. (2006) Plant telomere-binding proteins. *Biologia Plantarum* **50**, 1–7.

Kuchar, M. and Fajkus, J. 2004) Interactions of putative telomere-binding proteins in Arabidopsis thaliana: identification of functional TRF2 homolog in plants. *FEBS Letters* **578** (3), 311–315.

Labella, S., Woglar, A., Jantsch, V. and Zetka, M. (2011) Polo kinases establish links between meiotic chromosomes and cytoskeletal forces essential for homolog pairing. *Developmental Cell* **21**, 948–958.

Levan, A. (1939) The effect of colchicine on meiosis in Allium. *Hereditas* **25** (1), 9–26.

Linger, B.R. and Price, C.M. (2009) Conservation of telomere protein complexes: Shuffling through evolution. *Critical Reviews in Biochemistry and Molecular Biology* **44** (6), 434–466.

Liu, L., Franco, S., Spyropoulos, B. *et al.* (2004) Irregular telomeres impair meiotic synapsis and recombination in mice. *Proceedings of the National Academy of Sciences of the USA* **101** (17), 6496–6501.

Loidl, J. (1988) The effect of colchicine on synaptonemal complex formation in Allium ursinum. *Experimental Cell Research* **178** (1), 93–97.

Lopez, E., Pradillo, M., Romero, C. *et al.* (2008) Pairing and synapsis in wild-type Arabidopsis thaliana. *Chromosome Research* **16**, 701–708.

Lustig, A.J. (2001) Cdc13 subcomplexes regulate multiple telomere functions. *Nature Structural and Molecular Biology* **8**, 297–299.

Malone, C.J., Fixsen, W.D., Horvitz, H.R. and Han, M. (1999) UNC-84 localizes to the nuclear envelope and is required for nuclear migration and anchoring during C. elegans development. *Development* **126** (14), 3171–3181.

Martínez, P., Thanasoula, M., Carlos, A.R. *et al.* (2010) Mammalian Rap1 controls telomere function and gene expression through binding to telomeric and extratelomeric sites. *Nature Cell Biology* **12**, 768–770.

Martínez-Pérez, E., Shaw, P., Reader, S. *et al.* (1999) Homologous chromosome pairing in wheat. *Journal of Cell Science* **112**, 1761–1769.

McKnight, T.D. and Shippen, D.E. (2004) Plant telomere biology. *Plant Cell* **16** (4), 794–803.

Mitton-Fry, R.M., Anderson, E.M., Hughes, T.R. *et al.* (2002) Conserved structure for single-stranded telomeric DNA recognition. *Science* **5** 296 (5565), 145–147.

Moens, P.B. (1969) The fine structure of meiotic chromosome polarization and pairing in *Locusta migratoria* spermatocytes. *Chromosoma* **28**, 1–25.

Morimoto, A., Shibuya, H., Zhu, X. *et al.* (2012) A conserved KASH domain proteinassociates with telomeres, SUN1 and Dynactin during mammalian meiosis. *Journal of Cell Biology* **198**, 165–172.

Murphy, S.P., Simmons, C.R, and Bass, H.W. (2010) Structure and expression of the maize (Zea mays L.) SUN-domain protein gene family: evidence for the existence of two divergent classes of SUN proteins in plants. *BMC Plant Biology* **10**, 269–291.

Naranjo, T. and Corredor, E. (2008) Nuclear architecture and chromosome dynamics in the search of the pairing partner in meiosis in plants. *Cytogenetic Genome Research* **120**, 320–330.

Naito, T., Matsuura, A. and Ishikawa, F. (1998) Circular chromosome formation in fission yeast mutant defective in two ATM homologues. *Nature Genetics* **20**, 203–206.

Oda, Y. and Fukuda, H. (2011) Dynamics of Arabidopsis SUN proteins during mitosis and involvement in nuclear shapping. *The Plant Journal* **66** (4), 629–641.

Opresko, P.L., Mason, P.A., Podell, E.R. *et al.* (2005) POT1 Stimulates RecQ Helicases WRN and BLM to unwind telomeric DNA substrates. *The Journal of Biological Chemistry* **280** (37), 32069–32080.

Osman, K., Higgins, J.D., Sanchez-Moran, E. *et al.* (2011) Pathways to meiotic recombination in Arabidopsis thaliana. *New Phytologist* **190** (3), 523–544.

Paques, F. and Haber, J.E. (1999) Multiple pathways of recombination induced by double-strand breaks in Saccharomyces cerevisiae. *Microbiology and Molecular Biology Reviews* **63**, 349–404.

Penkner, A.M., Fridkin, A., Gioggnitzer, J. *et al.* (2009) Meiotic chromosome homologous search involves modifications of the nuclear envelope Matefin/SUN-1. *Cell* **139**, 920–933.

Penkner, A., Tang, L., Novatchkova, M. *et al.* (2007) The nuclear envelope protein Matefin/SUN-1 is required for homologous pairing in C. elegans meiosis. *Developmental Cell* **12** (6), 873–885.

Pennock, E., Buckley, K. and Lundblad, V. (2001) Cdc13 delivers separate complexes to the telomere for end protection and replication. *Cell* **104**, 387–396.

Phillips, C.M. and Dernburg, A.F. (2006) A family of zinc-finger proteins is required for chromosome-specific pairing and synapsis during meiosis in C. elegans. *Developmental Cell* **11** (6), 817–829.

Pinto, A.R., Nicholls, C. and Liu, J.P. (2011) Telomere protein complexes and interactions with telomerase in telomere maintenance. *Frontiers in Bioscience* **16**, 187–207.

Rabl, C. (1885) Über Zelltheilung. *Morphologisches Jahrbuk* **10**, 214–330.
Rasmussen, S.W. and Holm, P.B. (1980) Mechanics of meiosis. *Hereditas* **93**, 187–216.
Richards, E.J. and Ausubel, F.M. (1988) Isolation of a higher eukaryotic telomere from Arabidopsis thaliana. *Cell* **53**, 127–136.
Riha, K., Fajkus, J., Siroky, J. and Vyskot, B. (1998) Developmental control of telomere lengths and telomerase activity in plants. *The Plant Cell* **10**, 1691–1698.
Riha, K., McKnight, T.D., Griffing, L.R. and Shippen, D.E. (2001) Living with genome instability: plant responses to telomere dysfunction. *Science* **291** (5509), 1797–1800.
Riha, K. and Shippen, D.E. (2003) Telomere structure, function and maintenance in Arabidopsis. *Chromosome Research* **11** (3), 263–275.
Roberts, N.Y. (2009) Investigating the Control of Homologous Chromosome Pairing and Crossover Formation in Meiosis of *Arabidopsis thaliana*. Ph.D. thesis. University of Birmingham, UK.
Roberts, N.Y., Osman, K. and Armstrong, S.J. (2009) Telomere Distribution and dynamics in Somatic and Meiotic Nuclei of *Arabidopsis thaliana. Cytogenetic Genome Research* **124**, 193–201.
Rockmill, B. and Roeder, G.S. (1996) Telomere-mediated chromosome pairing during meiosis in budding yeast. *Genes and Development* **12**, 2574–2586.
Roeder, G.S. (1997) Meiotic chromosomes: it takes two to tango. *Genes and Development* **11** (20), 2600–2621.
Roig, I., Liebe, B., Egozcue, J. *et al.* (2004) Female-specific features of recombinational double-stranded DNA repair in relation to synapsis and telomere dynamics in human oocytes. *Chromosoma* **113**, 22–33.
Rossignol, P., Collier, S., Bush, M. et al. (2007) Arabidopsis POT1A interacts with TERT-V(I8), an N-terminal splicing variant of telomerase. *Journal of Cell Science* **120** (20), 3678–3687.
Salonen, K., Paranko, J. and Parvinen, M. (1982) A colcemid sensitive mechanism involved in regulation of chromosome movements during meiotic pairing. *Chromosoma* **85**, 611–618.
Sanchez-Moran, E., Mercier, R., Higgins, J.D., Armstrong, S.J., Jones G.H., Franklin, F.C. (2005) A strategy to investigate the plant meiotic proteome. *Cytogenetic and Genome Research*. 109, 181–189.
Santos, J.L., Jiménez, M.M. and Díez, M. (1994) Meiosis in haploid rye: extensive sinapsis and low chiasma frequency. *Heredity* **73**, 580–588.
Sato, A., Isaac, B., Phillips, C.M. *et al.* (2009) Cytoskeletal forces span the nuclear envelope to coordinate meiotic chromosome pairing and synapsis. *Cell* **139**, 907–919.
Scherthan, H. (2001) A bouquet makes ends meet. *Nature Reviews Molecular Cell Biology* **2** (8), 621–627.
Scherthan, H. (2007) Telomere attachment and clustering during meiosis. *Cellular and Molecular Life Sciences* **64** (2), 117–124.
Scherthan, H., Weich, S., Schwegler, H. *et al.* (1996) Centromere and telomere movements during early meiotic prophase of mouse and man are associated with the onset of chromosome pairing. *Journal of Cell Science* **134**, 1109–1125.
Scherthan, H., Sfeir, A. and de Lange, T. (2011) Rap-1 independent telomere attachment and bouquet formation in mammalian meiosis. *Chromosoma* **120**, 151–157.
Schmitt, J., Benavente, R., Hodzic, C. *et al.* (2007) Transmembrane protein Sun2 is involved in tethering mammalian meiotic telomeres to the nuclear envelope. *Proceedings of the National Academy of Sciences USA* **104** (18), 7426–7431.

Schrumpfova, P., Kuchar, M., Mikova, G. *et al.* (2004) Characterization of two Arabidopsis thaliana myb-like proteins showing affinity to telomeric DNA sequence. *Genome* **47**, 316–324.

Schrumpfová, P.P., Kuchar, M., Palecek, J. and Fajkus, J. (2008) Mapping of interaction domains of putative telomere-binding proteins AtTRB1 and AtPOT1b from *Arabidopsis thaliana. FEBs Letters* **582**, 1400–1406.

Schwarzacher, T. and Heslop-Harrison, J.S. (1991) *In situ* hybridization to plant telomeres using synthetic oligomers. *Genome* **34**, 317–323.

Sfeir, A., Kabir, S., van Overbeek, M. *et al.* (2010) Loss of Rap1 induces telomere recombination in the absence of NHEJ or a DNA damage signal. *Science* **327**, 1657–1661.

Shakirov, E.V., McKnight, T.D. and Shippen, D.E. (2009) POT1-independent single-strand telomeric DNA binding activities in Brassicaceae. *The Plant Journal* **58** (6), 1004–1015.

Shakirov, E.V., Surovtseva, Y.V., Osbun, N. and Shippen, D.E. (2005) The Arabidopsis Pot1 and Pot2 proteins function in telomere length homeostasis and chromosome end protection. *Molecular Cell Biology* **25** (17), 7725–7733.

Sheenan, M.J. and Pawlowski, W.P. (2009) Live imaging of rapid chromosome movements in meiotic prophase I in maize. *Proceedings of the National Academy of Sciences of the USA* **106**, 20989–20994.

Shimanuki, M., Miki, F., Ding, D.Q. *et al.* (1997) A novel fission yeast gene, kms1+, is required for the formation of meiotic prophase-specific nuclear architecture. *Molecular and General Genetics* **16**, 238–249.

Siroky, J., Zluvova, J., Rhia, K. *et al.* (2003) Rearrangement of ribosomal DNA clusters in late generation telomerase-deficient *Arabidopsis. Chromosoma* **112**, 116–123.

Smith, A. and Benevente, R. (1992) Identification of a short nuclear lamin protein selectively expressed during meiotic stages of rat spermatogenesis. *Differentiation* **52** (1), 55–60.

Song, X., Leehy, K., Warrington, R.T. et al. (2008) STN1 protects chromosome ends in Arabidopsis thaliana. *Proceedings of the National Academy of Sciences USA* **105** (50), 19815–19820.

Starr, D.A. and Fischer, J.A. (2005) KASH 'n' Karry: the KASH domain family of cargo specific cytoskeletal adapter proteins. *Bioessays*. **27**, 1136–1146.

Storlazzi, A., Tesse, S., Gargano, S. *et al.* (2003) Meiotic double-strand breaks at the interface of chromosome movement, chromosome remodeling, and reductional division. *Genes and Development* **17** (21), 2675–2687.

Surovtseva, Y.V., Shakirov, E.V., Vespa, L. et al. (2007) Arabidopsis POT1 associates with the telomerase RNP and is required for telomere maintenance. *The EMBO Journal* **26** (15), 3653–3661.

Surovtseva, Y.V., Churikov, D., Boltz, K.A. *et al.* (2009) Conserved telomere maintenance component 1 interacts with STN1 and maintains chromosome ends in higher eukaryotes. *Molecular Cell* **36** (2), 207–218.

Tamura, K., Adachi, Y., Chiba, K. et al. (2002) Identification of Ku70 and Ku80 homologues in Arabidopsis thaliana: evidence for a role in the repair of DNA double-strand breaks. *The Plant Journal* **29** (6), 771–781.

Tani, A. and Murata, M. (2005) Alternative splicing of Pot1 (Protection of telomere)-like genes in Arabidopsis thaliana. *Genes and Genetic Systems* **80** (1), 41–48.

Tepperberg, J.H., Moses, M.J. and Nath, J. (1997) Colchicine effects on meiosis in the male mouse. I. Meiotic prophase: synaptic arrest, univalents, loss of damaged spermatocytes and a possible checkpoint at pachytene. *Chromosoma* **106** (3), 183–192.

Thomas, J.B. and Kaltsikes, P.J. (1977) The effect of colchicine on chromosome pairing. *Canadian Journal of Genetics and Cytology* **19**, 231–249.

Tomita, K. and Cooper, J.P. (2006) The meiotic chromosomal bouquet: SUN collects flowers. *Cell* **125** (1), 19–21.

Trelles-Sticken, E., Adelfalk, C., Loidl, J. and Scherthan, H. (2005) Meiotic telomere clustering requires actin for its formation and cohesin for its resolution. *Journal of Cell Biology* **170** (2), 213–223.

Trelles-Sticken, E., Dresser, M.E. and Scherthan, H (2000) Meiotic telomere protein Ndj1p is required for meiosis-specific telomere distribution, bouquet formation and efficient homologue pairing. *Journal of Cell Biology* **151** (1), 95–106.

Trelles-Sticken, E., Loidl, J. and Scherthan, H. (1999) Bouquet formation in budding yeast: initiation of recombination is not required for meiotic telomere clustering. *Journal of Cell Science* **122**, 651–658.

Tzur, Y.B., Wilson, K.L. and Grumann, Y. (2006) SUN-domain proteins: 'Velcro' that links the nucleoskeleton to the cytoplasm. *Nature Reviews Molecular Cell Biology* **7**, 782–788.

Vespa, L., Warrington, R.T., Mokros, P. *et al.* (2007) ATM regulates the length of individual telomere tracts in *Arabidopsis*. *Proceedings of the National Academy of Sciences of the USA* **104**, 18145–18150.

Voet, T., Liebe, B. and Labaere, C. (2003) Telomere-independent homologue pairing and checkpoint escape of accessory ring chromosomes in male mouse meiosis. *Journal of Cell Biology* **162** (5), 795–808.

Wanat, J.J., Kim, K.P., Koszul, R. *et al.* (2008) Csm4, in collaboration with Ndj1, mediates telomere-led chromosome dynamics and recombination during yeast meiosis. *PLOS Genetics* Sep 26;**4**(9):e10001888.

Wang, Q., Du, X., Cai, Z. and Greene, M.I. (2006) Characterisation of the structures involved in localisation of the Sun proteins to the nuclear envelope and the centrosome. *DNA Cell Biology* **25**, 554–562.

Wang, S., Lapitan, N.L.V and Tsuchiya, T. (1992) Characterization of telomeres in *Hordeum vulgare* chromosomes by in situ hybridization. II. Healed broken chromosomes in telotrisomic 4L and acrotrisomic 4L4S lines. *Genome* **35**, 975–980.

Watson, J.M. and Riha, K. (2010) Comparative biology of telomeres: where plants stand. *FEBS Letters* **584**, 3752–3759.

Wu, H.Y. and Burgess, S.M. (2006) Ndj1, a telomere-associated protein, promotes meiotic recombination in budding yeast. *Molecular Cell Biology* **26** (10), 3683–3694.

Zhou, K., Rolls, M.M., Hall, D.H. *et al.* (2009) A ZYG-12-dynein interaction at the nuclear envelope defines cytoskeletal architecture in the C. elegans gonad. *Journal of Cell Biology* **186**, 229–241.

Zickler, D. (2006) From early homologue recognition to synaptonemal complex formation. *Chromosoma* **115**, 158–174.

Zickler, D. and Kleckner, N. (1998) The leptotene-zygotene transition of meiosis. *Annual Review of Genetics* **32**, 619–697.

Annual Plant Reviews (2013) **46**, 229–254 http://onlinelibrary.wiley.com
doi: 10.1002/9781118472507.ch8

Chapter 8

THE NUCLEAR PORE COMPLEX IN SYMBIOSIS AND PATHOGEN DEFENCE

Andreas Binder and Martin Parniske

Faculty of Biology, Genetics, University of Munich (LMU), Großhaderner Straße 2-4, 82152 Martinsried, Germany

Abstract: Nuclear pores mediate all transport in and out of the nucleus. While nuclear pore complexes (NPCs) have been intensively studied in yeast and animals, they are not well characterized in plants. Nevertheless, forward and reverse genetic screens have identified plant nucleoporins with functions in various plant processes, including plant-microbe interactions. This indicates that signalling between cytoplasm and nucleus is a key event in the corresponding pathways. Loss of function mutants of *Lotus japonicus NUCLEOPORIN85* (*NUP85*), *NUP133* and *NENA* are impaired in arbuscular mycorrhiza and root nodule symbiosis and unable to induce symbiotic calcium oscillations in the nucleus. *Arabidopsis thaliana NUP96* and *NUP88* are required in both pathogen-associated molecular pattern (PAMP)-triggered and disease-resistance (R) gene-mediated defence signalling. In comparison to the yeast and animal systems, this review discusses possible specific and general functions and targets of plant nucleoporins with a focus on the role of the NPC in plant-microbe interactions.

Keywords: plant-microbe interaction; root-nodule symbiosis; arbuscular mycorrhiza; plant defence; nuclear pore; nucleoporins; NUP107-160 sub-complex

8.1 Introduction

Nuclear pore complexes (NPCs) are large macromolecular protein assemblies connecting the cytosol with the nuclear lumen across the double membrane of the nuclear envelope (NE). The NPC is composed of approximately 30 nucleoporins (nups), of which many are structurally conserved in even distantly related eukaryotes (Bapteste *et al.*, 2005). Studies predominantly in vertebrates and yeast have revealed much information about the structure

Annual Plant Reviews Volume 46: Plant Nuclear Structure, Genome Architecture and Gene Regulation, First Edition. Edited by David E. Evans, Katja Graumann and John A. Bryant.

and function of the NPC (Alber *et al.*, 2007; Fernandez-Martinez and Rout, 2009; Hoelz *et al.*, 2011). Besides their classical role in facilitation of nucleo-cytoplasmic transport, NPCs and their associated components are involved in various cellular processes, including chromatin organization, regulation of gene expression and cell division (Capelson and Hetzer, 2009; Wente and Rout, 2010; Strambio-De-Castillia *et al.*, 2010,). With few exceptions (Fiserova *et al.*, 2009; Tamura *et al.*, 2010) there has been only a very limited directed effort to characterize the NPC in plants. Forward and reverse genetic studies have implicated components of the plant NPC in a variety of plant processes (Xu and Meier, 2008; Meier and Brkljacic, 2009), including plant-microbe symbiosis and plant defence signalling.

8.2 The nuclear pore and plant-microbe symbiosis

8.2.1 Common signalling in arbuscular mycorrhiza and root-nodule symbiosis

Arbuscular mycorrhiza (AM) is an ancient association between fungi of the phylum *Glomeromycota* and the majority of land plants (Schüßler *et al.*, 2001). The symbiosis is considered to be of great ecological and economic importance (Smith and Smith, 2011). The obligate symbiotic AM fungi provide the plants with phosphorus, sulfur, nitrogen, water and other inorganic nutrients from the soil. In return the fungus obtains an estimated 10–20% of fixed carbon from the host plants. Nutrient exchange is assumed to take place in specialized cells, harbouring the tree-shaped arbuscules (Parniske, 2008; Smith and Read, 2008). Root nodule symbiosis (RNS) between nitrogen-fixing bacteria and plant species is restricted to the eurosid I subclade of angiospermes (Soltis *et al.*, 2000). Legumes form RNS with diverse bacteria collectively referred to as rhizobia (Sprent, 2007), while plants of the orders Fagales, Curcubitales and Rosales interact with actinobacteria of the genus *Frankia* (Pawlowski and Sprent, 2008). Atmospheric nitrogen is fixed within root nodules by differentiated bacteria. In exchange for the supplied nitrogen, the plant provides the bacteria with a variety of nutrients including dicarboxylates and amino acids (White *et al.*, 2007).

Both RNS and AM share a common genetic programme, suggesting that symbiosis genes that evolved in the context of the older AM symbiosis were co-opted during the evolution of RNS (Kistner and Parniske, 2002). Rhizobia and AM fungi produce signalling molecules, which trigger the initiation of the plant's symbiotic programme. In case of rhizobia these include lipochito-oligosaccharide nodulation (nod) factors, which in *L. japonicus* are perceived by the lysin motif (LysM)-type receptor-like kinases NOD FACTOR RECEPTOR5 (NFR5) and NFR1 (Madsen *et al.*, 2003; Radutoiu *et al.*, 2003).

Equivalent signals from arbuscular mycorrhiza fungi, so called myc factors, include structurally similar lipochito-oligosaccharides secreted by the AM fungus *Rhizophagus irregularis* (Maillet *et al.*, 2011); *Rhizophagus irregularis* was previously classified as *Glomus intraradices* or *Glomus irregulare*

(Stockinger *et al.*, 2009; Schüßler and Walker, 2001). The corresponding receptors have not yet been identified in legumes, however in the tropical tree species of the genus *Parasponia,* which evolved root-nodule symbiosis independently from legumes, an NFR5-like receptor is required both for RNS and AM symbiosis (Op den Camp *et al.*, 2011). Signal transduction both in RNS and AM symbiosis is mediated by the SYMBIOSIS RECEPTOR-LIKE KINASE (SYMRK; *Medicago sativa* NORK or *M. truncatula* DMI2 – Endre *et al.*, 2002; Stracke *et al.*, 2002).

Morphological responses during early symbiotic signalling in RNS include root hair curling around entrapped rhizobia. Originating from the root hair curl, intracellular infection threads form and guide the rhizobia towards the root cortex and the developing nodule primordium, where the bacteria are eventually released into membrane enclosed compartments, the symbiosomes (Oldroyd *et al.*, 2011). In case of AM, entry of the fungal hyphae is associated with the formation of a specialized cytoplasmic assembly, the prepenetration apparatus (PPA) (Genre *et al.*, 2008). This plant-derived structure determines the path of fungal colonization through rhizodermal and cortical cells.

8.2.2 Symbiotic signalling at the nucleus

A so-far unidentified second messenger transduces the signal from the membrane receptors to the nucleus, which is the central compartment for further symbiotic signalling. Following rhizobial nod factor perception, periodic calcium oscillations (calcium spiking) occur in and around the nucleus (Ehrhardt *et al.*, 1996; Sieberer *et al.*, 2009). Similar calcium signals can be detected in response to AM fungal exudates and growing hyphopodia (Kosuta *et al.*, 2008; Chabaud *et al.*, 2011).

Mutations in either of the three *L. japonicus* nucleoporin genes, *LjNUP85*, *LjNUP133* and *NENA* abolish the calcium spiking response (Kanamori *et al.*, 2006; Saito *et al.*, 2007; Groth *et al.*, 2010). Other components of the nuclear envelope, the nuclear membrane localized ion channels CASTOR and POLLUX/DOES NOT MAKE INFECTIONS1 (DMI1) (Ané *et al.*, 2004; Imaizumi-Anraku *et al.*, 2005; Charpentier *et al.*, 2008) are also required for common symbiotic signalling and calcium spiking. The inner nuclear membrane localized SERCA-type calcium ATPase M. TRUNCATULA CALCIUM ATPASE8 (MCA8) also has an important role in generation of the calcium oscillations (Capoen *et al.*, 2011).

The calcium spikes are most likely interpreted by a calcium- and calmodulin-dependent protein kinase (CCaMK) (Lévy *et al.*, 2004; Mitra *et al.*, 2004; Tirichine *et al.*, 2006) that interacts with its phosphorylation substrate CYCLOPS inside the nucleus (Messinese *et al.*, 2007; Yano *et al.*, 2008). A gain of function mutation resulting in an autoactive version of CCaMK is sufficient to trigger rhizobial infection and nodule formation in calcium spiking-defective mutants, indicating that the primary role of the corresponding gene products is the activation of CCaMK (Hayashi *et al.*, 2010; Madsen *et al.*, 2010).

Downstream of calcium spiking and CCaMK/CYCLOPS several transcription factors, namely NODULATION SIGNALING PATHWAY1 (NSP1), NSP2, ETHYLENE-RESPONSIVE BINDING DOMAIN FACTOR REQUIRED FOR NODULATION1 (ERN1) and NODULE INCEPTION (NIN) regulate the expression of RNS-specific genes and thus coordinate ongoing rhizobial infection and nodule organogenesis (Oldroyd *et al.*, 2011).

8.2.3 Symbiotic defects in *ljnup85*, *ljnup133* and *nena* mutants

The sequence identity between the *L. japonicus* nucleoporins LjNUP133, LjNUP85 and NENA and their respective yeast/vertebrate counterparts NUP133, NUP85 and SEC13 HOMOLOG1 (SEH1) is low; however, their predicted domain structure is conserved. Fast evolution at the sequence level, with the retention of the overall structural appearance, is considered a typical feature of the NPC across all eukaryotic lineages (Bapteste *et al.*, 2005). Loss of function mutants of *ljnup85, ljnup133* and *nena* exhibit temperature-dependent defects in nodulation and AM colonization (Kistner *et al.*, 2005; Kanamori *et al.*, 2006; Saito *et al.*, 2007; Groth *et al.*, 2010). At high temperature (24 °C or 26 °C) the stronger mutant alleles form very few or no nodules and AM infection and colonization is severely reduced, while at lower temperature (18 °C) nodulation and AM colonization phenotypes are less severe. The residual nodulation in the mutants is delayed compared to wild-type plants and the large majority of nodules are not infected by rhizobia. While *ljnup85, ljnup133* and *nena* fail to initiate nod factor-induced calcium spiking, the earlier nod factor-induced calcium influx into the cytosol at the root hair tip (Felle *et al.*, 1998; Shaw and Long, 2003) is not affected in *nup85 and nup133* mutants (Miwa *et al.*, 2006; *nena* mutants were not tested). Cortical infection and development in mature nodules was indistinguishable between *nena* and wild type plants, indicating that the *L. japonicus* nucleoporins are required only for rhizodermal root hair infection, but not for cortical nodule development (Groth *et al.*, 2010). Waterlogging of *nena* plants significantly increased the percentage of infected nodules in an ethylene-dependent manner. This was attributed to an intercellular infection mode, which is independent of root-hair infection and still functions in *nena* (Groth *et al.*, 2010). Infection structures in *nena* mutants resemble those observed in the ethylene-dependent lateral root base (LRB) infection mode that is activated in waterlogging-tolerant legumes such as *Sesbania rostrata* under flooded conditions. In LRB infection, rhizobia can directly enter cortical cell layers through cracks in the lateral root bases, thus circumventing the epidermal stages of root-hair invasion (Goormachtig *et al.*, 2004).

In *L. japonicus* a large number of mutant alleles for *NUP85, NUP133* and *NENA* were identified in screens for symbiotic defects and via TILLING (Perry *et al.*, 2009; Groth *et al.*, 2010). Besides the symbiotic phenotype the mutation of a single nucleoporin has surprisingly mild effects on overall plant growth and development including shorter pollen tubes in *ljnup85* as well as seed yield reduction in *ljnup133 and ljnup85*, which was not observed in *nena* mutants.

8.2.4 How do nucleoporins function in plant-microbe symbiosis?

The mechanism by which LjNUP85, LjNUP133 and NENA facilitate calcium spiking in the nucleus is still obscure. One explanation is that the NPC itself has a role in the generation and maintenance of nuclear calcium signals. The observation that nuclear and cytosolic calcium changes are not necessarily linked and that nuclear calcium responses can be induced independently from the cytosol (Mazars *et al.*, 2009; Rodrigues *et al.*, 2009), indicated that the NPC acts as a diffusion barrier for ions. Consistent with this idea, in a patch-clamping setup, the NPC itself exhibited ion channel-like behaviour, partly existing in a closed confirmation that does not permit ion flow (Bustamante, 2006). On the other hand, studies that measured currents through individual nuclei held in an hourglass-shaped trap observed NPCs to be constitutively open to ions (Danker *et al.*, 2001). However, even in the later case, the conductance of the NPCs could be changed by application of factors such as calcium or ATP (Shahin *et al.*, 2001). In either case, it is possible that potential structural aberrations in the *ljnup85, ljnup133* and *nena* mutants could alter the electrophysiological properties of the nucleus in a way that prevents nod-factor-induced calcium spiking. Mutations in nucleoporins could change the conductance of individual NPCs or affect the distribution of NPCs in the nuclear envelope. For instance, several yeast nucleoporin mutants, including *nup133* and *nup85* suffer from aberrant clustering of NPCs (Siniossoglou *et al.*, 1996). It is currently not clear whether calcium spiking is initiated at the cytosolic or nucleoplasmic side and whether calcium is released from the nucleoplasmic reticulum, nuclear associated ER or both (Capoen *et al.*, 2011). Continuous calcium oscillations might require ion exchange through the NPC, such as movement of calcium originating from the nuclear associated ER into the nucleus. Ion flow through the pore could also influence the membrane potential of the inner or outer nuclear membrane in order to activate voltage-gated calcium channels. It would be interesting to determine whether other nuclear calcium responses induced by elicitors such as flagellin (flg22), harpin, jasmonic-acid-isoleucine or mechanical stresses (Xiong *et al.*, 2004; Lecourieux *et al.*, 2005; Walter *et al.*, 2007) are also affected in the nucleoporin mutants.

Alternatively to the direct involvement of NPC components in calcium signalling, LjNUP85, LjNUP133 and NENA may be necessary for properly targeting proteins required for calcium spiking. As demonstrated in case of *atnup88, atnup96* and *atnup160,* mutations in plant nucleoporins can lead to specific nuclear import or export defects (Parry *et al.*, 2006; Cheng *et al.*, 2009). Since LjNUP85, LjNUP133 and NENA are assumed to be part of the nuclear pore scaffold, current models predict that they do not directly interact with the nuclear transport machinery (Wente and Rout, 2010; Hoelz *et al.*, 2011). However, mutations in the scaffold nups could lead to structural changes in NPC that might disturb import and export by affecting peripheral nucleoporins. A large proportion of the peripheral nups contain phenylalanine-glycine (FG)-repeat domains, which mediate interactions with transport receptors. The intrinsically unfolded FG-repeats are also assumed to regulate the permeability of the NPC (Wente, 2000; Frey and Görlich, 2007; D'Angelo and Hetzer, 2008). In case of symbiosis, the transport of secondary messengers to the nucleus, which is assumed to activate the calcium oscillations, might be impaired in the nucleoporin mutants (Oldroyd and Downie, 2008). Another suggested target is the GRAS (GAI RGA SCR)-type transcription factor NSP2. NSP2 is required for nodule specific gene induction and was shown to change its localization from the nuclear envelope to the nucleus upon nod factor stimulation (Kaló *et al.*, 2005). The localization has not been assayed in the *ljnup85, ljnup133 or nena* mutants so far, however, as NSP2 is not required for calcium spiking it does not appear to be the most likely target. Additionally, while individual nucleoporins are dispensable for nodule organogenesis, NSP2 is not, as was shown by analyses of *nup snf1* double mutants expressing a deregulated CCaMK version (Madsen *et al.*, 2010).

Other obvious candidates to be affected by mutations in the nucleoporin genes are components of the nuclear calcium spiking machinery, including the nuclear membrane localized ion channels CASTOR and POLLUX/DMI1 (Ané *et al.*, 2004; Charpentier *et al.*, 2008). While these proteins do not appear to be calcium channels, they are required for symbiotic calcium signalling, potentially by balancing the ion charges across the nuclear membranes or by altering the membrane potential and thus activating voltage gated calcium channels (Charpentier *et al.*, 2008). The POLLUX homologue of *M. truncatula*, DMI1, was shown to preferentially localize to the inner nuclear membrane (INM), and this localization might be required for its symbiotic function (Capoen *et al.*, 2011).

There are still many open questions regarding the transport of membrane proteins from the outer nuclear membrane (ONM) to the INM. Smaller proteins seem to be able to passively diffuse from the ONM along the pore membrane (POM) into the INM, where they are retained by interaction with elements of the nuclear architecture, while larger ($> \sim$25 kDa) membrane proteins such as CASTOR and POLLUX require active transport, which implicates components of the classical nuclear import pathway (Ohba *et al.*, 2004; Lusk *et al.*, 2007; Zuleger *et al.*, 2008). A nuclear localization signal (NLS)-mediated transport mechanism for INM proteins that is dependent on karyopherin-α/β (Kap60/95) was recently elucidated in yeast (Meinema *et al.*, 2011). In this case a long unfolded linker is needed to translocate reporters based on the protein helix-extension-helix-2 (Heh2) into the INM. Accumulation of the Heh2 constructs in the INM is not achieved by trapping of the proteins in the IMN, but because Kap60/95 mediated import is faster than export (Meinema *et al.*, 2011). Efficient transport of Heh2 is dependent on the FG domains of nucleoporins, which indicates that the soluble part of the protein is likely to be transported through the central channel. Conclusively, during the active transport, part of the membrane protein must pass through the NPC scaffold, which implies a remodelling of nucleoporin interactions in order to create an opening for the cargo (Lusk *et al.*, 2007; Meinema *et al.*, 2011). It is therefore conceivable that structural defects in the *nup* mutants can impair the proper sub-cellular localization of INM proteins such as CASTOR and/or POLLUX.

Other membrane proteins of the nuclear calcium spiking machinery, such as calcium channels and pumps might also be affected by a potentially impaired ONM to INM transport. Nuclear localized calcium channels have so far not been identified. In animals nuclear calcium release is predominantly linked to the activation of Inositol(1,4,5)triphosphate receptors (InsP_3Rs) (Bootman *et al.*, 2009). However no sequences with overall homology to this type of calcium channel could be found in higher plants (Kudla *et al.*, 2010). The *M. truncatula* calcium ATPase MCA8 normally localizes to the ER, ONM as well as INM and RNA interference mediated silencing of MCA8 significantly reduces the observed frequency of nuclear calcium spiking (Capoen *et al.*, 2011). It is possible that INM protein levels of MCA8 might be reduced in the *nup* mutants and that this reduction could lead or contribute to impaired calcium spiking.

Independent from their functions in nucleo-cytoplasmic trafficking, NPC components can have direct roles in chromatin organization and gene regulation (Akhtar and Gasser, 2007; Brown *et al.*, 2008; Capelson and Hetzer, 2009). For instance, several Drosophila nups, including NUP88 and the NUP107-160 sub-complex member SEC13 bind to transcriptionally active genes. Interestingly the binding sites were localized in the nucleoplasm outside the NE. Downregulation of SEC13 inhibited the transcription of the associated target genes (Capelson *et al.*, 2010). It is therefore possible that genes required for calcium spiking might be affected in their transcriptional regulation in *ljnup85*, *ljnup133* and *nena*.

8.3 The nuclear pore and plant defence

8.3.1 Plant immune responses can be triggered by pathogen-associated molecular patterns and microbial effectors

Innate immunity in plants is mediated by transmembrane receptors that recognize conserved features of microbes, so called pathogen-associated molecular patterns (PAMPs), such as bacterial flagellin or fungal chitin (Zipfel and Felix, 2005). Many pathogens have evolved strategies to circumvent this first layer of defence, called PAMP-triggered immunity (PTI). By delivering effector molecules to, or into, plant cells, bacteria, fungi and oomycetes are able to suppress immune responses in the host cells. In turn plants have developed a second layer of defence, effector-triggered immunity (ETI) (Jones and Dangl, 2006). As part of ETI, plant disease resistance (R), proteins recognize avirulence signals from pathogens, which can often lead to a so called hypersensitive response (HR). Hypersensitive response is defined by localized cell death at the site of the infection and is considered a means to limiting the spread of the pathogen (Greenberg and Yao, 2004). Hypersensitive response is typically accompanied by several physiological responses including the production of reactive oxygen species (ROS), ion fluxes and the accumulation of the defence hormone salicylic acid (SA). The nucleus – including trafficking through the NPC – has emerged as a central compartment in defence signalling (Deslandes and Rivas, 2011).

Most identified R proteins are of the NB-LRR type, named after nucleotide binding site (NB) and leucin-rich repeat (LRRs) domains. There are three subgroups of NB-LRR proteins classified according to their N-terminal region: (i) Coiled-coil or leucine zipper group (CC-NB-LRRs or LZ-NB-LRRs), (ii) Toll/Interleukin-1 receptor domain group (TIR-NB-LRRs) and (iii) non-motif groups (Caplan *et al.*, 2008). Signalling via the majority of TIR-NB-LRR R proteins involves defence regulator ENHANCED DISEASE SUSCEPTIBILTY1 (EDS1), whereas CC-NB-LRR receptors often require the plasma membrane protein NON RACE SPECIFIC DISEASE RESISTANCE1 (NDR1) (Aarts *et al.*, 1998). However, CC-NB-LRR signalling can also depend on EDS1, which was previously masked by the fact that EDS1 and SA can function redundantly (Venugopal *et al.*, 2009). EDS1 forms both cytosolic and nuclear complexes with PHYTOALEXIN DEFICIENT4 (PAD4) and SENESCENCE-ASSOCIATED GENE101 (SAG101) and their combined activity is required to induce programmed cell death (Feys *et al.*, 2005; Zhu *et al.*, 2011). Moreover nuclear EDS1 accumulation is correlated with EDS1-dependent defence gene regulation (García *et al.*, 2010). In many cases R gene-induced HR initiates a secondary resistance response, known as systemic acquired resistance (SAR) (Durrant and Dong, 2004). SAR acts both locally and systemically and confers a longer protection against a broad range of pathogens. Induction of SAR requires SA (Gaffney *et al.*, 1993) and is associated with the expression of pathogenesis related (PR) genes (Durrant and Dong, 2004). NON-EXPRESSOR OF PR GENES1 (NPR1) is a positive regulator of SAR and mutations in *NPR1* completely prevent the induction of PR genes by SA (Cao *et al.*, 1997). Upon SA induction, NPR1 is translocated to the nucleus, where it interacts with TGA (TGACG motif binding) transcription factors in order to initialize defence gene expression (Zhang *et al.*, 1999; Després *et al.*, 2000; Zhou *et al.*, 2000).

8.3.2 AtNUP88 and AtNUP96 are required for basal and NB-LRR-mediated plant immunity

A screen for *npr1-1* suppressors revealed a gain of function mutation in a TIR-NB-LRR-type R protein, which was termed *suppressor of npr1-1, constitutive*

1 (*snc1*) (Li *et al.*, 2001). In *snc1*, a point mutation in the NB and LRR linker region causes constitutive resistance against the bacterial pathogen *Pseudomonas syringae maculicola (P.s.m.)* ES4326 and the oomycete *Hyaloperonospora arabidopsidis (H.a.)* Noco2 (formerly called *Perenospora parasitica*) (Zhang *et al.*, 2003). The *snc1* mutants exhibited high SA levels and a constitutive expression of PR genes without associated cell death (Li *et al.*, 2001). Morphologically homozygous *snc1* plants showed stunted growth and had dark curly leaves, which is often observed in mutants with increased SA levels (Scott *et al.*, 2004). A suppressor screen of *snc1* was aimed to identify further components of R gene mediated signalling. A variety of mutants, termed *modifiers of snc1 (mos)*, were found to suppress constitutive defence signalling in the *snc1* and *snc1 npr1-1* backgrounds. The characterized *MOS* genes code for proteins with predicted functions in RNA processing, protein modification and nucleo-cytoplasmic trafficking (Monaghan *et al.*, 2010).

Two of the MOS proteins, MOS3 and MOS7, are putative components of the nuclear pore (Zhang and Li, 2005; Cheng *et al.*, 2009). MOS3 is a homologue of vertebrate NUP96 and yeast NUP145, while MOS7 is related to vertebrate NUP88. In agreement with their predicted role, MOS3 (AtNUP96) and MOS7 (AtNUP88) localized to the nuclear envelope (Zhang and Li, 2005; Cheng *et al.*, 2009). Mutations in *MOS3* and *MOS7* abolish the enhanced disease resistance in the *snc1* background. Furthermore *snc1* dependent constitutive PR gene expression, as well as increased SA levels are suppressed. The *snc1 mos3* and *snc1 mos7* double mutants have an intermediate appearance regarding size, leaf colour and shape compared to wild type and *snc1* plants.

Independent of the *snc1* background, single *mos3* and *mos7* mutants are impaired in basal and R gene dependent immunity. Resistance mediated by multiple R proteins, both of the TIR- and CC-NB-LRR type is affected in the mutants, although to varying degrees (Zhang and Li, 2005; Cheng *et al.*, 2009). In addition *mos7* is defective in SAR and fails to induce PR genes in systemic tissues (Cheng *et al.*, 2009).[1]

8.3.3 Mechanisms of nucleoporin-mediated plant defence signalling

A specific mechanism in plant immunity has been demonstrated in case of AtNUP88. The *mos7* mutation is correlated with an increase in nuclear export signal (NES)-mediated nuclear export and a reduction of the nuclear level of SNC1 (Cheng *et al.*, 2009). A NES fusion to the mutant version of SNC1 demonstrated that increased export of the R protein is sufficient to

[1]A reverse genetics approach identified two other putative components of the NUP107-160 sub-complex with a role in defense signalling in Arabidopsis: T-DNA insertion lines of *AtNUP160* as well as of *AtSEH1* exhibit defects in basal resistance. Additionally AtNUP160 and to a lesser extend AtSEH1 are also required for TIR-NB-LRR R gene mediated resistance (Wiermer *et al.*, 2012).

suppress the *snc1* phenotype. The observed export defects are not ubiquitous, as the localization of several other nuclear proteins was unchanged (Cheng *et al.*, 2009). The nuclear levels of NPR1 as well as EDS1 are also reduced in the *mos7* mutants. A decrease in the level of nuclear NPR1 could be the cause for the observed defects in SAR (Kinkema *et al.*, 2000), and enhanced EDS1 export from the nucleus was previously correlated with impaired pathogen resistance (García *et al.*, 2010). EDS1 interacts with the TIR-NB-LRR R protein RESISTANCE TO P. SYRINGAE4 (RPS4) and the bacterial effector AvrRps4 and coordinated nuclear and cytoplasmic activities are required for a complete immune response (García *et al.*, 2010; Heidrich *et al.*, 2011). While total SNC1 protein levels are not altered in *mos7*, the overall amount of NPR1 and EDS1 protein is decreased. This is not caused by changes in gene expression, as *EDS1* and *NPR1* transcript levels are not affected in the *mos7* mutant (Cheng *et al.*, 2009). Elevated nuclear export of NPR1 and EDS1 in *mos7* might lead to an increased degradation of the proteins. Alternatively the reduction in protein abundance could be an indirect result of lowered basal resistance (Cheng *et al.*, 2009). While it was previously assumed that the majority of R proteins is cytosolic, accumulating evidence suggests that nuclear localization is important for the function of many NB-LRR proteins (Burch-Smith *et al.*, 2007; Shen *et al.*, 2007; Wirthmueller *et al.*, 2007; Caplan *et al.*, 2008). Since both CC-NBS-LRR- and TIR-NBS-LRR-dependent defence signalling was strongly impaired in the *mos7* mutants, the nuclear export of R proteins other than SNC1 could also be affected (Cheng *et al.*, 2009).

The described mechanism fits well with the characterization of NUP88 homologues in other systems. Mammalian and Drosophila NUP88 are part of the of the asymmetric cytoplasm-oriented NUP214-NUP88 complex, which anchors the export receptor CHROMOSOME REGION MAINTAINANCE1 (CRM1)/EXPORTIN1 (XPO1) to the nuclear pore and mediates NES-dependent nuclear export (Roth *et al.*, 2003; Bernad *et al.*, 2004). Inhibition of the Arabidopsis CRM1/XPO1 homologue led to an increase in nuclear accumulation of EDS1 (García *et al.*, 2010), which argues for a direct connection between the *mos7* mutation and the decrease in nuclear EDS1 levels. Defects in the Drosophila gene *DmNUP88/MBO* (Members Only) are also associated with an impaired immune response (Uv *et al.*, 2000). It was demonstrated that DmNUP88 together with DmNUP214 are required for the nuclear accumulation of the NFκB-like transcription factors Dorsal and Dif, a response that is associated with Toll receptor signalling upon bacterial infection (Xylourgidis *et al.*, 2006).

Independent of their defence related phenotype, the *mos7-1* mutant plants have no obvious pleiotropic defects with regard to development, salt tolerance or hormone signalling. *mos7-1* encodes an AtNUP88 mutant version lacking four amino acids at the N terminus of the protein. It was not possible to obtain homozygous plants of other *mos7* mutant alleles (Cheng *et al.*, 2009). This indicates that more severe loss of function mutations in *AtNUP88* are lethal, potentially by causing a general impairment in NES mediated nuclear export. Consistent with this, null alleles of *MBO*, the *NUP88* homologue of Drosophila, were also not viable (Uv *et al.*, 2000).

The mechanism of AtNUP96 function in basal and R gene-mediated defence signalling is still unknown. However, both the effect of the *mos7* mutant and the fact that another *MOS* gene, *MOS6*, codes for Arabidopsis importin α3 (Palma *et al.*, 2005) suggest that the observed defects in *mos3* may also depend on nucleo-cytoplasmic trafficking. There are several potential targets whose nuclear function could be affected by impaired protein import or increased export in *mos3* mutants, including SNC1 or other R proteins, EDS1 and NPR1, defence-related transcription factors such as BASIC DOMAIN-LEUCINE ZIPPER10 (bZIP10) or components of the mitogen-activated protein kinase (MAPK) signalling cascades (Wiermer *et al.*, 2007). Further candidates are nuclear localized transcriptional co-repressors, such as TOPLESS RELATED1 (TPR1) and TOPLESS (TPL). TPR1 interacts with SNC1 and likely activates TIR-NB-LRR R protein-mediated resistance by repression of negative immune regulators (Zhu *et al.*, 2010). Mutations in *TPR1* and *TPL* lead to impaired basal and R gene-mediated immunity and suppress the constitutive active defence of *snc1*, while overexpression of *TPR1* constitutively activates SNC1-mediated immune signalling (Zhu *et al.*, 2010).

Interestingly, mouse NUP96 also has a role in innate and adaptive immunity. In mice, interferon induces *NUP96* gene expression and increased NUP96 levels can reverse a virus-induced block of mRNA export (Enninga *et al.*, 2002). Animals expressing low levels of NUP96 are more susceptible to viral infections, which is correlated with a nuclear retention of certain mRNAs, whose expression is induced by interferon (Faria *et al.*, 2006). Given these findings the phenotype of *mos3* may also depend on impaired export of certain defence-related transcripts.

The symbiotic defects of *L. japonicus ljnup85, ljnup133* and *nena* mutants are associated with a loss of symbiotic calcium spiking. Calcium is also an important second messenger in plant defence responses (Lecourieux *et al.*, 2006). There is growing evidence that nuclear calcium homeostasis is at least partially independent from the cytosol (Mazars *et al.*, 2009). While nuclear calcium signalling in plant immunity is not well understood, certain defence-related elicitors such as harpin, flg22 and cryptogein were shown to trigger a nuclear calcium increase that is most likely independent of cytosolic diffusion (Lecourieux *et al.*, 2005). It is therefore possible that nuclear defects in the *mos3* and less likely in the *mos7* mutants could affect calcium responses required for defence signalling.

8.4 Specificity, redundancy and general functions of plant nucleoporins

8.4.1 The NUP107-160 sub-complex

As nucleoporins have been implicated in plant defence and symbiosis signalling, it is possible that they share a common function relevant for both types of plant-microbe interactions. The homologous proteins of Arabidospis MOS3 and *L. japonicus* LjNUP85, LjNUP133 and NENA are part of the same sub-structure of the nuclear pore, the vertebrate NUP107-160/ yeast NUP84 sub-complex (Lutzmann *et al.*, 2002; Walther *et al.*, 2003; Loïodice *et al.*, 2004). An additional component of this complex, AtNUP160, was characterized with a role in auxin and stress signalling (Parry *et al.*, 2006; Dong *et al.*, 2006).

While the existence of the complex has not been fully demonstrated in plants, there is evidence that it is conserved: single homologues of all yeast NUP84 and most vertebrate NUP107-160 complex constituents have been identified in Arabidopsis (Tamura *et al.*, 2010). Additionally LjNUP85 interacts with NENA (LjSEH1), while neither protein interacts directly with LjNUP133 (Groth *et al.*, 2010), which fits with the published assembly of the nucleoporins in the Y-shaped NUP84 sub-complex structure (Figure 8.1) (Lutzmann *et al.*, 2002). The NUP107-160 complex is the largest of the NPC subunits and present both on the cytosolic and nuclear side in a total of 16 copies per pore (Alber *et al.*, 2007). In vertebrates it is composed of the nucleoporins NUP37, NUP43, NUP85, NUP96, NUP107, NUP133, NUP160, SEH1 and SEC13 (Belgareh *et al.*, 2001; Vasu *et al.*, 2001; Harel *et al.*, 2003; Loïodice *et al.*, 2004). It is an essential component of the NPC scaffold and required for nuclear pore assembly both at the end of mitoses as well as during interphase, albeit via different mechanisms (Doucet *et al.*, 2010). The complete loss of the complex results in nuclei without nuclear pores (Harel *et al.*, 2003; Walther *et al.*, 2003). The complex also has cell cycle specific roles; during mitosis it is recruited to kinetochores and contributes to normal kinotochore functions including spindle assembly (Belgareh *et al.*, 2001; Orjalo *et al.*, 2006; Zuccolo *et al.*, 2007).

Single mutant alleles of NUP107-160 complex components in plants are viable. This indicates that the sub-complex as a whole can tolerate the omission of individual elements, while at the same time particular functions can no longer be fulfilled. Mutations in Arabidopsis *AtNUP96* and *AtNUP160* have seemingly more pronounced defects than in Lotus *LjNUP85, LjNUP133* and *NENA* (*LjSEH1*). At present it is not clear if these differences are species-specific or depend on the particular nucleoporins. The observed temperature sensitivity in the Lotus and yeast mutants could be related to the stability of the sub-complex. At lower temperatures the NUP107-160 structure may tolerate the total or partial loss of individual components, while elevated temperature could increase instability of the remaining nucleoporin complex.

In general a broad range of defects is associated with the loss or down-regulation of NUP107-160 nucleoporins. Yeast *nup133* and *nup85* mutants show reduction in growth at permissive temperatures (23 °C) and growth arrest or even lethality at non-permissive temperatures (30 °C or 37 °C) (Li *et al.*, 1995; Siniossoglou *et al.*, 1996). SEH1 disruption on the other hand still permits growth at 37 °C, but leads to reduced growth rates particularly at lower temperatures. Double mutants of yeast NUP84 complex components are usually lethal (Siniossoglou *et al.*, 1996), while double mutants of Arabidopsis *atnup96* and *atnup160* are viable, albeit severely affected in

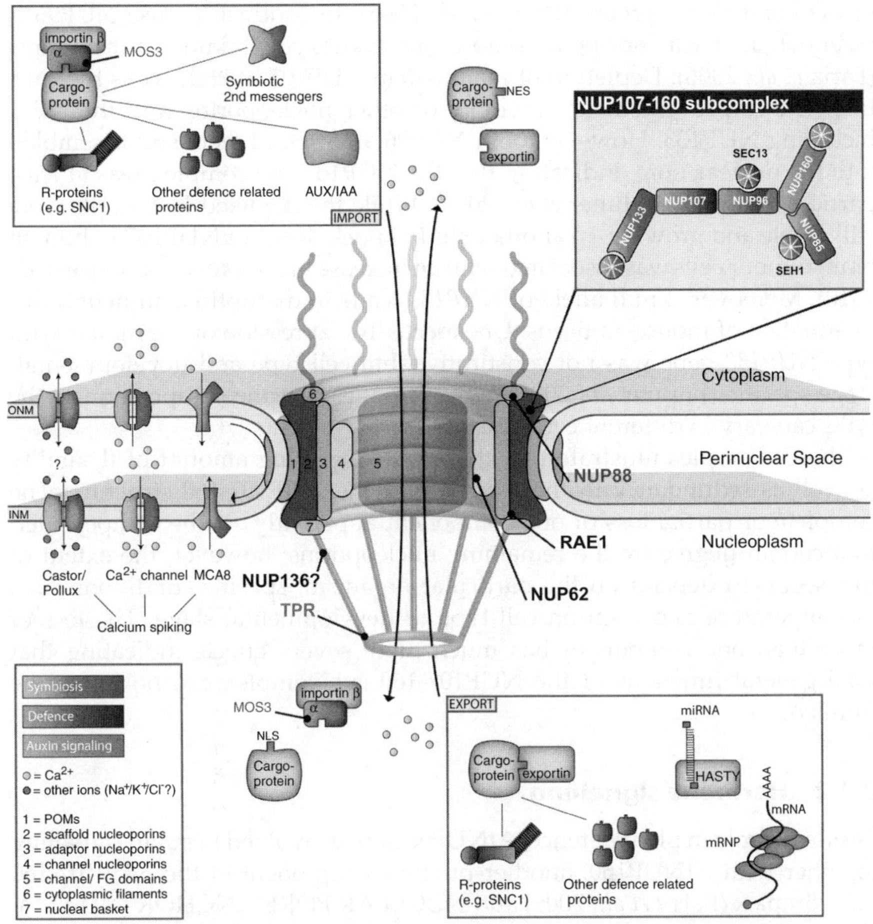

Figure 8.1 Model of the plant nuclear pore complex and its putative functions in plant signalling pathways. Plant nucleoporins are represented according to the position of the known homologues in the vertebrate NPC. The model of the NUP107-160 sub-complex in the top left panel is based on the structural characterization of the yeast NUP84 complex (Hoelz, Debler and Blobel, 2011). Mutations in plant nucleoporins affect various plant processes, including symbiotic signalling, plant defence and auxin signalling. The mutant phenotypes could be caused by deficient nucleo-cytoplasmic transport of various molecules. Potential targets in plant-defence and auxin signalling include R-proteins such as SNC1, other defense related proteins, AUX/IAA repressors, as well as certain mRNAs or miRNAs. In case of plant-microbe symbiosis, nuclear calcium spiking is abolished in the *ljnup85*, *ljnup133* and *nena/ljseh1* mutants. This might be due to structural changes in the NPC that either directly affect calcium homeostasis or indirectly impair proper transport of symbiotic components upstream of calcium spiking, such as calcium channels, the ion channels CASTOR and POLLUX, the calcium ATPase MAC8 or symbiotic second messengers. mRNP = messenger ribonucleoprotein; NLS = Nuclear localization signal; NES = Nuclear export signal; INM = Inner nuclear membrane; ONM = Outer nuclear membrane; POM = pore membrane protein. (*For colour details please see colour plate section.*)

growth and development (Parry *et al.*, 2006). In contrast, mouse NUP96 is essential, as it was not possible to create homozygous knock-out mutants (Faria *et al.*, 2006). Depletion of mammalian NUP107 in HeLa cells by RNA interference prevented the assembly of other nucleoporins into the NPC, including NUP133. However, both NUP96 and SEC13 were still assembled into the nuclear pore, indicating that the NUP107-160 complex was at least partially present (Boehmer *et al.*, 2003). While the depleted HeLa cells were still viable and growing, in another study knock down of NUP107 in human brain tumour cells was associated with an increase in apoptosis (Banerjee *et al.*, 2010). Moreover, a null allele of *NUP133* leads to disruptions in neural differentiation of mouse embryos. Unexpectedly, expression of the mouse wild type *NUP133* gene was not constitutive, but cell type and developmental-stage specific (Lupu *et al.*, 2008), which implies that the composition of the NPC can vary in different cell types.

These examples illustrate that there is a surprising amount of flexibility as well as redundancy in the assembly of the NUP107-160 structure. The complete or partial loss of one component apparently can be compensated to a certain degree by the remaining nucleoporins; however, the extent of this seems to depend on the particular organism, specific conditions such as temperature and even on cell type or developmental stage. The loss of more than one component has much more severe effects indicating that vital general functions of the NUP107-160 sub-complex can no longer be fulfilled.

8.4.2 Hormone signalling

Besides its role in plant defence, AtNUP96 is also involved in auxin signalling together with AtNUP160, another putative component of the NUP107-160 sub-complex (Parry *et al.*, 2006) and NUCLEAR PORE ANCHOR (NUA)/*A. thaliana* TRANSLOCATED PROMOTER REGION (AtTPR) (Jacob *et al.*, 2007; Xu *et al.*, 2007), the Arabidopsis homologue of the vertebrate inner basket nucleoporin TPR (see Figure 8.1). Mutations in SUPPRESSOR OF AUXIN RESISTANCE3 (*SAR3/AtNUP96*), *SAR1* (*AtNUP160*) and *AtTPR* suppress many of the phenotypes of the *auxin resistant1* (*axr1*) mutant. AXR1 functions in the RELATED TO UBIQUITIN (RUB) pathway, which is required for the auxin-dependent degradation of the nuclear Auxin/Indole-3-Acetic Acid (Aux/IAA) proteins that act as transcriptional repressors in auxin signalling (Dharmasiri and Estelle, 2004). The *axr1* mutation leads to accumulation of Aux/IAA proteins in the nucleus and thereby represses the normal auxin response. An altered localization of the Aux/IAA protein IAA17 was demonstrated in the *sar1* and *sar3* mutant background (Parry *et al.*, 2006). While an IAA17-β-Glucuronidase (GUS) fusion showed the expected nuclear accumulation in the wild type, in the nucleoporin mutants GUS staining was observed throughout the cell, indicating either decreased nuclear import or increased export. Thus a decrease in nuclear accumulation of Aux/IAA proteins in the *sar* mutants could potentially compensate the effect of *axr1*.

Another possible mechanism might depend on the function of particular micro (mi) RNAs. Unlike animals, plants cells may have the potential to fully process miRNA in the nucleus (Hohn and Vazquez, 2011). Levels of several miRNAs are reduced in the *attpr* mutant (Jacob *et al.*, 2007), among them miR165 and miR393. miR393 was shown to target TRANSPORT INHIBITOR RESPONSE1 (TIR1) (Sunkar and Zhu, 2004), a positive regulator of auxin signalling. A reduction in miR393 levels should therefore result in an increased auxin response that could potentially counteract the repressed auxin response in *axr1*. However, *TIR1* transcript levels were unchanged in the *sar1* and *sar3* mutants (Parry *et al.*, 2006). miR165 overexpression was shown to change the transcript level of several auxin related genes (Zhou *et al.*, 2007), its reduction in *attpr* might therefore also affect the *axr1* phenotype.

Auxin signalling could be particularly susceptible to defects in nucleocytolasmic trafficking as suppression of the Arabidopsis Ran binding protein AtRanBP1c leads to auxin hypersensitivity (Kim *et al.*, 2001). It has to be determined if the observed defects in auxin signalling in *atnup96*, *atnup160* and *attrp* are related to the defence phenotype in *atnup96*.

8.4.3 Development, flowering time, stress tolerance and RNA transport

Pleiotropic phenotypes with regard to growth, development and flowering time are associated with several Arabidopsis nucleoporin mutants, including *atnup96*, *atnup160* and *attpr* (Zhang and Li, 2005; Dong *et al.*, 2006; Jacob *et al.*, 2007; Xu *et al.*, 2007). In addition, developmental defects are also observed in mutant or knock-down plants of Arabidopsis *AtNUP62* (Zhao and Meier, 2011), *AtNUP136/AtNUP1* (Tamura *et al.*, 2010; Lu *et al.*, 2010) and tobacco *RNA EXPORT1* (*RAE1*) (Lee *et al.*, 2009), whose gene products all have potential functions in nucleo-cytoplasmic trafficking. AtNUP62 is an FG-nucleoporin that interacts with an Arabidopsis homologue of NUCLEAR TRANSPORT FACTOR2 (NTF2), which is required for Ran-dependent nucleo-cytoplasmic transport (Ribbeck *et al.*, 1998; Zhao *et al.*, 2006; Zhao *et al.*, 2008). AtNUP136 was identified as a plant-specific nucleoporin with functional similarity to vertebrate NUP153. Both AtNUP136 and NUP153 interact with importins (Kodiha *et al.*, 2008; Tamura *et al.*, 2010). In addition AtNUP136 anchors the Arabidopsis TRANSCRIPTION-COUPLED EXPORT2 (TREX-2) complex to the NPC, which has roles in both mRNA export and transcription (Iglesias and Stutz, 2008). AtNUP136 was also found to regulate nuclear morphology (Tamura *et al.*, 2010). RAE1 probably has a role in mRNA export during interphase and is required for mitotic spindle pole formation (Lee *et al.*, 2009).

The majority of the Arabidopsis mutants mentioned above (Table 8.1), as well as several yeast and vertebrate nucleoporin mutants, exhibit accumulation of poly(A)+ RNA in the nucleus, indicating a potential disruption of RNA export. There are other links between nucleoporin function, flowering time, stress tolerance, developmental defects and RNA levels in plants. The

Table 8.1 Plant nucleoporins with functions in plant signalling processes and development

Plant Nucleoporin	Vertebrate/Yeast Homolog	Mutant/Silencing Phenotypes: *Molecular level*	Mutant/Silencing Phenotypes: *Plant level*	Reference
LjNUP85	NUP85 (NUP75)/Nup85	Impaired calcium spiking	Temperature dependent defect in RNS and AM symbiosis; reduction in seed yield	Saito *et al.* (2007)
LjNUP133	NUP133/Nup133	Impaired calcium spiking	Temperature dependent defect in RNS and AM symbiosis; reduction in seed yield	Kanamori *et al.* (2006)
NENA	SEH1(SEC13?)/ Seh1(Sec13?)[a]	Impaired calcium spiking	Temperature dependent defect in RNS and AM symbiosis;	Groth *et al.* (2010)
AtNUP88/ MOS7	NUP88/Nup82	Increased NES-mediated export; decreased nuclear levels of snc1, NPR1 and EDS1	Suppression of *snc1*; impairment in basal and R-gene mediated immunity and SAR; null mutations potentially lethal	Cheng *et al.* (2009)
AtNUP96/ MOS3/ SAR3	NUP96/Nup145C	Accumulation of nuclear poly(A)+ RNA; altered localization of AUX/IAA protein IAA17	Suppression of *snc1;* impairment in basal and R-gene mediated immunity; suppression of *axr1;* developmental defects; early flowering	Zhang and Li (2005); Parry *et al.* (2006)
AtNUP160/ SAR1	NUP160/Nup120	Accumulation of nuclear poly(A)+ RNA; altered localization of AUX/IAA protein IAA17; reduced levels of cold induced *CBF* transcripts	Suppression of *axr1*; impaired cold stress and acquired freezing tolerance; developmental defects; early flowering	Parry *et al.* (2006)
NUA/AtTPR	TPR/Mlp1, Mlp2	Accumulation of nuclear poly(A)+ RNA; reduced levels of miRNAs; increased abundance of SUMO conjugates; altered expression of flowering regulators;	Suppression of *axr1*; developmental defects; reduction in seed yield; early flowering;	Xu *et al.* (2007); Jacob *et al.* (2007)
NUP136/ AtNUP1	NUP153/Nup1[b]	Accumulation of nuclear poly(A)+ RNA; altered nuclear morphology	Developmental defects; impaired pollen development and seed production; early flowering	Tamura *et al.* (2010); Lu *et al.* (2010)
NbRAE1	RAE1/Gle2	Accumulation of nuclear poly(A)+ RNA; altered nuclear morphology; defects in spindle formation and chromosome segregation; cell-cycle delay; increased ploidy	Growth arrest; developmental defects	Lee *et al.* (2009)
AtNUP62	NUP62/Nsp1	Unknown	Developmental defects; early flowering	Zhao and Meier (2011)

[a]homologous to both SEH1 and SEC13, with slightly higher homology to SEH1
[b]potentially functional analogs and not orthologous

Arabidopsis EARLY IN SHORT DAYS4 (ESD4) protein interacts with nucleoporin AtTPR and mutations in *ESD4* and *AtTPR* cause similar pleiotropic defects, as well as changes in mRNA levels of the floral regulators *FLOWERING LOCUS C* (*FLC*), *SUPPRESSOR OF OVEREXPRESSION OF CO1* (*SOC1*) and *FLOWERING LOCUS T* (*FT*) (Reeves *et al.*, 2002; Xu *et al.*, 2007). Mutations in the DEAD (Asp-Glu-Ala-Asp) box RNA helicase *LOW EXPRESSION OF OSMOTICALLY RESPONSIVE GENES4* (*LOS4*) also result in accumulation of nuclear poly(A)+ RNA, early flowering and developmental defects (Gong *et al.*, 2005). Moreover *LOS4* is important for chilling and freezing tolerance and *los4* mutants are affected in the expression levels of *C-REPEAT BINDING FACTOR* (*CBF*) genes, which code for transcription factors involved in low-temperature-dependent gene induction (Gong *et al.*, 2002; Gong *et al.*, 2005). A mutation in *AtNUP160* also renders the plants susceptible to chilling stress and reduces CBF gene expression (Dong *et al.*, 2006).

It is unclear if the nuclear accumulation of poly(A)+ RNA is directly correlated with changes in mRNA levels in general or of specific targets. In case of the *attpr* mutant it was demonstrated that there was no preferential accumulation of a subset of transcripts and that the overall transcriptome was relatively similar to the wild type (Jacob *et al.*, 2007). In addition, microarray analysis in the *atnup160* background revealed only very limited global changes in gene expression (Dong *et al.*, 2006).

It was previously proposed by Meier and Brkljacic (2009) that the observed phenotypes could be caused by changes in miRNA homeostasis, which might be more susceptible to alterations than general mRNA levels. For instance, double mutants of the miRNAs miR159a and miR159b exhibit pleiotropic morphological defects (Allen *et al.*, 2007). Therefore, the reduction of miR159 levels in *attpr* (Jacob *et al.*, 2007) might be the cause or at least contribute to the observed developmental phenotype in the *attpr* mutant. In addition, mutations in *HASTY* (*HST*), the Arabidopsis homologue of mammalian *EXPORTIN5*, also cause an early flowering and developmental phenotype as well as a reduction in miRNA levels (Bollman *et al.*, 2003, Park *et al.*, 2005).

In the *L. japonicus nup* mutants no significant pleiotropic defects were described beyond a reduction in seed yield and pollen tube growth (Saito *et al.*, 2007; Kanamori *et al.*, 2006). However, this should be confirmed in more detail, possible by comparing mutant and wild type plants under certain stress conditions. While there is currently little first-hand information about impaired RNA transport or miRNA homeostasis in *ljnup85, ljnup133* and *nena*, the homologous proteins in vertebrates and yeast as well as the whole NUP107-160 sub-complex have been implicated to play a role in mRNA export (Li *et al.*, 1995; Vasu *et al.*, 2001; Boehmer *et al.*, 2003; Bai *et al.*, 2004), making it likely that the Lotus mutants could also be affected.

8.5 Challenges and conclusion

While much progress has been made in elucidating the NPC architecture in yeast and vertebrates, the large number of components as well as an inherent plasticity and flexibility in its composition complicates the analysis of the whole structure on the atomic level (Hoelz *et al.*, 2011). Knowledge about the plant NPC in particular is still incomplete and many open questions regarding its role in plant signalling pathways remain.

While some of the previously described defects in the plant nucleoporin mutants could be the results of specific mechanisms, it is also possible that certain processes are particularly susceptible to more general disruptions in NPC function, such as an overall decrease in the level of nuclear import or export. Moreover, secondary effects, which are not directly related to the observed phenotypes, will have to be distinguished from the actual targets that are affected in a particular pathway. Correlating specific functions conclusively with individual nucleoporins will be challenging, as changes in structural proteins such as the scaffold Nups of the NUP107-160 complex can potentially have widespread effects on the overall topology of NPC and on associated elements. Added to this complexity is the fact that nucleoporins have roles outside the NPC during mitosis (Chatel and Fahrenkrog, 2011) and in regulation of gene expression in association with chromatin in the nucleoplasm (Capelson *et al.*, 2010; Kalverda *et al.*, 2010).

In case of AtNUP88 it has been demonstrated that a mutation in a nucleoporin can lead to specific defects in plant immunity, without causing obvious pleiotropic effects. The particular mode of action of many other plant Nups still remains to be investigated in detail. The specific phenotype correlated with a loss of nuclear calcium spiking in the *L. japonicus nup* mutants potentially increases the chance to elucidate the role of the NPC in plant-microbe symbiosis. To distinguish between general and specific mechanisms, it will be helpful to determine if individual nucleoporins share common roles in plant symbiosis, defence and other plant processes. A continued study of the plant NPC will elucidate similarities and differences to the yeast and animal systems and contribute to the understanding of NPC function in general.

References

Aarts, N., Metz, M., Holub, E. *et al.* (1998) Different requirements for *EDS1* and *NDR1* by disease resistance genes define at least two R gene-mediated signalling pathways in *Arabidopsis*. *Proceedings of the National Academy of Sciences* **95**, 10306–10311.

Akhtar, A. and Gasser, S.M. (2007) The nuclear envelope and transcriptional control. *Nature Review Genetics* **8**, 507–517.

Alber, F., Dokudovskaya, S., Veenhoff, L.M. *et al.* (2007) The molecular architecture of the nuclear pore complex. *Nature* **450**, 695–701.

Allen, R.S., Li, J., Stahle, M.I. *et al.* (2007) Genetic analysis reveals functional redundancy and the major target genes of the *Arabidopsis* miR159 family. *Proceedings of the National Academy of Sciences* **104**, 16371–16376.

Ané, J.-M., Kiss, G.B., Riely, B.K. *et al.* (2004) *Medicago truncatula DMI1* required for bacterial and fungal symbioses in legumes. *Science* **303**, 1364–1367.

Bai, S.W., Rouquette, J., Umeda, M. *et al.* (2004) The fission yeast Nup107-120 complex functionally interacts with the small GTPase Ran/Spi1 and is required for mRNA export, nuclear pore distribution, and proper cell division. *Molecular and Cellular Biology* **24**, 6379–6392.

Banerjee, H., Gibbs, J., Jordan, T. and Blackshear, M. (2010) Depletion of a single nucleoporin, Nup107, induces apoptosis in eukaryotic cells. *Molecular and Cellular Biochemistry* **343**, 21–25.

Bapteste, E., Charlebois, R., Macleod, D. and Brochier, C. (2005) The two tempos of nuclear pore complex evolution: highly adapting proteins in an ancient frozen structure. *Genome Biology* **6**, R85.

Belgareh, N., Rabut, G., Baï, S.W. *et al.* (2001) An evolutionarily conserved NPC subcomplex, which redistributes in part to kinetochores in mammalian cells. *The Journal of Cell Biology* **154**, 1147–1160.

Bernad, R., Van Der Velde, H., Fornerod, M. and Pickersgill, H. (2004) Nup358/RanBP2 attaches to the nuclear pore complex via association with Nup88 and Nup214/CAN and plays a supporting role in CRM1-mediated nuclear protein export. *Molecular and Cellular Biology* **24**, 2373–2384.

Boehmer, T., Enninga, J., Dales, S. *et al.* (2003) Depletion of a single nucleoporin, Nup107, prevents the assembly of a subset of nucleoporins into the nuclear pore complex. *Proceedings of the National Academy of Sciences* **100**, 981–985.

Bollman, K.M., Aukerman, M.J., Park, M.-Y. *et al.* (2003) HASTY, the *Arabidopsis* ortholog of exportin 5/MSN5, regulates phase change and morphogenesis. *Development* **130**, 1493–1504.

Bootman, M.D., Fearnley, C., Smyrnias, I. *et al.* (2009) An update on nuclear calcium signalling. *Journal of Cell Science* **122**, 2337–2350.

Brown, C.R., Kennedy, C.J., Delmar, V.A. *et al.* (2008) Global histone acetylation induces functional genomic reorganization at mammalian nuclear pore complexes. *Genes and Development* **22**, 627–639.

Burch-Smith, T.M., Schiff, M., Caplan, J.L. *et al.* (2007) A novel role for the TIR domain in association with pathogen-derived elicitors. *PloS Biology* **5**, e68.

Bustamante, J.O. (2006) Current concepts in nuclear pore electrophysiology. *Canadian Journal of Physiology and Pharmacology* **84**, 347–365.

Cao, H., Glazebrook, J., Clarke, J.D. *et al.* (1997) The *Arabidopsis NPR1* gene that controls systemic acquired resistance encodes a novel protein containing ankyrin repeats. *Cell* **88**, 57–63.

Capelson, M. and Hetzer, M.W. (2009) The role of nuclear pores in gene regulation, development and disease. *EMBO Reports* **10**, 697–705.

Capelson, M., Liang, Y., Schulte, R. *et al.* (2010) Chromatin-bound nuclear pore components regulate gene expression in higher eukaryotes. *Cell* **140**, 372–383.

Caplan, J., Padmanabhan, M. and Dinesh-Kumar, S.P. (2008) Plant NB-LRR Immune receptors: from recognition to transcriptional reprogramming. *Cell Host and Microbe* **3**, 126–135.

Capoen, W., Sun, J., Wysham, D. *et al.* (2011) Nuclear membranes control symbiotic calcium signalling of legumes. *Proceedings of the National Academy of Sciences* **108**, 14348–14353.

Chabaud, M., Genre, A., Sieberer, B.J. *et al.* (2011) Arbuscular mycorrhizal hyphopodia and germinated spore exudates trigger Ca2+ spiking in the legume and nonlegume root epidermis. *New Phytologist* **189**, 347–355.

Charpentier, M., Bredemeier, R., Wanner, G. *et al.* (2008) *Lotus japonicus* CASTOR and POLLUX are ion channels essential for perinuclear calcium spiking in legume root endosymbiosis. *Plant Cell* **20**, 3467–3479.

Chatel, G. and Fahrenkrog, B. (2011) Nucleoporins: Leaving the nuclear pore complex for a successful mitosis. *Cellular Signalling* **23**, 1555–1562.

Cheng, Y.T., Germain, H., Wiermer, M. *et al.* (2009) Nuclear pore complex component MOS7/Nup88 is required for innate immunity and nuclear accumulation of defense regulators in *Arabidopsis*. *Plant Cell* **21**, 2503–2516.

D'Angelo, M.A. and Hetzer, M.W. (2008) Structure, dynamics and function of nuclear pore complexes. *Trends in Cell Biology* **18**, 456–466.

Danker, T., Shahin, V., Schlune, A. *et al.* (2001) Electrophoretic plugging of nuclear pores by using the nuclear hourglass technique. *Journal of Membrane Biology* **184**, 91–99.

Deslandes, L. and Rivas, S. (2011) The plant cell nucleus: a true arena for the fight between plants and pathogens. *Plant Signaling and Behavior* **6**, 42.

Després, C., Delong, C., Glaze, S. *et al.* (2000) The *Arabidopsis* NPR1/NIM1 protein enhances the DNA binding activity of a subgroup of the TGA family of bZIP transcription factors. *Plant Cell* **12**, 279–290.

Dharmasiri, N. and Estelle, M. (2004) Auxin signalling and regulated protein degradation. *Trends in Plant Science* **9**, 302–308.

Dong, C.-H., Hu, X., Tang, W. *et al.* (2006) A putative *Arabidopsis* nucleoporin, AtNUP160, is critical for RNA export and required for plant tolerance to cold stress. *Molecular and Cellular Biology* **26**, 9533–9543.

Doucet, C.M., Talamas, J.A. and Hetzer, M.W. (2010) Cell cycle-dependent differences in nuclear pore complex assembly in metazoa. *Cell* **141**, 1030–1041.

Durrant, W.E. and Dong, X. (2004) Systemic acquired resistance. *Annual Review of Phytopathology* **42**, 185–209.

Ehrhardt, D.W., Wais, R. and Long, S.R. (1996) Calcium spiking in plant root hairs responding to rhizobium nodulation signals. *Cell* **85**, 673–681.

Endre, G., Kereszt, A., Kevei, Z. *et al.* (2002) A receptor kinase gene regulating symbiotic nodule development. *Nature* **417**, 962–966.

Enninga, J., Levy, D.E., Blobel, G.N. and Fontoura, B.M.A. (2002) Role of nucleoporin induction in releasing an mRNA nuclear export block. *Science* **295**, 1523–1525.

Faria, A.M.C., Levay, A., Wang, Y. *et al.* (2006) The nucleoporin Nup96 is required for proper expression of interferon-regulated proteins and functions. *Immunity* **24**, 295–304.

Felle, H.H., Kondorosi, É., Kondorosi, Á. and Schultze, M. (1998) The role of ion fluxes in Nod factor signalling in *Medicago sativa*. *The Plant Journal* **13**, 455–463.

Fernandez-Martinez, J. and Rout, M.P. (2009) Nuclear pore complex biogenesis. *Current Opinion in Cell Biology* **21**, 603–612.

Feys, B.J., Wiermer, M., Bhat, R.A. *et al.* (2005) *Arabidopsis* SENESCENCE-ASSOCIATED GENE101 stabilizes and signals within an ENHANCED DISEASE SUSCEPTIBILITY1 complex in plant innate immunity. *Plant Cell* **17**, 2601–2613.

Fiserova, J., Kiseleva, E. and Goldberg, M.W. (2009) Nuclear envelope and nuclear pore complex structure and organization in tobacco BY-2 cells. *The Plant Journal* **59**, 243–255.

Frey, S. and Görlich, D. (2007) A saturated FG-repeat hydrogel can reproduce the permeability properties of nuclear pore complexes. *Cell* **130**, 512–523.

Gaffney, T., Friedrich, L., Vernooij, B. *et al.* (1993) Requirement of salicylic acid for the induction of systemic acquired resistance. *Science* **261**, 754–756.

García, A.V., Blanvillain-Baufumé, S., Huibers, R.P. *et al.* (2010) Balanced nuclear and cytoplasmic activities of EDS1 are required for a complete plant innate immune response. *PLoS Pathogens* **6**, e1000970.

Genre, A., Chabaud, M., Faccio, A. *et al.* (2008) Prepenetration apparatus assembly precedes and predicts the colonization patterns of arbuscular mycorrhizal fungi within the root cortex of both *Medicago truncatula* and *Daucus carota*. *Plant Cell* **20**, 1407–1420.

Gong, Z., Dong, C.-H., Lee, H. *et al.* (2005) A DEAD box RNA helicase is essential for mRNA export and important for development and stress responses in *Arabidopsis*. *Plant Cell* **17**, 256–267.

Gong, Z., Lee, H., Xiong, L. *et al.* (2002) RNA helicase-like protein as an early regulator of transcription factors for plant chilling and freezing tolerance. *Proceedings of the National Academy of Sciences* **99**, 11507–11512.

Goormachtig, S., Capoen, W. and Holsters, M. (2004) Rhizobium infection: lessons from the versatile nodulation behaviour of water-tolerant legumes. *Trends in Plant Science* **9**, 518–522.

Greenberg, J.T. and Yao, N. (2004) The role and regulation of programmed cell death in plant–pathogen interactions. *Cellular Microbiology* **6**, 201–211.

Groth, M., Takeda, N., Perry, J. *et al.* (2010) *NENA*, a *Lotus japonicus* homolog of *Sec13*, is required for rhizodermal infection by arbuscular mycorrhiza fungi and rhizobia but dispensable for cortical endosymbiotic development. *Plant Cell* **22**, 2509.

Harel, A., Orjalo, A.V., Vincent, T. *et al.* (2003) Removal of a single pore subcomplex results in vertebrate nuclei devoid of nuclear pores. *Molecular Cell* **11**, 853–864.

Hayashi, T., Banba, M., Shimoda, Y. *et al.* (2010) A dominant function of CCaMK in intracellular accommodation of bacterial and fungal endosymbionts. *The Plant Journal* **63**, 141–154.

Heidrich, K., Wirthmueller, L., Tasset, C. *et al.* (2011) *Arabidopsis* EDS1 connects pathogen effector recognition to cell compartment-specific immune responses. *Science* **334**, 1401–1404.

Hoelz, A., Debler, E.W. and Blobel, G. (2011) The structure of the nuclear pore complex. *Annual Review of Biochemistry* **80**, 613–643.

Hohn, T. and Vazquez, F. (2011) RNA silencing pathways of plants: silencing and its suppression by plant DNA viruses. *Biochimica et Biophysica Acta (BBA) – Gene Regulatory Mechanisms* **1809**, 588–600.

Iglesias, N. and Stutz, F.O. (2008) Regulation of mRNP dynamics along the export pathway. *FEBS Letters* **582**, 1987–1996.

Imaizumi-Anraku, H., Takeda, N., Charpentier, M. *et al.* (2005) Plastid proteins crucial for symbiotic fungal and bacterial entry into plant roots. *Nature*, **433**, 527–531.

Jacob, Y., Mongkolsiriwatana, C., Veley, K.M. *et al.* (2007) The nuclear pore protein AtTPR is required for RNA homeostasis, flowering time, and auxin signaling. *Plant Physiology* **144**, 1383–1390.

Jones, J.D.G. and Dangl, J.L. (2006) The plant immune system. *Nature* **444**, 323–329.

Kaló, P., Gleason, C., Edwards, A. *et al.* (2005) Nodulation signaling in legumes requires NSP2, a member of the GRAS family of transcriptional regulators. *Science* **308**, 1786–1789.

Kalverda, B., Pickersgill, H., Shloma, V.V. and Fornerod, M. (2010) Nucleoporins directly stimulate expression of developmental and cell-cycle genes inside the nucleoplasm. *Cell* **140**, 360–371.

Kanamori, N., Madsen, L., Radutoiu, S. *et al.* (2006) A nucleoporin is required for induction of Ca2+ spiking in legume nodule development and essential for rhizobial and fungal symbiosis. *Proceedings of the National Academy of Sciences* **103**, 359.

Kim, S.-H., Arnold, D., Lloyd, A. and Roux, S.J. (2001) Antisense expression of an *Arabidopsis* Ran binding protein renders transgenic roots hypersensitive to auxin and alters auxin-induced root growth and development by arresting mitotic progress. *Plant Cell* **13**, 2619–2630.

Kinkema, M., Fan, W. and Dong, X. (2000) Nuclear localization of NPR1 is required for activation of PR gene expression. *Plant Cell* **12**, 2339.

Kistner, C. and Parniske, M. (2002) Evolution of signal transduction in intracellular symbiosis. *Trends in Plant Science* **7**, 511–518.

Kistner, C., Winzer, T., Pitzschke, A. *et al.* (2005) Seven *Lotus japonicus* genes required for transcriptional reprogramming of the root during fungal and bacterial symbiosis. *Plant Cell* **17**, 2217–2229.

Kodiha, M., Tran, D., Qian, C. *et al.* (2008) Oxidative stress mislocalizes and retains transport factor importin-α and nucleoporins Nup153 and Nup88 in nuclei where they generate high molecular mass complexes. *Biochimica et Biophysica Acta (BBA) – Molecular Cell Research* **1783**, 405–418.

Kosuta, S., Hazledine, S., Sun, J. *et al.* (2008) Differential and chaotic calcium signatures in the symbiosis signalling pathway of legumes. *Proceedings of the National Academy of Sciences* **105**, 9823–9828.

Kudla, J., Batistic, O. and Hashimoto, K. (2010) Calcium signals: the lead currency of plant information processing. *Plant Cell* **22**, 541–563.

Lecourieux, D., Lamotte, O., Bourque, S.P. *et al.* (2005) Proteinaceous and oligosaccharidic elicitors induce different calcium signatures in the nucleus of tobacco cells. *Cell Calcium* **38**, 527–538.

Lecourieux, D., Ranjeva, R. and Pugin, A. (2006) Calcium in plant defence-signalling pathways. *New Phytologist* **171**, 249–269.

Lee, J.-Y., Lee, H.-S., Wi, S.-J. *et al.* (2009) Dual functions of *Nicotiana benthamiana* Rae1 in interphase and mitosis. *The Plant Journal* **59**, 278–291.

Lévy, J., Bres, C., Geurts, R. *et al.* (2004) A putative Ca2+ and calmodulin-dependent protein kinase required for bacterial and fungal symbioses. *Science* **303**, 1361–1364.

Li, O., Heath, C.V., Amberg, D.C. *et al.* (1995) Mutation or deletion of the *Saccharomyces cerevisiae* RAT3/NUP133 gene causes temperature-dependent nuclear accumulation of poly(A)+ RNA and constitutive clustering of nuclear pore complexes. *Molecular Biology of the Cell* **6**, 401–417.

Li, X., Clarke, J.D., Zhang, Y. and Dong, X. (2001) Activation of an EDS1-mediated R-gene pathway in the *snc1* mutant leads to constitutive, NPR1-independent pathogen resistance. *Molecular Plant–Microbe Interactions* **14**, 1131–1139.

Loïodice, I., Alves, A., Rabut, G. *et al.* (2004) The entire Nup107-160 complex, including three new members, is targeted as one entity to kinetochores in mitosis. *Molecular Biology of the Cell* **15**, 3333–3344.

Lu, Q., Tang, X., Tian, G. *et al.* (2010) *Arabidopsis* homolog of the yeast TREX-2 mRNA export complex: components and anchoring nucleoporin. *The Plant Journal* **61**, 259–270.

Lupu, F., Alves, A., Anderson, K. *et al.* (2008) Nuclear pore composition regulates neural stem/progenitor cell differentiation in the mouse embryo. *Developmental Cell* **14**, 831–842.

Lusk, C.P., Blobel, G. and King, M.C. (2007) Highway to the inner nuclear membrane: rules for the road. *Nature Reviews Molecular Cell Biology*, **8**, 414–420.

Lutzmann, M., Kunze, R., Buerer, A. *et al.* (2002) Modular self-assembly of a Y-shaped multiprotein complex from seven nucleoporins. *EMBO Journal* **21**, 387–397.

Madsen, E.B., Madsen, L.H., Radutoiu, S. *et al.* (2003) A receptor kinase gene of the LysM type is involved in legumeperception of rhizobial signals. *Nature* **425**, 637–640.

Madsen, L.H., Tirichine, L., Jurkiewicz, A. *et al.* (2010) The molecular network governing nodule organogenesis and infection in the model legume *Lotus japonicus. Nature Communications* **1**, 10.

Maillet, F., Poinsot, V., Andre, O. *et al.* (2011) Fungal lipochitooligosaccharide symbiotic signals in arbuscular mycorrhiza. *Nature* **469**, 58–63.

Mazars, C., Bourque, S., Mithöfer, A. *et al.* (2009) Calcium homeostasis in plant cell nuclei. *New Phytologist* **181**, 261–274.

Meier, I. and Brkljacic, J. (2009) The nuclear pore and plant development. *Current Opinion in Plant Biology* **12**, 87–95.

Meinema, A.C., Laba, J.K., Hapsari, R.A. *et al.* (2011) Long unfolded linkers facilitate membrane protein import through the nuclear pore complex. *Science* **333**, 90–93.

Messinese, E., Mun, J.-H., Yeun, L.H., *et al.* (2007) A novel nuclear protein interacts with the symbiotic DMI3 calcium- and calmodulin-dependent protein kinase of *Medicago truncatula. Molecular Plant-Microbe Interactions* **20**, 912–921.

Mitra, R.M., Gleason, C.A., Edwards, A. *et al.* (2004) A Ca2+/calmodulin-dependent protein kinase required for symbiotic nodule development: Gene identification by transcript-based cloning. *Proceedings of the National Academy of Sciences of the United States of America* **101**, 4701–4705.

Miwa, H., Sun, J., Oldroyd, G.E.D. and Downie, J.A. (2006) Analysis of nod-factor-induced calcium signaling in root hairs of symbiotically defective mutants of *Lotus japonicus. Molecular Plant-Microbe Interactions* **19**, 914–923.

Monaghan, J., Germain, H., Weihmann, T. and Li, X. (2010) Dissecting plant defence signal transduction: modifiers of *snc1* in *Arabidopsis. Canadian Journal of Plant Pathology* **32**, 35–42.

Ohba, T., Schirmer, E.C., Nishimoto, T. and Gerace, L. (2004) Energy- and temperature-dependent transport of integral proteins to the inner nuclear membrane via the nuclear pore. *The Journal of Cell Biology* **167**, 1051–1062.

Oldroyd, G.E.D. and Downie, J.A. (2008) Coordinating nodule morphogenesis with rhizobial infection in legumes. *Annual Review of Plant Biology* **59**, 519–546.

Oldroyd, G.E.D., Murray, J.D., Poole, P.S. and Downie, J.A. (2011) The rules of engagement in the legume-rhizobial symbiosis. *Annual Review of Genetics* **45**, 119–145.

Op Den Camp, R., Streng, A., De Mita, S. *et al.* (2011) LysM-type mycorrhizal receptor recruited for rhizobium symbiosis in nonlegume *Parasponia. Science* **331**, 909–912.

Orjalo, A.V., Arnaoutov, A., Shen, Z. *et al.* (2006) The Nup107-160 nucleoporin complex is required for correct bipolar spindle assembly. *Molecular Biology of the Cell* **17**, 3806–3818.

Palma, K., Zhang, Y. and Li, X. (2005) An importin alpha homolog, MOS6, plays an important role in plant innate immunity. *Current Biology* **15**, 1129–1135.

Park, M.Y., Wu, G., Gonzalez-Sulser, A. *et al.* (2005) Nuclear processing and export of microRNAs in *Arabidopsis. Proceedings of the National Academy of Sciences of the United States of America* **102**, 3691–3696.

Parniske, M. (2008) Arbuscular mycorrhiza: the mother of plant root endosymbioses. *Nature Reviews Microbiology* **6**, 763–775.

Parry, G., Ward, S., Cernac, A. *et al.* (2006) The *Arabidopsis* SUPPRESSOR OF AUXIN RESISTANCE proteins are nucleoporins with an important role in hormone signalling and development. *Plant Cell* **18**, 1590–1603.

Pawlowski, K. and Sprent, J.I. (2008) Comparison between actinorhizal and legume symbiosis nitrogen-fixing actinorhizal symbioses. In: Pawlowski, K. and Newton, W.E. (eds) *Nitrogen-Fixing Actinorhizal Symbioses*, Springer, Dordrecht.

Perry, J., Brachmann, A., Welham, T. *et al.* (2009) TILLING in *Lotus japonicus* identified large allelic series for symbiosis genes and revealed a bias in functionally defective ethyl methanesulfonate alleles toward glycine replacements. *Plant Physiology* **151**, 1281–1291.

Radutoiu, S., Madsen, L.H., Madsen, E.B., *et al.* (2003) Plant recognition of symbiotic bacteria requires two LysM receptor-like kinases. *Nature* **425**, 585–592.

Reeves, P.H., Murtas, G., Dash, S. and Coupland, G. (2002) *Early in short days 4*, a mutation in *Arabidopsis* that causes early flowering and reduces the mRNA abundance of the floral repressor *FLC*. *Development* **129**, 5349–5361.

Ribbeck, K., Lipowsky, G., Kent, H.M. *et al.* (1998) NTF2 mediates nuclear import of Ran. *EMBO Journal* **17**, 6587–6598.

Rodrigues, M.A., Gomes, D.A., Nathanson, M.H. and Leite, M.F. (2009) Nuclear calcium signalling: a cell within a cell. *Brazilian Journal of Medical and Biological Research* **42**, 17–20.

Roth, P., Xylourgidis, N., Sabri, N. *et al.* (2003) The *Drosophila* nucleoporin DNup88 localizes DNup214 and CRM1 on the nuclear envelope and attenuates NES-mediated nuclear export. *The Journal of Cell Biology* **163**, 701–706.

Saito, K., Yoshikawa, M., Yano, K. *et al.* (2007) NUCLEOPORIN85 is required for calcium spiking, fungal and bacterial symbioses, and seed production in *Lotus japonicus*. *Plant Cell* **19**, 610–624.

Schüßler, A., Schwarzott, D. and Walker, C. (2001) A new fungal phylum, the *Glomeromycota*: phylogeny and evolution. *Mycological Research* **105**, 1413–1421.

Schüßler, A. and Walker, C. (2010) The *Glomeromycota*: A species list with new families and genera. Edinburgh & Kew, UK: The Royal Botanic Garden; Munich, Germany: Botanische Staatssammlung Munich; Oregon, USA: Oregon State University. URL: http://www.amf-phylogeny.com. ISBN-13: 9781466388048; ISBN-10: 1466388048.

Scott, I.M., Clarke, S.M., Wood, J.E. and Mur, L. (2004) Salicylate accumulation inhibits growth at chilling temperature in *Arabidopsis*. *Plant Physiology* **135**, 1040–1049.

Shahin, V., Danker, T., Enss, K. *et al.* (2001) Evidence for Ca2+- and ATP-sensitive peripheral channels in nuclear pore complexes. *The FASEB Journal* **15**, 1895–1901.

Shaw, S.L. and Long, S.R. (2003) Nod factor elicits two separable calcium responses in *Medicago truncatula* root hair cells. *Plant Physiology* **131**, 976–984.

Shen, Q.-H., Saijo, Y., Mauch, S. *et al.* (2007) Nuclear activity of MLA immune receptors links isolate-specific and basal disease-resistance responses. *Science* **315**, 1098–1103.

Sieberer, B.J., Chabaud, M., Timmers, A.C. *et al.* (2009) A nuclear-targeted cameleon demonstrates intranuclear Ca2+ spiking in *Medicago truncatula* root hairs in response to rhizobial nodulation factors. *Plant Physiology* **151**, 1197–1206.

Siniossoglou, S., Wimmer, C., Rieger, M. *et al.* (1996) A novel complex of nucleoporins, which includes Sec13p and a Sec13p homolog, is essential for normal nuclear pores. *Cell* **84**, 265–275.

Smith, S.E. and Read, D.J. (2008) *Mycorrhizal symbiosis*, London, Academic Press.

Smith, S.E. and Smith, F.A. (2011) Roles of arbuscular mycorrhizas in plant nutrition and growth: new paradigms from cellular to ecosystem scales. *Annual Review of Plant Biology* **62**, 227–250.

Soltis, D.E., Soltis, P.S., Chase, M.W., *et al.* (2000) Angiosperm phylogeny inferred from 18S rDNA, *rbcL*, and *atpB* sequences. *Botanical Journal of the Linnean Society* **133**, 381–461.

Sprent, J.I. (2007) Evolving ideas of legume evolution and diversity: a taxonomic perspective on the occurrence of nodulation. *New Phytologist* **174**, 11–25.

Stockinger, H., Walker, C. and Schüßler, A. (2009) '*Glomus intraradices* DAOM197198', a model fungus in arbuscular mycorrhiza research, is not *Glomus intraradices*. *New Phytologist* **183**, 1176–1187.

Stracke, S., Kistner, C., Yoshida, S. *et al.* (2002) A plant receptor-like kinase required for both bacterial and fungal symbiosis. *Nature* **417**, 959–962.

Strambio-De-Castillia, C., Niepel, M. and Rout, M.P. (2010) The nuclear pore complex: bridging nuclear transport and gene regulation. *Nature Reviews Molecular Cell Biology* **11**, 490–501.

Sunkar, R. and Zhu, J.-K. (2004) Novel and stress-regulated microRNAs and other small RNAs from *Arabidopsis*. *Plant Cell* **16**, 2001–2019.

Tamura, K., Fukao, Y., Iwamoto, M. *et al.* (2010) Identification and characterization of nuclear pore complex components in *Arabidopsis thaliana*. *Plant Cell* **22**, 4084–4097.

Tirichine, L., Imaizumi-Anraku, H., Yoshida, S., *et al.* (2006) Deregulation of a Ca2+/calmodulin-dependent kinase leads to spontaneous nodule development. *Nature* **441**, 1153–1156.

Uv, A.E., Roth, P., Xylourgidis, N. *et al.* (2000) *Members only* encodes a *Drosophila* nucleoporin required for Rel protein import and immune response activation. *Genes and Development* **14**, 1945–1957.

Vasu, S., Shah, S., Orjalo, A. *et al.* (2001) Novel vertebrate nucleoporins Nup133 and Nup160 play a role in mRNA export. *The Journal of Cell Biology* **155**, 339–354.

Venugopal, S.C., Jeong, R.-D., Mandal, M.K. *et al.* (2009) Enhanced disease susceptibility 1 and salicylic acid act redundantly to regulate resistance gene-mediated signaling. *PlOS Genetics* **5**, e1000545.

Walter, A., Mazars, C., Maitrejean, M. *et al.* (2007) Structural requirements of jasmonates and synthetic analogues as inducers of Ca2+ signals in the nucleus and the cytosol of plant cells. *Angewandte Chemie International Edition* **46**, 4783–4785.

Walther, T.C., Alves, A., Pickersgill, H. *et al.* (2003) The conserved Nup107-160 complex is critical for nuclear pore complex assembly. *Cell* **113**, 195–206.

Wente, S.R. (2000) Gatekeepers of the nucleus. *Science* **288**, 1374–1377.

Wente, S.R. and Rout, M.P. (2010) The nuclear pore complex and nuclear transport. *Cold Spring Harbor Perspectives in Biology* **2**.

White, J., Prell, J., James, E. and Poole, P. (2007) Nutrient sharing between symbionts. *Plant Physiology* **144**, 604–614.

Wiermer, M., Cheng, Y.T., Imkampe, J. *et al.* (2012) Putative members of the *Arabidopsis* Nup107-160 nuclear pore sub-complex contribute to pathogen defense. *The Plant Journal* **70**, 796–808.

Wiermer, M., Palma, K., Zhang, Y. and Li, X. (2007) Should I stay or should I go? Nucleocytoplasmic trafficking in plant innate immunity. *Cellular Microbiology* **9**, 1880–1890.

Wirthmueller, L., Zhang, Y., Jones, J.D.G. and Parker, J.E. (2007) Nuclear accumulation of the *Arabidopsis* immune receptor RPS4 is necessary for triggering EDS1-dependent defense. *Current Biology* **17**, 2023–2029.

Xiong, T.C., Jauneau, A., Ranjeva, R. and Mazars, C. (2004) Isolated plant nuclei as mechanical and thermal sensors involved in calcium signalling. *The Plant Journal* **40**, 12–21.

Xu, X.M. and Meier, I. (2008) The nuclear pore comes to the fore. *Trends in Plant Science* **13**, 20–27.

Xu, X.M., Rose, A., Muthuswamy, S. *et al.* (2007) NUCLEAR PORE ANCHOR, the *Arabidopsis* homolog of Tpr/Mlp1/Mlp2/megator, is involved in mRNA export and SUMO homeostasis and affects diverse aspects of plant development. *Plant Cell* **19**, 1537.

Xylourgidis, N., Roth, P., Sabri, N., *et al.* (2006) The nucleoporin Nup214 sequesters CRM1 at the nuclear rim and modulates NFκB activation in *Drosophila. Journal of Cell Science* **119**, 4409–4419.

Yano, K., Yoshida, S., Müller, J. *et al.* (2008) CYCLOPS, a mediator of symbiotic intracellular accommodation. *Proceedings of the National Academy of Sciences* **105**, 20540–20545.

Zhang, Y., Fan, W., Kinkema, M. *et al.* (1999) Interaction of NPR1 with basic leucine zipper protein transcription factors that bind sequences required for salicylic acid induction of the *PR-1* gene. *Proceedings of the National Academy of Sciences* **96**, 6523–6528.

Zhang, Y., Goritschnig, S., Dong, X. and Li, X. (2003) A gain-of-function mutation in a plant disease resistance gene leads to constitutive activation of downstream signal transduction pathways in *suppressor of npr1-1, constitutive 1. Plant Cell* **15**, 2636–2646.

Zhang, Y. and Li, X. (2005) A putative Nucleoporin 96 is required for both basal defense and constitutive resistance responses mediated by *suppressor of npr1-1,constitutive 1. Plant Cell* **17**, 1306–1316.

Zhao, Q., Brkljacic, J. and Meier, I. (2008) Two distinct interacting classes of nuclear envelope-associated coiled-coil proteins are required for the tissue-specific nuclear envelope targeting of *Arabidopsis* RanGAP. *Plant Cell* **20**, 1639–1651.

Zhao, Q., Leung, S., Corbett, A.H. and Meier, I. (2006) Identification and characterization of the *Arabidopsis* orthologs of nuclear transport Factor 2, the nuclear import factor of Ran. *Plant Physiology* **140**, 869–878.

Zhao, Q. and Meier, I. (2011) Identification and characterization of the *Arabidopsis* FG-repeat nucleoporin Nup62. *Plant Signaling and Behavior* **6**, 330–334.

Zhou, G.-K., Kubo, M., Zhong, R. *et al.* (2007) Overexpression of miR165 affects apical Meristem formation, organ polarity establishment and vascular development in *Arabidopsis. Plant and Cell Physiology* **48**, 391–404.

Zhou, J.-M., Trifa, Y., Silva, H. *et al.* (2000) NPR1 differentially interacts with members of the TGA/OBF family of transcription factors that bind an element of the *PR-1* gene required for induction by salicylic acid. *Molecular Plant-Microbe Interactions* **13**, 191–202.

Zhu, S., Jeong, R.-D., Venugopal, S.C. *et al.* (2011) SAG101 forms a ternary complex with EDS1 and PAD4 and is required for resistance signaling against turnip crinkle virus. *PLOS Pathogens* **7**, e1002318.

Zhu, Z., Xu, F., Zhang, Y. *et al.* (2010) *Arabidopsis* resistance protein SNC1 activates immune responses through association with a transcriptional corepressor. *Proceedings of the National Academy of Sciences of the United States of America* **107**, 13960–13965.

Zipfel, C. and Felix, G. (2005) Plants and animals: a different taste for microbes? *Current Opinion in Plant Biology*, **8**, 353–360.

Zuccolo, M., Alves, A., Galy, V. *et al.* (2007) The human Nup107-160 nuclear pore subcomplex contributes to proper kinetochore functions. *EMBO Journal* **26**, 1853–1864.

Zuleger, N., Korfali, N. and Schirmer, E. (2008) Inner nuclear membrane protein transport is mediated by multiple mechanisms. *Biochemical Society Transactions* **36**, 1373–1377.

INDEX

Annual Plant Reviews Volume 46: Plant Nuclear Structure, Genome Architecture and Gene Regulation, First Edition. Edited by David E. Evans, Katja Graumann and John A. Bryant.